D1208806

European Communication Council Report

E-CONOMICS

Strategies for the Digital Marketplace

Axel Zerdick, Arnold Picot, Klaus Schrape,

Alexander Artopé, Klaus Goldhammer, Ulrich T. Lange, Eckart Vierkant

Esteban López-Escobar, Roger Silverstone

with an introduction by Carl Shapiro and Hal R. Varian

Springer
Berlin
Heidelberg
New York
Barcelona
Hong Kong
London
Milan
Paris
Singapore
Tokyo

E–CONOMICS

Strategies for the Digital Marketplace

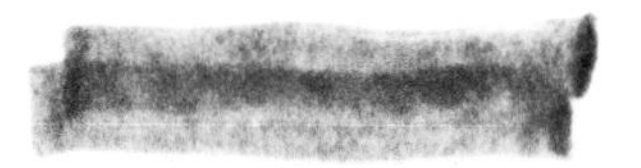

European Communication Council Report

E-CONOMICS

Strategies for the Digital Marketplace

The authors:

Axel Zerdick (1941), Prof. Dr., Institut für Publizistik- und Kommunikationswissenschaft, Freie Universität Berlin, E-mail: zerdick@attglobal.net

Arnold Picot (1944), Prof. Dr. Dr. h.c., Institut für Organisation, Seminar für Betriebswirtschaftliche Informations- und Kommunikationsforschung, Ludwig-Maximilians-Universität München, E-mail: picot@bwl.uni-muenchen.de

Klaus Schrape (1946), Prof. Dr., Vicedirector and Chair of the Department for Media and Communication, Prognos AG, Basel, a.o. Professor, Universität Basel, E-mail: klaus.schrape@prognos.com

Alexander Artopé (1969), Dipl.-Kfm. M.A., Managing Editor, European Communication Council, E-mail: alexander@datango.de

Klaus Goldhammer (1967), Dr., Managing Editor, European Commmunication Council, E-mail: klaus.goldhammer@goldmedia.de

Ulrich T. Lange (1953), Dr., Director of "institut für medienarchitektur und -gestaltung", Berlin, and CEO, dynavions AG, Berlin, E-mail: ulilange@compuserve.com

Eckart Vierkant (1972), Dipl.-Kfm., Institut für Publizistik- und Kommunikationswissenschaft, Freie Universität Berlin, E-mail: evierkant@aol.com

Esteban López-Escobar (1941), Prof. Dr.; Departamento de Comunicación Pública, Universidad de Navarra, Pamplona, E-mail: elef@unav.es

Roger Silverstone (1945), Prof., PhD; Chair of Media and Communications, Department of Sociology, London School of Economics, E-mail: R.Silverstone@lse.ac.uk

Supported by: MEDIAGRUPPE MÜNCHEN
Webpage: www.mgmuc.de

ISBN 3-540-64943-3 Springer-Verlag Berlin Heidelberg New York

CIP-Data applied for

E-conomics : strategies for the digital marketplace/ed : ECC
European Communication Council By Axel Zerdick _– Berlin ; Heidelberg ; New York ;
Barcelona ; Hong Kong ; London ; Milan ; Paris ; Singapore ; Tokyo : Springer, 2000
(Report ... / European Communication Council ; 2000)
ISBN 3-540-64943-3

Springer-Verlag Berlin Heidelberg New York
a member of BertelsmannSpringer Science+Business Media GmbH
Printed in Germany

Cover design and setting: SchömannCorporate/Berlin
SPIN: 10789753 68/3111 – 5 4 3 2 1 – Printed on acid free paper

Workshop Participants and Interview Partners

The ECC is particularly grateful to those experts who gave their valuable time for interviews with us and for discussions in our workshops:

Name	Function	Organisation
Alfieri, Thomas	Manager Interactive Research	NBC
Bachem, Christian	Strategic Planning, Marketing Services	Pixelpark
Beer, Joachim	Associate Director	ICSI
Bonnaure, Patrick	Manager Business Development	WebTV Networks
Bunger, Stan	Producer New Media News	KRON-TV, Channel 4
Carlton, Jim	Technology Reporter	The Wall Street Journal
Catchings, John	Station Operations Director	KPIX Television - CBS
Cavagnaro, Ed	Director News and Programming	KCBS NewsRadio
Cringely, Robert X.	Author	
Dennis, Everette E.	Distinguished Visiting Professor	Fordham University
Diamond, Michael	Associate Director, Corporate Strategy and Planning	Time Warner
Dixon, Christopher P.	Managing Director, Entertainment Research	PaineWebber
Eliashberg, Jehoshua	Professor of Marketing	University of Pennsylvania
Fischer, Claude S.	Professor of Sociology, College of Letters and Science	University of California at Berkeley
Henkin, Michael	Director of Operations Planning	Loral Space & Communications
Kellog, Nelson R.	Professor of Interdisciplinary Studies	Sonoma State University
Kim, Jasmin	Marketing Manager	Yahoo!
Kounalakis, Markos	Executive Communications Strategist	Silicon Graphics
Kürble, Peter	Lecturer	Fontys Hogescholen, Venlo
McGowan, David	Attorney at Law	Howard, Rice, Nemerovski, Canady, Falk & Rabkin
Meluso, Peter	Chief Service Officer	I-Traffic
Neuert, Ulrich	Executive Vice President	Bertelsmann New Media
Noam, Eli M.	Professor of Finance and Economics, Director of the Columbia Institute for Tele-Information	Columbia University
Poliza, Andreas	Chairman and CEO	GoLive Systems
Reiss, Spencer	Senior Editor	Wired Magazine
Riquier, David	Associate Director	MIT Media Lab
Schambach, Stefan	President and CEO	Intershop Communications
Schöneberger, Markus	Manager Movies, TV, Media Economics	RTL Television
Shirk, George	News Director	The Gate - Online Service
Stipp, Horst	Director, Social and Development Research	Research Department, NBC
Swank, Gabriel	Corporate Strategist	Intershop Communications
Tannenbaum, Percy	Professor	University of California at Berkeley
Varian, Hal R.	Dean of the School of Information Managament and Systems	University of California at Berkeley
Weiber, Rolf	Professor of Marketing	Universität Trier
Weinberg, Ulrich	General Manager, Professor for Computer Animation	Terratools, HFF Potsdam / Babelsberg
Woroch, Glenn A.	Professor of Economics, Consortium for Research on Telecommunications Policy	University of California at Berkeley
Worthman, Susan	Managing Director	MDG.ORG, Multimedia Development Group

The European Communication Council (ECC) is an independent group of scholars and scientists from different European countries and the USA. For each new report, communication experts from diverse academic backgrounds are invited to participate in research and production as either Fellows of the Council or authors. The ECC's objective is to discuss trends and issues in European communications for leaders and visionaries in communication companies and for leading policymakers.

ECC Reports concentrate on key trends and issues in media, telecommunications and information technologies, which are expected to be predominant for future development of communication industries, and which deserve higher profile in future debates.

The European Communication Council is organised as an independent research project at Freie Universität Berlin. It is fully funded by non-government resources from MEDIAGRUPPE MÜNCHEN.

Fellows, editors and authors who contributed to this and previous reports:

Name	ECC Function	Home Organisation
Artopé, Alexander	Managing Editor 1999, 2000	datango.de, Berlin
Carlton, Jim	Author 1999, 2000	The Wall Street Journal, San Francisco
Claisse, Gérard	Author 1997	Laboratoire d'Economie des Transports, Lyon
Cringely, Robert X.	Author 1999, 2000	Journalist and Author, San Mateo
Gaster, Jens	Author 1997	European Commission, Brussels
Goldhammer, Klaus	Co-ordinating Author 1997, Managing Editor 1999, 2000	Goldhammer Medienberatung, Berlin
Kelly, Kevin	Author 1999, 2000	Wired Magazine, San Francisco
Kleinsteuber, Hans J.	Author 1997	Universität Hamburg
Lange, Ulrich T.	Co-ordinating Author 1997, Author 1999, 2000	dynavisions AG, Berlin
López-Escobar, Esteban	Fellow 1999, 2000	Universidad de Navarra, Pamplona
Martinoli, Mario	Author 1999, 2000	Databank Consulting, Milano
Negroponte, Nicholas	Author 1999, 2000	MIT Media Lab, Boston
Noam, Eli M.	Author 1997	Columbia University, New York
Paterson, Richard	Author 1997	British Film Institute, London
Picot, Arnold	Fellow 1999, 2000	Ludwig-Maximilians-Universität, München
Pilati, Antonio	Author 1997	Instituto di Economia dei Media, Milano
Richeri, Giuseppe	Author 1997	University Bologna / University Lugano
Rosenbach, Marcel	Author 1997	Universität Hamburg
Schlesinger, Philip	Fellow 1997	University of Stirling, Scotland
Schrape, Klaus	Author 1997, Fellow 1999, 2000	Universität Basel / Prognos AG
Seufert, Wolfgang	Author 1997	Deutsches Institut für Wirtschaftsforschung
Shapiro, Carl	Author 1999, 2000	University of California at Berkeley
Silj, Alessandro	Fellow 1997	ROMA - Research on Media Associates / Consiglio Italiano per le Scienze Sociali, Roma
Silverstone, Roger	Author 1997, Fellow 1999, 2000	London School of Economics
Tannenbaum, Percy	Fellow 1997	University of California at Berkeley
Varian, Hal	Author 1999, 2000	University of California at Berkeley
Vierkant, Eckart	Author 1999, 2000	Freie Universität Berlin
Wattenberg, Ulrich	Author 1997	GMD Forschungszentrum Informationstechnik, Berlin
Zerdick, Axel	Speaker 1997, 1999, 2000	Freie Universität Berlin

The Authors

back row (left to right): Axel Zerdick, Ulrich T. Lange, Esteban López-Escobar, Alexander Artopé
front row (left to right): Klaus Goldhammer, Roger Silverstone, Klaus Schrape, Arnold Picot
not in the picture (here, that is – but very much so otherwise): Eckart Vierkant

ECC
EUROPEAN COMMUNICATION COUNCIL

Table of Content

ECC
EUROPEAN COMMUNICATION COUNCIL

Introduction

By Carl Shapiro and Hal R. Varian, University of California, Berkeley

Authors of: Information Rules:
A Strategic Guide to the Network Economy, Boston, MA: Harvard Business School Press, 1999

As the century closed, the world became smaller. The public rapidly gained access to new and dramatically faster communication technologies. Entrepreneurs, able to draw on unprecedented scale economies, built vast empires. Great fortunes were made. The government demanded that these powerful new monopolists be held accountable under antitrust law. Every day brought forth new technological advances to which the old business models seemed no longer to apply. Yet, somehow, the basic laws of economics asserted themselves. Those who mastered these laws survived in the new environment. Those who did not, failed.

Basic laws of economics in a new environment

A prophecy for the next decade? No. You have just read a description of what happened a century ago when the twentieth century industrial giants emerged. Using the infrastructure of the emerging electricity and telephone networks, these industrialists transformed the U.S. economy, just as today's Silicon Valley entrepreneurs are drawing on computer and communications infrastructure to transform the world's economy.

Systematic and thorough approach, European emphasis

E-conomics – Strategies for the Digital Marketplace is a welcome attempt to understand the New Economy we are living in. It distinguishes itself from other such attempts by its systematic and thorough approach, as well as by its European emphasis. Too often, Americans (especially those who live in Silicon Valley) see only what is happening in our own economy. This makes a systematic understanding of European developments in "digital economics" particularly useful.

The information revolution, like the preceding industrial revolutions, rests on basic infrastructure. Just as transportation networks had to be built in order to facilitate trade in physical goods, communications and information networks have to be built to facilitate trade in digital goods. But it is not just the optical fiber and the telecommunications equipment that forms the infrastructure for the network economy. Of equal or greater importance are the communications standards and protocols that allow for coherent communication. A string of bits sent between two computers is meaningless unless both computers can agree on the interpretation of those bits. That requires standardisation.

One of the great challenges facing the world in the next phase of the network economy is developing standards setting processes that work quickly enough

"Europe and the U.S. tend to take different approaches to standards setting."

and well enough to facilitate rather than impede the growth of the digital economy. This requires not only technical expertise, but also the development of a legal and policy infrastructure to develop, support, and evolve these standards. Streamlined means of dispute resolution are particularly important.

Europe and the U.S. tend to take different approaches to standards setting. In the U.S. "the market decides" while in Europe there is generally a greater degree of government involvement. Each approach has its advantages and disadvantages and there is much to be said for a diverse approach. However, if we are to realize the full fruits of the digital revolution, ultimately most standards need to be worldwide. This, in turn, requires mutual understanding and appreciation of the issues facing the participants in the standards-setting processes, including the fact that standards set in one region can greatly influence other regions, and the critical role of intellectual property rights in many standard-setting efforts.

Interweaving of European and American perspectives: mutual understanding can facilitate a truly worldwide digital marketplace

Through their skillful interweaving of European and American perspectives, the authors of E-conomics contribute to the mutual understanding necessary to facilitate the standards and interconnection policies that can lead to a truly worldwide digital marketplace.

Foreword

The European Communication Council

Objectives of the European Communication Council ECC

The idea of the "European Communication Council" was born in 1996 on the initiative of Michael Wölfle and Hans Lauber from MEDIAGRUPPE MÜNCHEN. It was conceived as a small group of predominantly European communication scientists faced with an unusually appealing task. This was to bring contributions from European (and international) communication science on the development of the media and communications industries to the notice of the decision-makers identifying, in particular, themes of special relevance for future development. Various scientific fields concerned with developments in the communications industries and countless research institutions in many countries have recently witnessed the emergence of new strands of discussion in the area, some of which are highly specialized, many of which are exceptionally interesting. It is a tough job even for specialists to keep an overview of all the contributions arising in these discussions. It is a well-nigh impossible one for the decision-makers and visionaries in business and politics, people whose work leaves them no time to read and digest such a wealth and diversity of scientific texts. This is where the European Communication Council steps in. Its task is to sound out the broad field of current scientific discussion, pinpointing those trends and themes that seem of special relevance for the future evolution of the media and communication industries and which might serve to stimulate the visionaries and decision-makers in their discussions.

European focus and European perspective, rather than narrower national points of view

Calling the European Communication Council "European" is in certain respects something of a challenge. On a pragmatic level things are clear-cut enough. "European" here means that the focus will be on the key communications markets in Europe, but also on other European countries (both inside and outside the European Union) where interesting and significant developments can be discerned. The term is also meant to denote a European viewpoint, both in contrast to narrower national perspectives and with a view to examining developments in other countries (principally the United States and Japan) from a specifically European angle of vision.

A complementary relationship to European Institutions ...

The relation of the European Communication Council to other European institutions is in three respects a complementary one. Firstly, we have built upon the important and successful work performed by European institutions such as the European Commission, the European Audiovisual Observatory, the European Information Technology Observatory (EITO) and the European Institute for the Media (EIM). Our activity is in turn intended to direct more attention to these institutions. Secondly, the work of the ECC is complementary to the European Union in that the ECC is financed entirely by private funding – not a single EU-ECU has been spent on this volume. Visible independence from economic and

political interests is thus a central objective in the work of the ECC. Thirdly, the ECC reports are complementary to official EU reports in that the ECC is not obliged to reflect Europe as a political entity – neither the Europe of the EU in its entirety, nor the range of interests and perspectives of the member states.The European Communication Council treats its subject matter independently of these considerations, thus giving it the opportunity to produce stimulating thought that is international, interdisciplinary and rooted in an intellectual exchange between theory and practice.

... but an independent academic process and an intellectual exchange between theory and practice

The Internet Economy and E–conomics

The first ECC report, "Exploring the Limits – Europe's Changing Communication Environment," contained a number of views, analyses and suggestions relating to the Internet. Yet the future of the Internet was by no means clearly marked out. "Internet years," as the rule of thumb in the communications industry has it, "are like dogs' years: they pass seven times as quickly." Today, no area of the economy can escape the pull exerted by networking. The significance of the Internet is becoming apparent everywhere, not just in the predictions of research institutes on e-commerce but in new work procedures in a whole range of industries as well as in everyday life.

The significance of the Internet

Yet companies, it seems, are finding it quite a challenge. The potential inherent in the Internet is clearly something that is particularly difficult to tap, and it appears to entail higher risks than is normal. Frequently it is the unknown newcomers in the Internet who open up new markets at breathtaking speed or become feared rivals to the long-established giants. It seems as though the traditional laws of classical economics no longer apply or have only limited validity, as though the rules, strategies and perspectives in the Internet business are completely different from the ones we have known up to now. As these new rules will prove decisive to competitiveness in virtually all areas of the economy, but especially in the media, information and telecommunications industries, the European Communication Council has geared this second report entirely to the question of which economic laws will determine the Internet's future development.

New developments in the economy and in economics

This report was written in cooperation with numerous scientists who were integrated into the work process in a variety of ways. First of all, the speaker of the ECC (Axel Zerdick, Freie Universität Berlin) invited four independent scientists to participate as fellows of the European Communication Council, determining the basic core of the discussion. The collaboration of Arnold Picot (Ludwig-Maximilians-Universität München), Klaus Schrape (Prognos AG and Universität Basel), Esteban López-Escobar (Universidad de Navarra) and Roger Silverstone (London School of Economics) meant that the report has benefited from exceptionally interesting specialists who have won international recognition for their contributions to the media economy. On the basis of the joint discussions held, a second group then wrote the various texts in the report, in the process bringing in many new ideas of their own. As managing editors and authors,

The team of authors

Alexander Artopé, Klaus Goldhammer, Ulrich T. Lange and Eckart Vierkant also went way beyond their duties as writers, providing crucial support to the development and realisation of the ECC report at every point and in every way.

Selected texts by invited contributors to shed particular light on specific aspects

A third group to contribute to this report are those further specialists whose thought-provoking ideas have left their mark on our work in the most diverse of ways. In some cases the collaboration took the form of personal conversations and shared discussions which brought substantial and exceedingly valuable new dimensions to our understanding of the area. Selected texts from Robert X. Cringely (journalist and author of the book "Accidental Empires"), Bill Gates (with a memorandum published by Jim Carlton of the Wall Street Journal), Kevin Kelly (Executive Editor, Wired Magazine), Mario Martinoli (Senior Consultant, Databank Consulting), Nicholas Negroponte (Founder of the MIT Media Lab, Cambridge), as well as Carl Shapiro and Hal Varian (University of California at Berkeley) are intended to shed particular light on certain aspects of the theme.

Additional intellectual input

The resulting report was marked substantially by the intensive discussions that came as the fruit of a (for a scientific project) relatively elaborate working method. Along with a series of work meetings between fellows and managing editors in which the broad outlines of the project and the form of its gradual realization were drawn up, the ECC in 1998 organized two larger workshops in Berlin and one in Berkeley. Many sincere thanks go once again to all the participants (listed on page 6) as well as to everyone else who took part in the discussions. A special thank you goes to Uta Hartleb and Armin Doll (Schömann Corporate) as well as to Thomas Lehnert (Springer-Verlag), responsible for the report's perfect professional appearance. The English version of our report has gained considerably from additional suggestions by Takashi Nakayama (NTT New York). The inspired translation by Rupert Glasgow (Barcelona) and the congenial reading by Jess Thacker (London) will help you enjoy the different European flavours of the English language.

Our basic data are relevant and reliable ... for updated information we recommend the Internet

Chapter 6 ("Facts and Figures") was produced by Prognos AG. This represents the ambitious attempt to provide a self-consistent base of data bringing together in fifty pages the most essential European statistics. Here too the focus is upon the larger European states, the EU as a whole, as well as the United States and Japan. In a fast moving field like the Internet, a book must supply those facts and figures relevant for analysis. Updating them in print is a sisyphean task, however, and any reader inspired by our analysis to go beyond should use those sources on the Internet we have been benefiting from.

Chapter 1

The Internet Economy as a Strategic Challenge

1.1 Ten Theses Relating to the Internet Economy

1. The Digitization of Value as a Strategic Challenge

Reality is changing. The Internet economy is spreading to more and more areas of the national economy. This is based on a new electronic infrastructure that is speeding up the change from physical atoms to digital bits and making tried and trusted strategies increasingly ineffectual. Though only coming to light gradually, the consequences are radical.

Bits can be sold and kept at the same time: original and copy cannot be distinguished from one another. Marginal costs for the production of further copies tend towards zero. Storage space is unneeded: bits have no weight and move at the speed of light. They know no boundaries, and in a networked economy it is practically impossible to control or restrain their movements. The marketplace for bits is a global one (Negroponte).

2. Critical Mass as a Key Factor in the Networked Economy

The new network economy means the increasing networking of participants, infrastructures and objects both within and outside the media and communications sectors. The result is a new set of special economic rules (relating to critical mass, the setting of standards) that are essential knowledge if media and communications companies are to be successful. Otherwise they will find themselves navigating the network economy without a clue where they are heading!

Positive feedback from network effects turn previous market models on their heads. It is not scarcity but surplus that determines the value of a good. Only by achieving a critical mass within the shortest possible time can standards be set: the customer lock-in that is produced in this way is the prerequisite for market leadership.

3. Cannibalize Yourself, before Someone Else Does it to You!

The increasing networking of media and communications sectors leads to the erosion of traditional value chains. The motto "Cannibalize yourself, before someone else does it for you!" is a call for early action not just by media and communications enterprises. Digital markets deal in information products and information-intensive goods and services subject to extremely high network effects and economies of scale. Delays in market entrance do not go unpunished.

The reason for the shift to multimedia value added processes is not only lower transaction costs but also increased product diversity and greater media richness. Traditional value chains are being steadily undermined: it will be those quickest to gain access to the growing cash flows of the multimedia value chains involving the Internet who survive. The precondition for success here remains added value in the eyes of customers.

4. Giving Products Away as a Recipe for Success

The "Follow the Free" price strategy is a tenet particularly characteristic of the network economy. The first step involves giving a product away for free as a means of using network effects to build up a critical mass of users within the shortest possible time. In a second step revenue is then generated through the sale of complementary products such as upgrades or more powerful programs.

As this makes clear, moreover, competition in a network economy is first and foremost a matter of speed. A prime example of this is the policy followed by Netscape, who in 1995 grabbed a worldwide share of some eighty per cent of the Web browser market within six months by giving its Netscape Navigator away for free. What makes the "Follow the Free" price strategy possible is not only the chance of setting a standard but also the advantageous cost structure of digital products: marginal costs for production and distribution tend towards zero.

5. Competition and Cooperation through Value networks

The competition strategy of business webs demands a change of strategic perspective for media and communications companies. The focus becomes both narrower and broader than before: narrower, because the tendency is for firms to limit themselves to their core competencies, and broader, in that the formation of alliances comes to be seen as a key strategic weapon.

The business web strategy stems from the systemic world of information technology. The best-known example is the "Wintel" partnership: since the early nineties, Microsoft and Intel have been dominating their value network and thus also the profit margins in the computer industry. The basis for this success is the positive feedback enjoyed by a complementary system architecture able with increasing networking to expand as a standard. To learn from the computer industry is to learn to be successful.

6. Simultaneous Price Cuts and Increased Differentiation

Extreme specialization – i.e. concentration on core competencies – and worldwide networking mutually reinforce one another to form the basis for configurations of goods and services that are low-cost and at the same time geared to the individual customer's needs. This applies not only to information, media and communications products, but to goods and services of all types.

Global networking opens up new possibilities for specialization, for as is well known the size of the market is a key factor in determining the degree of specialization. At the same time, the networking of specialized suppliers allows the customer-oriented intermediary to provide a flexible configuration of highly differentiated or individual market products. In this way it becomes possible to combine two basic strategies up to now regarded as mutually contradictory: cost leadership (through specialization) and customer-oriented differentiation (through flexible configurations).

7. Product Differentiation through Versioning

Versioning is a new way of marketing digital content that applies the principle of product differentiation to media and information products. Their digital form makes it possible to vary this content in their particular characteristics (such as comprehensiveness, how up-to-date they are, their presentation form), thus conceiving and marketing them as a product line. This is easier and less expensive than ever before: product design at your fingertips.

An example of this is "PAWWS Financial Network," which offered real-time stock market information at a cost of $50 per month but charges only $8.95 if customers are prepared to accept a twenty-minute delay in the information they are given. The prerequisite for this is to identify the product characteristic that has a high value for some consumers but is relatively unimportant to others, in this case how up-to-the-minute it is. A product line is then designed by providing gradations in this characteristic.

8. Individualization of Mass Markets

New communication strategies aimed at specific market segments and individuals have become possible, as the Internet makes the exchange of information between suppliers and consumers quicker, more intensive and more international. The use of intelligent agents and collaborative filters in particular, as well as customer-oriented opportunities for interaction, are crucial for gaining a competitive edge.

In the Internet economy, effective one-to-one marketing will be feasible even for suppliers of mass media, in the form of individualized mass communication. This requires the combination of three instruments: a potential for interaction with customers, economical yet powerful databases, and the capacity to individualize goods and services (mass customization).

9. Traditional Regulatory Models Become Obsolete

Today's regulatory framework is becoming less and less adequate for the relentless processes of convergence and networking in the media and communications markets. Current regulations have thus become at best ineffectual, at worst counterproductive.

Decision-making competences are too diffuse and lacking in coordination. It is too complicated and takes too long to implement what is decided upon. In many cases this makes it impossible to achieve what needs to be achieved. In Germany in particular, the regulatory framework for digital broadcasting is splintered both by sector and vertically: the European Union, Federal Government, sixteen state chancelleries; the regulatory authority for post and telecommunications, fifteen Landesmedienanstalten (regional licensing authorities), the KEK (the Commission of Enquiry into Media Concentration), the Supervisory Committee for the Public Broadcasting Service, the KEF (the Commission of Enquiry into the Financial Requirements of the Public Broadcasting Service). This splintering has led to a lack of strategic orientation in the market participants and if anything has put a brake on market development. One of the tasks faced by regulatory policies of the future will be to make the jurisdiction simpler, more transparent and better integrated.

10. Main Challenges to Future Regulation

The globalization and convergence of the media and communications markets mean that the demands placed on the level(s) of regulation are more likely to increase than decrease in the future. In order to avoid creating a permanent strain, regulation must succeed in becoming at the same time both more abstract and more concrete and firmly based in experience.

The technological possibilities and market conditions in the digital world are changing far too fast for regulations to keep up with, let alone stay ahead. The impression is one of perpetual political failure. A way out of the situation would perhaps be a withdrawal from this intensity of regulation (i.e. deregulation), at the same time reinforcing a "strong" and effective regulatory framework as protection against infringements in fundamental matters, e.g. going beyond preannounced intervention thresholds. Strong regulation of this kind is indispensable for European and national fair trading law, questions of copyright, the protection of children and young people, data protection and consumer protection.

1.2 Perspectives

Work on this report is finished now, and we feel sure that we have been able to provide some new ideas for discussion. What, then, are the perspectives for the Internet economy, and what will the role of the European economy be in bringing them to fruition?

After a phase of intense Internet hype, expectations at present seem to be experiencing a slight downward turn. The gold-digger mood is slowly giving way to the rather obvious realization that not all investment in the Internet will pay off in the short run. For the time being at least, the sober truth asserting itself is that the markets connected with the Internet have to be developed like any other market with care and well thought-out strategies. Even so, the rules of the Internet economy are spreading ever more rapidly to other – in principle all – areas of the national economy. The question is not whether they will catch on, but how soon.

The rules of the Internet economy are spreading to all areas of the economy

One possible exception in this sobering-up phase comes in the form of so-called "concept IPOs," which are Internet companies based on promising concepts and high profit expectations whose shares are floated on the public stock exchange. These promise founders and capital investors in particular a high short-term return on investment, but here too the risk of failure cannot be ruled out. It remains to be seen how far the caution currently being shown by investors will establish itself as a trend, and to what extent it will continue to be possible for enterprising founders to convert their brilliant (or brilliant-seeming) concepts into stock market profits even when turnovers are slight and real profits negative.

The winners of the Internet rush may be similar to those of the gold rush

The analogy with the gold-digger mood is perhaps not wholly inappropriate: it was not just a few lucky prospectors who made a mint from the American gold rush last century but also the many traders able to sell the necessary tools and equipment. Most interesting in this context, however, were the profits made by those able to sell their goldmine shares to hopeful investors following the fantastic (and on occasion perfectly accurate) reports about them.

Preconditions for the Internet Economy to be Successful

To the detriment of the European economy in general, the development of the Internet economy in Europe will continue to lag behind that in America unless it finally succeeds in meeting a few basic requirements. In the competition with the United States, the main thresholds are infrastructure development, prices for access to communication services, and political insight into the all-embracing nature of the changes taking place. These make it imperative to establish Internet competences in all areas of society.

Internet access must be available everywhere, conveniently and cheaply

The realization that universal on-line access is to everyone's benefit is gaining acceptance too slowly and with insufficient practical effect. A number of central demands are at issue here: Internet access must be available everywhere, conveniently and cheaply. This is the responsibility of all institutions capable of supporting or putting these demands into practice, whether it be infrastructure enterprises taking a more forward-looking, customer-oriented stance or the State taking the comprehensive measures that are necessary. State projects up to now are here just the first (groping) attempts, pointing in the right direction but without yet going far enough to be effective. In German schools, not even the running costs for Internet access are covered at present, let alone expenditure on hardware and software updates. While testifying to a growing awareness of the problem on the part of those responsible, such programs at the same time betray a lack of realistic concepts and solutions for dealing with the situation as it currently presents itself.

The responsibility also lies with the "Rulers of the Waves." The key field of competition in the on-line world at present is infrastructure. The availability of bandwidth will become the decisive criterion in global competition, regardless of the technological means by which these bandwidths are realized. The rapid expansion of bandwidth in the telephone network is not an alternative to the use of cable modems, but in the long run must be seen as a complement to it. For both, the basic conditions have to be improved not just in the form of pilot projects but with concrete offers that function and can be afforded. Access possibilities to the mobile networks likewise require rapid increases in bandwidth.

To tap the substantial potential represented by smaller business enterprises and private on-line custom it is above all how the costs of access evolve that will prove decisive. What is essential here is not so much the one-off costs for technical equipment as the regular burden on the customer's budget that comes from paying for access to the Internet. The cost of on-line use for the private consumer in the USA is now insignificant, covered as it is by standing charges or advertising. In a number of European countries, not least Germany, however, telephone call charges still constitute a burden for a large part of the population. A different pricing policy in charges for on-line sessions would be of advantage to both parties: customers would benefit through the lower costs of Internet use, while suppliers would gain from the huge increase in time spent on-line and the greater use made of their digitalized networks.

The Future of the Internet

In the future the use of global infrastructures will be taken for granted as something no user will have to pay for and no supplier will be able to charge for. The focus of multimedia strategies will thus shift to attractive content, which coupled with on-line access is what generates and justifies cash flow. Up to now too much attention has been directed at technical developments, riveting for technophiles but of relatively little interest to the majority of consumers. Few people care, for example, what sort of technology their TV set is equipped with. It is easy, by contrast, to get people interested in additional functions and applications.

Consumers don't care about technology

The task thus faced by producers of content is to realize their core competency in content-creation and conceptually and technically to transfer it to a new medium. The willingness of customers to make optional use of the interactive potential of the Internet offers should not be underestimated here: interactivity and passivity are not inevitably opposed to one another, but distinct manifestations of a more general behaviour pattern that varies for each consumer according to the situation, in much the same way as the choice between information and entertainment.

Interactivity and rational passivity

As a consequence, the future will see not only a lot of different (fixed and mobile) transmission channels, but also a number of distinct receiving appliances (with completely different screen sizes), whose use will depend on the purpose and situation in question. In all such appliances multifunctional technical possibilities and above all their situation-specific usage are in principle conceivable. The use of teletext in Germany, for example, thus shows that even in television interactivity has already gained acceptance. New projects like Web-TV can then be viewed as "enhanced TV" – i.e. as television supplemented and augmented by improved facilities for interaction.

Enhanced TV

It nonetheless seems to make sense in the long term to draw a functional distinction between the television set and the PC and between their different spatial collocations. The distinct spheres of usage (relaxation or work) also determine the design and alignment of the equipment, and in each case

supplementary uses can then be integrated according to class and price range: TV use on the PC, Internet use on the television set. The provision of signals via telephone, cable, satellite or terrestrial broadcasting reception is here perfectly conceivable in different combinations. Similar functional mixtures between the original sphere of usage and optional supplementary uses may also prove of value in other fields, especially in the mobile sector: with car radios and navigation systems, mobile telephones, pocket organizers and games consoles. What all these developments have in common is that the question of one type of appliance or use being superseded by another is now beside the point. What have come to the fore instead are the opportunities opened up by the greater number of different appliances, their enhancement through supplementary functions, and above all the different versions of information and entertainment content.

The US have made a head start, but Europe is in a strong position

We are currently at an early stage in the Internet revolution. The greater part is yet to come. In spite of the USA's head start, Europe is in a strong position, though it has up to now failed to take advantage of it. In comparison with the United States or Japan, Europe can be seen to have formed good connections with the worldwide Internet development, which it should be able to maintain. An encouraging indicator here is the fact that European software companies are more and more frequently opting to develop their programs in Europe but then have the marketing and distribution for the world market carried out in the United States.

It is thus up to the European communications industry to spur on the development of the Internet economy with a new self-confidence and the requisite measures and activities in order to gain a lasting advantage in the global market. This applies to business enterprises and regulatory authorities alike. With this report and the organizational and strategic recommendations it contains, the ECC hopes to make a (modest) contribution to this process.

Chapter 2

Fundamental Principles of the Network Economy

The focus of this report is an economic analysis of the media and communications sectors, or in other words media, telecommunications and information technology (IT). The following remarks are more than a mere description of the market reality in these sectors, but also an attempt to provide an insight into the most fundamental aspects of the way they function. For this purpose, three specific concepts are introduced that will facilitate a brief overview of the key economic features of the media and communications sectors.

Three basic concepts

2.1 systematization of possible forms in which revenue flow can be generated;

2.2 presentation of the notion of value chains as a tool for the structural analysis of the media and communications sectors;

2.3 problems associated with attempts to characterize revenue sources and potential.

2.1 A Classification of Revenue Types and Revenue Models

Among the most important decisions faced by a business enterprise are the questions of how to raise the revenue to finance business activities and how much to aim for. These decisions can as a rule be subdivided into two categories: decisions of general principle concerning revenue types and models, and decisions concerning pricing policy. The field of pricing policy is treated exhaustively in the management and marketing literature.

> *"Pricing policy comprises…all measures relating to sales strategy that serve to fix and impose the monetary remuneration paid by buyers for the products and services offered by a company"1*

Pricing policy decisions

Pricing policy decisions thus relate fundamentally to the fixing of prices for a product that is being offered. This setting of prices can be based on production costs, on demand, on the behaviour of competitors and/or on interdependencies within product lines. When the prices fixed differ at distinct points in time or in distinct market segments, then it is a price strategy that is being devised. Price strategies concerned with the development of prices over a period of time are known as penetration pricing and skimming strategies and are further elucidated in section 5.1.3. Price strategies concerned with fixing different prices for distinct market segments are known as price differentiation strategies. The differen-

1 Diller 1991, p. 20f.

tiations may here be in terms of temporal criteria (as with telephone rates that vary according to the time of day), spatial criteria (as with theatre seats that vary according to row), the customer group in question (as with magazine subscriptions for students or working people) or other considerations.

Revenue types and revenue models

Yet this broad field of pricing policy is logically preceded by an area of decision-making that has so far received little attention either in theory or practice: the area of decisions concerning the revenue types and revenue models to be applied. For a media concern, for example, the question thus arises whether revenue should be generated in the form of subscriptions, individual transactions or through advertising. Only once this decision has been taken can the respective prices for subscriptions, individual products or advertising times be fixed. A magazine publisher is thus faced at the outset with the decision of whether to produce a freesheet financed exclusively by advertising revenue or to fall back instead on a mixture of advertising and sales revenues. Not until this choice has been made can the price for advertising pages or the retail price fetched by the magazine at the kiosk be determined. The diagram below provides an illustration of this two-stage decision process.

Fig. 2.1: The two areas of revenue decisions

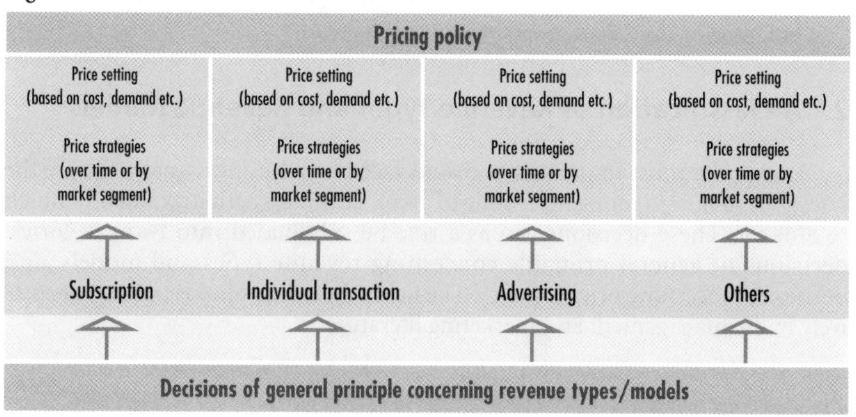

In spite of the relevance of these considerations, no one has thus far provided a structural classification of revenue forms for the media and communications sector which could be used by companies in the financial planning of their range of offers. The reasons for this are clear enough: no such classification has been necessary, as the market itself has provided each and every media product with a limited "corset" of possible revenue forms. There has been little scope, and just as little need, for decision-making in the field.

The onset of the Internet has brought drastic changes to the scope for decision-making, with the addition not only of another profit window of content but also of further revenue models available to the producer of content. Whereas

SOURCE
Fig. 2.1: ECC 1999

revenue forms for media products were formerly determined as constants once and for all, it is now possible for different forms to be combined as variables irrespective of the earlier constants.

"Construction kit" of revenue types

Given this context, the following diagram is an attempt to identify the various revenue forms in the media and communications sector and to analyse the connections and divergences that exist between them. A systematic overview is thus presented which can be viewed as a sort of "construction kit" for putting together the distinct revenue forms.

Fig. 2.2: Systematic overview of revenue forms

Revenue forms				
Direct			Indirect	
Usage-related	Non-usage-related		via company	via state
	One-off	Recurring regularly		
Individual transactions *by* quantity *by* duration	connection fees licence fees special receivers (e.g. decoders)	**subscription** **licence fees for public broadcasters** other standing charges	**advertising** datamining commissions other forms	**via state**

The diagram contains four revenue types to be systematically differentiated from one another, and which are described in more detail in the following section: direct usage-related revenue; direct non-usage-related revenue; indirect revenue from companies; and indirect revenue from the state. This approach further reveals five basic forms of revenue types specifically in the media and communications sectors, highlighted in bold type in our diagram: transaction-based revenue, subscription fees, licence fees for public broadcasters, advertising revenue and subsidies. Essential to these five revenue types as basic forms are not systematic distinctions but rather their frequent use in the media and communications sectors and the underlying importance that is thus attached to them.

Five basic types

This analysis of the relations between the individual revenue forms is based first and foremost upon a systematic distinction between direct and indirect forms of payment. While direct revenue types are obtained straight from the user of the product that is provided, indirect revenue types take into account that the use of media and communications services is often not paid for directly by the consumer. The revenue, instead, stems from third parties with some sort of interest in the consumer making use of the media and communications service in question.[2]

2 Underlying the concept of the indirect revenue form is the notion that the indirect refinancing of media and communications services by companies or the state means that the resulting costs are transferred back to the end consumer in the form of higher product prices or taxes. Even though it is here important to take into consideration divergences between the aggregated and the individual perspective, this understanding seems suitable to the conceptual framework here.

Source

Fig. 2.2: ECC 1999

Transaction-based revenues

In the second classificatory stage, the direct revenue forms are subdivided into three further categories. Firstly, payment can be based on the quantity or duration of the service used by the customer. Examples of this kind of usage-related revenue forms include the charges for database services (where payment depends on quantity) or the charges made for telephone calls (which depend on call duration). In both cases the payments are determined by a specific usage parameter relating to the transaction in question – quantity or duration – and together they comprise the first basic form of transaction-related revenue type.

One-off connection fees

This usage-related revenue form can be contrasted with one-off and regularly recurring non-usage-related revenue forms, characterized by a flat-rate payment method instead of an individual transaction charge. One-off non-usage-related payments include connection and licence fees as well as purchase prices for special reception devices (such as decoders) which are required for the services in question but do not have any original value in themselves.

Subscription

Subscription payments and standing charges on the other hand are payments that are repeated at regular intervals. Subscriptions play a major role in the refinancing of media and communications services and can thus be characterized as a further basic form. The remarkable feature of this revenue form is the optionality of actual usage: regardless of whether and how frequently a consumer makes use of content, the same flat rate is charged. In the print sector subscription fees are charged for taking a newspaper and in the TV sector for taking a pay-TV channel. They are founded on a voluntary purchase decision which then entails regularly recurring payments.

Licence fees a special form of subscription

Like subscription fees, licence fees for public broadcasters are a form of regularly recurring non-usage-related payment. Where they differ from subscription fees, however, is in the concrete obligation to pay them, making them seem more like a tax or a duty. Given their considerable significance in the media sector, they can be viewed as a further basic form.

Indirect forms of payment

With the indirect revenue types, the distinguishing factor is who actually makes the payment to the media and communications suppliers. The most common alternatives are payments by other companies or by the state. The reasons that companies may have an interest in financing or subsidizing media and communications services for the user can be subdivided into three categories. With the first of these, financing by advertising, the interest of the company whose product is being advertised is in the user's attention. Suppliers from the print or electronic media sector use their services to generate this attention in the form of circulation or reach, and in this way obtain the revenue required.

Datamining

Companies also have a particular interest in possessing more or less detailed data concerning consumers. Media and communications suppliers use their services to generate these data, which they then sell to third parties. These data can also serve to optimize their own media and communications services. This

revenue form is known as datamining and has gained particular import in connection with the possibilities provided by the Internet.

Commissions

A third incentive for companies to finance the media and communications sector may further arise through the direct mediation of transactions. This is the case, for example, when the user of a particular media and communications service in the Internet is guided by a banner ad to the on-line bookshop Amazon.com and then proceeds to buy a book there. Whenever this happens, Amazon.com pays a share of 15 per cent of the transaction proceeds to the media and communications supplier who placed the banner ad. This revenue form is known as a commission.[3]

Advertising

On account of the essential role it plays – in the media industry in particular – advertising constitutes the fourth basic form. The possibilities here range from a "subsidised" sale, as for example with periodicals (except freesheets), to a transmission of content that is apparently free for the customer, as with broadcasters financed solely by advertising.

Subsidies

Indirect income that comes from the state is designated subsidisation. As a rule this is offered for services considered worthy or deserving of the support. An example of this indirect revenue form is the provision of free Internet access to students by universities. The refinancing will in this case take the form of state subsidies to the universities. This method of refinancing media and communications services is distinct from the rest in that the payments in question depend on the existence of a subsidy claim and are thus not dependent on the product meeting with consumer approval. They consequently run the risk of supporting services with low market utility. In principle, income from subsidies does not constitute revenue. Even so, it is one of the basic forms for financing business activity and plays a crucial role in the media and communications sector.

The significance enjoyed by the five basic forms described above varies considerably from one segment of the media and communications sector to another. Financing by advertising is the dominant revenue form in the private broadcasting sector, for example, whereas it plays no part in information technology.

Combining different types of revenue

The deployment of a single revenue form on its own is a comparatively rare phenomenon within the media and communications sectors. Transaction-based payment depending on quantity bought has established itself as the norm with books for example,[4] but in contrast to newspapers and periodicals, three distinct revenue forms are in use. Some users pay subscription fees, others pay with each individual transaction, and on top of this comes the industry's advertising income.

The combination of revenue forms and the weighting given to them contribute greatly to the success or failure of media and information goods. In

3 Datamining and commissions as sources of revenue will be explained in more detail in section 4.2.3.
4 Other forms only play a subordinate role (subscriptions for book clubs, for instance) or can be regarded as unsuccessful attempts (advertising in paperback books).

"Step two: synthesise basic revenue types into combined revenue models."

Two target dimensions

fixing an appropriate mixture of revenue forms for refinancing individual media and information goods, there are two target dimensions to be taken into account.

Cost structure

From the companies' point of view, the structure of the costs incurred is a primary consideration in the organization of revenue forms. In the case of periodically recurring services for example, subscription fees tend to be preferable. With technical reception devices on the other hand, the high development costs make it desirable to refinance and amortise the investments in the shortest possible time, so in this case it makes more sense to use one-off payments.

Consumer perception of value

The second dimension in the fixing of revenue forms consists in the consumers' perception of product value and their patterns of usage. In general the connection is that the more concrete the perception of value, the greater the willingness to pay. For this reason transaction-based revenue forms, where the relation to usage is most clearly in evidence, seem appropriate for media and information products. An example here is the transaction-based payment for films, whether in the cinema, on video or as pay-per-view.

Patterns of usage

Patterns of usage relate above all to consumers' established habits. Consumers are willing, though with varying degrees of acceptance, to allow their TV programmes to be interrupted by commercials if this means that the service is provided free of charge. With books it is difficult to envisage this revenue form being successfully applied except in special circumstances.[5] In general there are few consumers willing to let the relaxing pleasure of reading a book be interrupted by an advertisement every two sides.

The choice of appropriate revenue forms thus takes place in a field of conflict between financial cost constraints and consumer notions of product value. The structuring of the revenue mixture requires a balanced consideration of these two dimensions. For the product to sell well there must always be acceptance and an accompanying willingness to pay on the part of the consumer. The choice of appropriate revenue forms must thus take its orientation first and foremost from the consumers' perception of product value, but covering the costs incurred can only be neglected in the short term, not the medium or long term.

The result, on the whole, is that one-off non-usage-related revenue forms have a restricted field of applicability, limited for example to telephone connection fees or the purchase of certain reception devices. The choice of this revenue form seems straightforward for products that do not have any value in their own right but are necessary for the use of services with a high and lasting subjective value. This explains the success of games consoles with young people or satellite receivers with TV viewers who do not have access to cable networks.

The situation is substantially different with products where the difficulty of assessing the offer in question results in vague perceptions of product value. It was for this reason that the one-off-payment revenue form that was chosen by the German digital Pay-TV DF1 for its set-top box, the d-box, proved so problematic.

5 Advertising is perhaps feasible (and acceptable to students) in the case of textbooks, if could be substantially reduced.

The high novelty level made it difficult for potential buyers to judge the programme offer. The DF1 offer thus lacked a clear concrete value that would have justified a payment of the equivalent of $ 400 for the acquisition of the d-box.

This example provides a clear illustration of the field of conflict between the two dimensions. According to press reports, considerations of cost were at the fore of DF1's pricing policy, which was striving to refinance the high investment costs incurred by the d-box as quickly as possible. The consumer perspective by contrast was not paid sufficient attention. One possibility in such cases is the cross-subsidisation of the device's development and production costs by higher monthly subscription fees, as consumers associate these with a concrete concept of product value. DF1 has now taken this step.

Conflict between cost structure and consumer value perception

2.2 The Concept of Value Chains

The concept of value added stems originally from the field of national accounting. The gross value added refers to the value of that part of total production by all sectors of the economy which in the period in question is provided for so-called final use (as a consumer good, item of capital expenditure or export) and has thus definitively left the domestic production cycle. If all the production values identified for all sectors of the economy were to be added up, some of the figures would be duplicated, since a part of total production enters into the production process of other companies as preliminary products and services. The gross value added of a sector of the economy can thus be calculated by subtracting preliminary products and services from the production value. The sum of the gross value added for all sectors of the economy is the gross domestic product (GDP).

Gross value added: a concept from national accounting

The value added rate indicates the ratio of gross value added to production value. It serves as an indicator of the division of labour in an economy or the number of production stages passed through by the goods in the production process. At the enterprise level, it corresponds to the concept of production depth.

Value added rate

From the microeconomic or business-management perspective of the individual company, the "value chain" is an instrument of competition-oriented business analysis that serves the development of strategies. A company's value chain comprises its value activities together with the profit margin.

Value chain

Fig. 2.3: Porter's model of the value chain

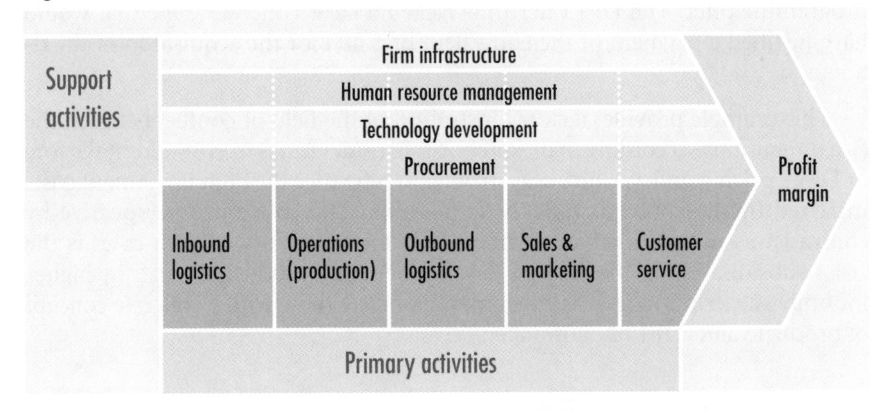

Value activities are processes that create value for the customer. The profit margin is the difference between the total value or return and the costs of carrying out the value activities in terms of the purchase of inputs, machinery, human resources, technology and information.

Value chain system

It is important to emphasise that any particular company's value chain is connected up with the preceding and subsequent value chains of suppliers and customers. Together these constitute the value chain system of a product, market or branch. Only the "frictionless" interplay of the individual links in the value chain guarantees that sufficient value is created for customers to regard it as worth making the purchase. If this fails to come off, then no value added accrues, or only an insufficient one, and the process of market creation has been unsuccessful. In the extreme case of a failed market introduction, the individual company may even be faced with a negative value added.[6]

Links to other economic concepts

One advantage of the concept of the value chain is that it opens up a number of possible links with economic concepts that have up to now co-existed unconnected with one another. Examples include:

* innovation-diffusion theory with its distinction between system and network goods;
* the concepts of product cycles and market life cycles;
* classificatory distinctions between different types of goods (such as search goods, experience goods and credence goods);
* various production, sales and pricing concepts (such as skimming and penetration pricing strategies), approaches to transaction costs or the analysis of competition and cooperation strategies.

6 See also the Web page of the Dead Media Project: www.eff.org/pub/Net_culture/Folklore/Dead_Media_Project/

Source

Fig. 2.3: ECC 1999

The following basic model of the value chain will be our starting point for the analysis of markets:

Fig. 2.4: Basic model of the value chain

Investment → Production → Sales → Billing/collection → **Customer**

This representation has the advantage of identifying the customer as the target of all the supplier's value added activities. The model originally stems from the field of material goods, where business activities are portrayed as a simplified sequence of investment, production, sales and billing or collection.[7] Starting from this basic model, the following sections[8] will present adaptations of it showing the market structures particular to the individual media and communications sectors. This should ensure that despite the considerable variations between the respective media and communications value chains there is an underlying structure that they will be seen to have in common.[9]

Variations of the basic model

2.3 New Markets – Old Data: The Problem of Assessing Revenue Potential

The development of the media and communications markets depends in large measure upon how far companies are able to finance their value added activities. For this it is essential that the total product being offered should be accepted, purchased and utilized in sufficient measure by the end consumers. Only thus will there be financial returns for the suppliers, allowing investments to be refinanced and profits made. Revenue sources and potentials consequently play a central role in the development of the media and communications markets.

Demand from three economic sectors

There are in principle three segments that count as potential revenue sources. Expenditure on media and communications services is decided on in private households (consumer sector), in companies (business sector) and in public or state institutions (public sector). In all three demand sectors the scarce resource of spending power – which is the biggest possible revenue potential – can be employed for a whole variety of products, and competition thus arises between media and communications services and other possible uses such as accommodation, clothing or travel. For the decision-makers, this expenditure choice is coupled in varying degrees to three other equally scarce resources: time, acceptance and competence.

7 In general it should be pointed out that the schematic representation of a value chain fails to take into account that in reality the production and consumption of a product may often take place simultaneously. This is especially the case with the integrated provision of services (see Kleinaltenkamp 1997). In section 4.2.4 these problems relating to the order of value chains are discussed in more detail.
8 See sections 3.1.3, 3.2.3 and 3.3.3
9 Another set of concepts relating to the idea of value chains is discussed in section 4.2.4.

Source

Fig. 2.4: ECC 1999

In all three demand sectors, the level of revenue potential and how it develops over time is intricately bound up with the economic situation as a whole:

Private households

* Media and communications expenditure by private households constitutes a part of private consumption in general and as such is directly dependent upon the standard of living in the economy as a whole. Its absolute level as well as the share of disposable income it represents vary according to the income bracket concerned. Comparisons on an international scale also reveal considerable differences from country to country.

Business

* Media and communications expenditure by businesses relates in particular to the goals of internal and external communication, for example in the form of advertising. There is a close connection here between expenditure of this sort and the overall business climate. It has thus far not proved possible to detect the anticyclical uncoupling of advertising expenditure from the economic situation that has frequently been proposed in management theory. What can be ascertained in the long run is that an increase in the intensity of competition is linked to a steady rise in the intensity of communication (which is the proportion of a company's total spending that goes on communication expenditure). Communication expenditure in the business sector can only be comprehensively assessed at the level of individual countries in particular years. Figures allowing international comparisons are only available for advertising expenditure.

Government

* Media and communications expenditure by state organisations varies with the revenue from taxation and thus likewise depends on the progress of the economy as a whole. Such expenditure is as a rule neither systematically nor comprehensively recorded.

Estimates and their underlying problems

Estimates of future developments in revenue potential generally take figures from the past and present as their point of departure. There are several problems connected with this.[10]

* National accounting still lacks a standardised definition of the media and communications sector as a branch of the economy.
* This makes it hard to draw comparisons on an international level, as the individual countries use different definitions in making their studies.
* Depending on whether a narrower or a broader definition is favoured, divergences thus emerge in the basic figures used, leading in turn to discrepancies in the predictions for future development.
* In addition to this, it is not always possible to make the necessary statistical distinctions between the media, telecommunications and information technology sectors on the supply side or between the consumer, business or public sectors on the demand side.

10 See Seufert/Schrape 1997, pp. 69-109

* Finally, there is not an adequate statistical allowance for areas of substitution. Substitution relationships within sectors exist, for example, between expenditure for print media, video and letter post. An example of substitution relationships between sectors is provided by expenditure for seminars, business trips, parcel post and trade.

Differing models, different sources

Predictions relating to revenue potentials have up to now tended to be turnover predictions pertaining either to one of the three partial media and communications markets (media, telecommunications or information technology) or to a multimedia sector still lacking a precise definition. The statistical basis has often been unclear, moreover, making it difficult to compare results.

Large estimates, small empirical base

What all predictions nonetheless have in common is that there will be great growth in the entire media and communications sector. What remains uncertain, by contrast, is what this growth will feed off. If all the growth predictions for the individual media and communications segments are added together, the total arrived at far exceeds any realistic order of magnitude justified by the expected growth in private consumption or in the economy as a whole. For these predictions to be fulfilled there would have to be a substantial shift in expenditure from areas of substitution in other sectors to the media and communications sector.

Statistical groundwork has to be standardised

If any such redistribution is to be reflected in the statistics, however, a corresponding change in the conventions of registering them is required. As these considerations make clear, a standardised definition and method of registration must be worked out as quickly as possible to trace the development of the media and communications sector and the segments within it, one that can be used both by national statistics authorities and by European Union institutions. Only thus will it be possible to ensure that studies between sectors and between countries will provide a valid source of comparisons.

Three types of statistical sources

Germany serves as a concrete illustration of the statistical obstacles to be overcome. There are in principle three sources of statistics available for estimates of revenue potentials from the private household segment.

National accounting

* National accounting: the official statistics utilised for national accounting, however, do not include an individual media and communications sector. Companies concentrating principally on the manufacture and sale of media, telecommunications and information technology products are assigned to different areas of the economy. The definition and statistical assessment of a separate media and communications sector in its own right is thus extremely difficult to structure. This applies to the supply side and the demand side alike. For parts of certain sectors only rough estimates are feasible. Changes in the systematization of the branches of industry (such as the rearrangements carried out in 1979 and 1993) have additionally entailed new categories and time divisions.

Microcensus

* Microcensus data: the microcensus records the total spending of private households by means of random sample surveys analysed in terms of three distinct classes of household. This rules out an assessment of media and communications expenditure representative of all households. The statistical methodology thus puts limits on the reliability and completeness of the figures reached.

Realised demand

* Revenue potentials can in the end best be measured by the demand realised for media and communications products. The data in question are provided by trade associations as well as a number of individual studies. It is worth noting, however, that the three demand segments (the consumer, business and public sectors) are as a rule lumped together, making a differentiated analysis of the individual segments impossible. As well as this, there are variations in the underlying principles of the survey. On some occasions it is the consumer prices that are used, while on others it is the trade prices.

Such are the main problems thrown up by attempts to estimate the revenue potentials of private households.

A model for consistent estimates of the revenue potential

In 1995, Prognos together with the German Institute for Economic Research (Deutsches Institut für Wirtschaftsforschung) worked out a definition for the media and communications sector.[11] According to this definition, media and communications spending by private households in Germany in 1992 amounted to a total volume of 103 billion DM. Using the same terms, this volume had by 1996 nominally risen 26 per cent to a total of 130 billion DM.

By comparison with the 1995 figures, somewhat weaker growth to a level of 235 billion DM is expected for the year 2010. This takes into account the slowing down in economic growth in the last few years. The table below presents the figures for the private household sector and also shows the structural changes in media and communications spending.

11 See Seufert/Schrape 1997, pp. 69ff.

Fig. 2.5: Media and communications spending by private households in Germany

	1992		1996		2010		Index 2010
	abs.	%	abs.	%	abs.	%	(1996 = 100)
Total in billion DM							
Media spending	76	74	86	66	136	58	158
Communications spending	27	26	44	34	99	42	225
M&C spending household	103	100	130	100	235	100	181
Per household in DM							
in DM/year in DM/month	2.886 240		3.523 294		6.152 513		175 175
Private households (in millions)	35,7		36,7		37,0		

Given the problems of drawing distinctions between individual segments, estimates of financial potentials are only possible at the aggregated level of the media and communications sector as a whole. As estimates of business and public sector demand can likewise only be approximate, moreover, the following analysis of the three media and communications sectors – media, telecommunications and information technology – will dispense with a more detailed consideration of revenue potentials.

Source

Fig.: ECC 1999

Chapter 3

3. The Media and Communications Sectors: A Review Looking Forward

3.1 Kings of Content – An Analysis of the Media Sector

The following section concentrates on an investigation into the media sector as the most conspicuous area of the media and communication industry. Along with the classic print media (newspapers, magazines and books), this includes the field of cinema and video films, the sound-carrier branch, as well as the broadcasting sector with radio and television.

Broadcasting on center stage

Ever since the day in 1906 when Reginald A. Fessenden in Brant Rock, Massachusetts, first succeeded in using radio waves to transmit music and human voices, the advance of the broadcasting industry throughout the world has been unstoppable. With television since the fifties coming to assume the place it now has in the mass of private households, broadcasting has today developed into what is economically and socially the most significant and probably also most influential medium in modern industrial societies. For this reason the focus of the following section is on the broadcasting sector.

While the other M&C (media and communications) sectors, telecommunications[12] and information technology,[13] are centred more on infrastructural aspects, the media sector has come to be synonymous with the dissemination and creation of content. The manufacture and marketing of content is difficult, because they can rarely be integrated into industrial production processes, often stemming as they do from the creative inspiration of individual artists. The success of such content, measured by the attention they attract, is almost impossible to predict. But what is unusual about media products shows itself not only on the production side but also in their consumption. Although used on a private basis, media content has the character of a "public good." Especially in film and television production, therefore, it is characterized by a perfect nonrivalry in consumption. This means that, regardless of the number of acts of consumption, a specific media content such as a cinema film undergoes no reduction in physical value. No matter how often and in how many different countries a media product is consumed, the producers are not faced with any additional distribution costs apart from the one-off production expenditure.[14]

Content: nonrival good

Even so, or perhaps for this very reason, the media are typified by an almost insatiable need for interesting content, as this is what underlies the value chain. It is for content and not technology as such that people use media infrastructures and that enterprises can set up value chains in the media sector in the first place.[15] So content continues to be king, not only for the media but for the entire M&C sector.

12 See section 3.2
13 See section 3.25
14 Kruse 1994, p. 185; Owen/Wildman 1992, p. 25.
15 See section 3.1.3

3.1.1 Structure and Development of European Media Markets

The most important determining factor in an analysis of the media market is without doubt the number of suppliers in Europe. In the case of the European TV market what is striking is that the number of channels in the market as a whole is continuing to grow. While the markets in Western Europe are beginning to reach their saturation point, the East European states offer further growth potential.

Fig. 3.1: Number of national TV channels in Europe

	1993	1994	1995	1996	1999
France	14	14	16	18	83
Germany	14	14	18	19	52
Italy	12	12	13	13	47
Spain	13	13	13	13	57
UK	52	52	64	80	113
EU5*	105	105	124	143	352
EU15**	159	162	189	214	442

Substitute Products and Substitution Potential

What can also be ascertained is a threat to the traditional broadcasting sector from substitute products.[16] The field of computer and video games in particular shows great potential, as does the Internet and especially today's WorldWideWeb with the individualized mass communication it offers. Private on-line use is already substituting above all prime-time television use. Although technical standards like the Real Player have only officially been on the market since February 1997,[17] more and more Websites are offering audiovisual content as a streaming process: by mid-1998 more than 5000 radio stations worldwide were already on-line.

Print Media vs. Electronic Media

According to the traditional classification, which draws a distinction between print media and electronic media,[18] the following categories can be assigned to the media sector:

Fig 3.2: Traditional distinction between print media and electronic media

Print media	Electronic Media
Newspapers Magazines Freesheets Books	TV programmes Radio programmes Feature films Video programmes Sound carriers

16 See section 5.1.5
17 Preliminary versions like Real Audio were available on the market as early as 1995.
18 In this overview, all new media such as video and computer games as well as all other multimedia products are left out of the account, as these will be treated separately in chapters 4 and 5.

Source

Fig. 3.1: 1993 to 1996 data see European Commission/Screen Digest, April 1998, p. 96; 1999 data ECC 1999 based on European Audiovisual Observatory 1999, p. 144f.

* National channels (terrestrial, cable and/or satellite); not including digital bouquets, local or regional channels and windows, or channels for foreign markets

** EU5 = France, Germany, Italy, Spain, Great Britain (the large EU markets)
EU 15 = the 15 EU member countries

Fig. 3.2: ECC 1999

Five main types of revenue

As was shown in chapter 2, the media refinance themselves predominantly from three sources: advertisements, subscriptions and fees, and sales or transactions. Numerous combinations and special forms of the above are possible. In the following section the market structures of the media sector will be analysed by focusing on three sample areas: private broadcasting as an example of a media market financed by advertising; the pay-TV market as an example of financing by subscription; the pay-per-view sector as an example of a transaction -based media market.

Market Structures of Media Financed by Advertising: Commercial TV

Advertising: the triangular relationship

All the media financed by advertising are characterised by a triangular relationship between the producer of content suppliers or content-packager[19] (the medium), the advertising customer and the recipient. Within this configuration both the recipients and the content themselves form only a secondary point of reference for what is in fact the crucial business relationship between the medium and the advertising customer. Any medium financed by advertising develops its performance primarily with respect to the interests of the advertising customer, selling its audience reach as a form of potential attention. Media financed by advertising are thus typified by a systemically inherent double address, which means that performance is directed primarily at the advertising customer and secondarily at the recipient. For the spectator the secondary product, the programme, is seemingly free of charge.[20]

Fig. 3.3: Triangular relationship in media financed by advertising

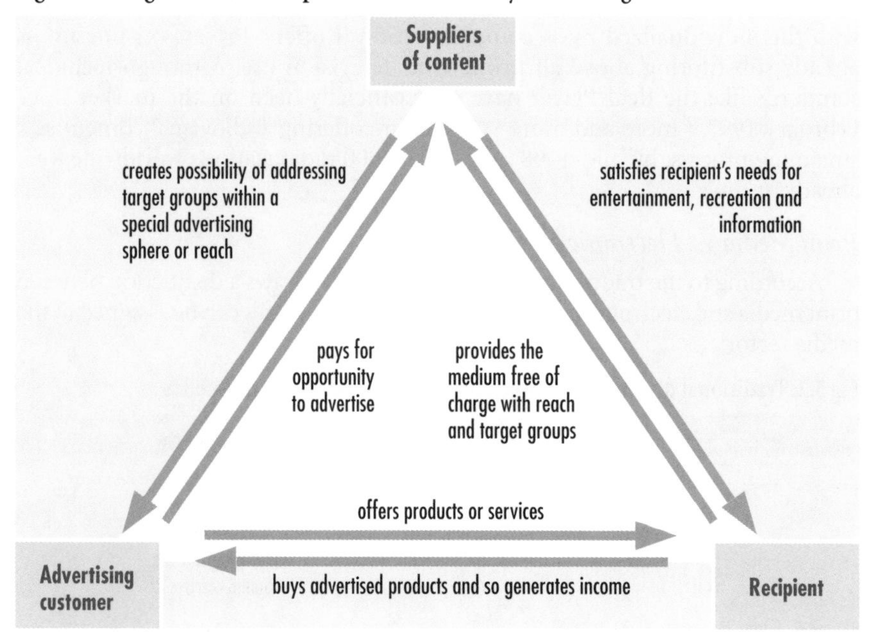

19 Primary producers can be understood to include, for example, film studios or musicians and their record companies who create content independently. TV or radio stations that use already existing films and music for their programmes can be said to be using available content rather than producing their own, fitting what is to hand into their programming schedule according to the interests of the recipients. These can thus be called repackagers or aggregators. It goes without saying that radio and TV stations can equally well become producers of content in their own right, as soon as they start producing their own programmes or, for example, TV movies.

20 The programmes are only seemingly free of charge because the enterprises that refinance the programme-suppliers through their advertising will then transfer the costs to their products. At the same time, it is often the case that the purchaser of a product (i.e the programme-refinancer) is not a user of the corresponding programme.

Source
Fig. 3.3: Goldhammer 1998, p. 149; Hass/Frigge/Zimmer 1991, p. 692

As the finance flow between the advertising customer and the content-supplier gradually reaches the limits of its growth potential, it makes sense for the media to open up new, direct financing possibilities, this time between the recipient and the content-supplier, for example through pay-TV or pay-per-view. In this way it is possible to tap new, direct finance flows, which – by means of versioning concepts[21] within the value network (see Sec. 5.2.3) – present the consumer with what is perceived as sufficient added value for it to count as an additional offer.

Pay-TV opens new (direct) revenue sources

A further step in a consideration of the structure of the media market also incorporates the acquisition market, since most media-suppliers function less as producers of content than as repackagers or aggregators, buying a variety of content which they then assemble and arrange in accordance with the interests of their recipients. Seen in this light, the media-suppliers function as mediators between the interests of the three relevant markets in which they operate.

Media as intermediaries

Fig. 3.4: Relevant constituents of a media market financed by advertising

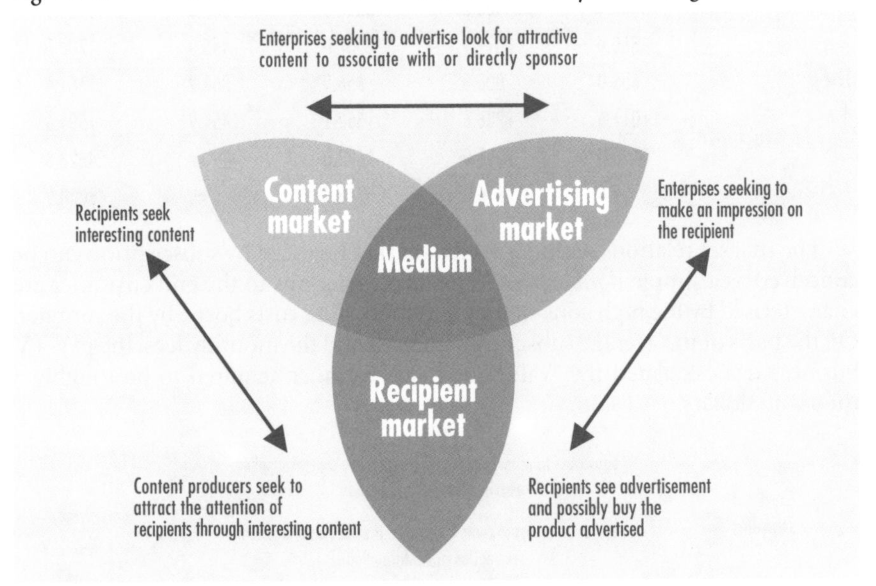

Admittedly, the individual relations vary in intensity. The deep connection between the advertising and the recipient markets is clear. But in the acquisition markets too (for content, i.e. commissioned productions, broadcasting rights, licences for merchandising, as well as staff, transmission licences, equipment and capital), the quantity and quality of the recipient market attained is important because of its direct effect on potential revenue. The interrelation between the advertising and the acquisition markets only seems to be insignificant. On the one hand, there is the effect of indirect forces in the form of attractive rights from the

21 Versioning refers to the supply of an information product within one specific stage of value added, which is then differentiated according to various criteria (such as how up-to-date it is, how broad its relevance).

Source

Fig. 3.4: ECC 1999

acquisition market, which in turn give rise to attractive advertising areas in the advertising economy. However, the dimensions of the relationship via the content-supplier between the acquisition market and the advertising market come to light most revealingly when it is taken into account that it is the proceeds from advertising or sponsoring that make the purchase of photo and film rights possible in the first place.

Market Structures of Media Financed by Subscription: Pay-TV

Great Britain and France leading in pay-tv development

Media areas financed by subscription, such as the pay-TV market, have undergone the most significant development in Europe, particularly in France and Great Britain, in recent years. This is true for the number of subscribers and the corresponding turnover.

Fig. 3.5: Number of subscribers to pay-TV channels in Europe (in thousands)

	1994	1995	1996	1997	1998
France	4437.0	5423.3*	5300.9*	10831.0*	18925.9*
Germany	860.0	1011.9	1337.0	1455.0	1700.0
Italy	655.0	800.0*	856.9*	868.2*	1161.4
Spain	1011.6	1286.6	1562.1	1464.9*	1593.6*
UK	6796.0*	9095.0*	1623.0*	na	24628.0*
EU5	13759.6*	17616.8*	25286.9*	na	48008.9*

The market relations within a media market financed by subscription can be conceived as a simple *double bond*. The direct relations to the end customer are characterised by the high consultancy and marketing costs borne by the supplier. On the basis of the average subscription length and the monthly fees, the pay-TV business has calculated the "value" of each customer acquired to be roughly a thousand dollars.

Although the use of such direct financial sources may seem attractive, along with the high marketing costs there is also a substitution problem. The higher the number of available channels financed by advertising and providing content seemingly free of charge, the less willing customers are to reward the added value of an additional service or bouquet with a subscription (see fig. 3.6).

Source

Fig. 3.5: ECC 1999 based on European Audiovisual Observatory 1999, p. 175f.
* data not available for some channels
Fig. 3.6: ECC 1999

Fig. 3.7: Number of TV channels financed by advertising vs. pay-TV penetration in Europe

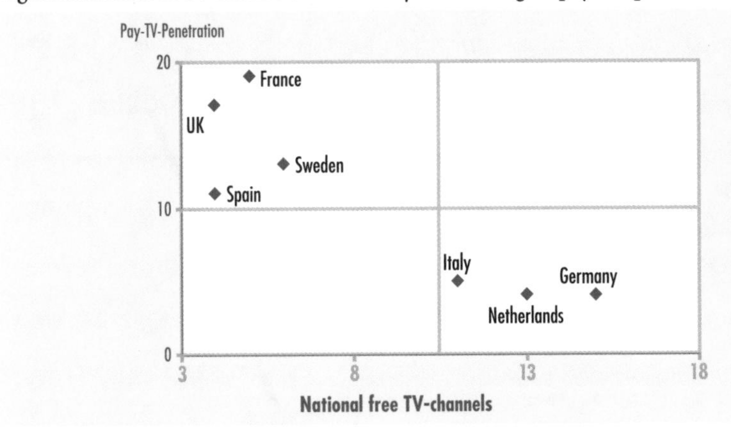

It will be difficult to resolve these problems without drawing distinctions of time and quality between the various profit windows in such a way that attractive content can first be seen exclusively on pay-TV, and only becomes available for broadcasting by advertisement-funded channels after a significant time-lapse.

High number of "free" tv channels – low attractiveness of pay-tv

Market Structures of Media Financed by Transactions: Pay-per-View (PPV)

Pay-per-view (PPV) broadcasters, which are the focus of this section, are not the only media to be financed by transactions. In particular the cinema and records markets come into this category. Here the revenue stems more or less solely from direct transactions between customer and supplier. In the print market too there is a high proportion of transaction-based financing. Books are almost entirely refinanced through their sales revenue. In the case of newspapers and magazines the percentage of sales within total revenue varies considerably, depending on the type of media and on the country.[22]

Skilful packaging is the challenge

Rather than offering access to a continuously running service for a fixed monthly subscription price, PPV-TV sells its customers individual programmes or items at constantly varying prices.

Its content generally consists of cinema films, erotic films, concerts or sporting events. Access to a PPV service is sold for a particular period of time (a film, a "night" or 24 hours), with several single events sometimes being packaged together in a "season ticket" in the case of sporting events. While sporting events are considered by most recipients to be obsolete as soon as they are over, certain films have a very much longer "life-span", on occasion up to 50 years and beyond. It is therefore in the interests of the pay-per-view station to generate a maximum of attention through a skilful packaging of content with varying life-spans.

22 For daily newspapers in Europe, sales revenue is between 33 and 66 per cent of total revenue. For magazines, sales revenue varies between 0 and 100 per cent. See ECC 1997, p. 289

Source

Fig. 3.7: Künstner 1997

Fig. 3.8: Development of pay-per-view channels and services in Europe

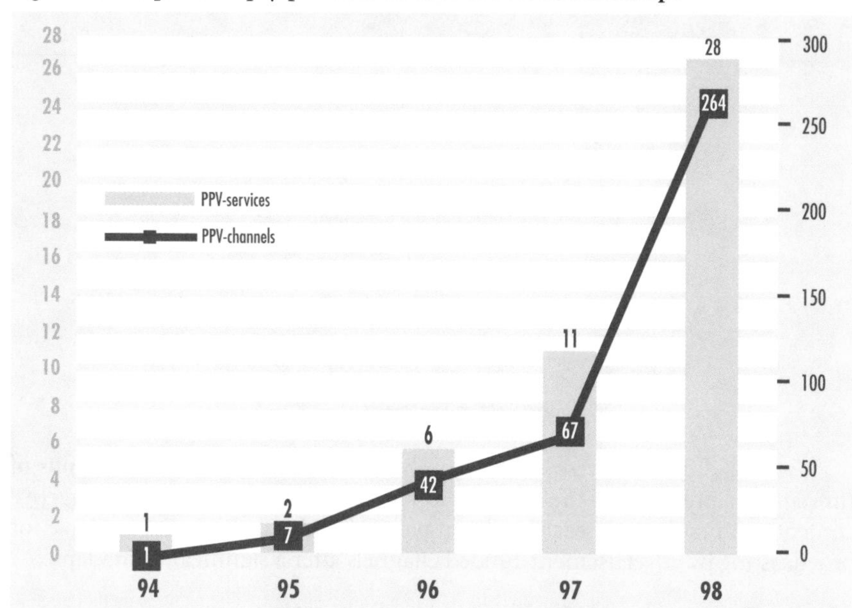

Different developments

Since the inception of the French broadcaster Multivision in 1994, the number of pay-per-view services and of channels offered has grown constantly. In Europe in 1998 there is now a total of 17 pay-per-view services offering roughly 200 channels. There are PPV broadcasters operating in most European countries, including Denmark, Germany, England, France, Holland, Spain and Sweden. A lot of the suppliers are already transmitting digitally. In 1999 there will be PPV services starting in Italy, Norway and Finland. In terms of PPV turnover, French broadcasters are currently responsible for around three quarters of the European market volume. The French share is expected to fall to about 30 per cent by 2002, however, as other European states each reach a market share of around 10 per cent, or even as much as 25 per cent in the case of England.[23]

As with TV financed by subscription, the PPV market is characterised by a double bond between customer and supplier. The difference is that instead of a one-off decision to buy a subscription to the pay-TV channel the customer is perpetually having to make the choice of whether or not to go for a particular offer (except in the case of season tickets). While this can be an advantage for the customer who does not have to make a long-term commitment, it may at first seem a drawback from the perspective of the supplier, as there is a high need for marketing in order to mobilise purchasers. The situation is made more difficult by the fact that it is often not possible to make a purchase on impulse, because up to now the customers of most PPV services have had to go to the trouble of ordering their programmes through a hotline. Even so, PPV has proved attractive

23 See Screen Digest, Jan. 1998, pp. 9 - 16

Source
Fig. 3.8: European Audiovisual Observatory 1998, p. 179; data for 1998 corrected, see European Audiovisual Observatory 1999, p.182

to suppliers too, allowing them to maximise income from top events, since the consumer is generally willing to pay more for pay-per-view than for borrowing a video owing to the topicality and/or attractiveness of a particular event as well as the convenience.

As (apart from the feature film sector) there are not enough high quality events to fill a PPV programme schedule completely, the great majority of suppliers in Europe have to fall back on Hollywood films. By means of output deals, the studios secure a share of 40 to 50 per cent of the final sales price, making this additional profit window financially interesting to the studios too: "In other words, the studios get a bigger piece of a bigger pie when consumers see their films on PPV instead of renting a video."[24]

Market Development: Digital Television

It is very frequently the case that the development of a market is determined by the success of its technical innovations and their availability. The media sector, like other M&C sectors, is at present dominated by a trend towards digitisation, both in production and distribution and in the reception of content. It is not only the book and music trade whose evolution has manifested itself through the Internet and the digital "books"[25] that have been appearing since mid-1998. Digitisation is the decisive factor in the market development of the media sector in general and above all the television sector. Digital television will entail a transformation in the technology of the entire range of production, contribution, distribution and reception, enabling potential innovations and cost reductions to be realised on all levels. Digital television allows a four to sixfold increase[26] of channels as compared to analogue satellite, cable and terrestrial transmission channels. For this reason and others, the possibilities opened up by digital broadcasting will result in radical changes in the television sector, at least provided they can be widely carried through in the market.

Digitisation is the most important development in the tv sector, and on all steps of the value chain

In Europe the era of digital television began as early as 1996 with the Italian station Telepiù, just one year after the US station DirecTV. Although by the end of 1997 there were in total around one hundred channels being digitally transmitted via satellite, digital stations have so far found it difficult to make a widespread impact except in France, especially since digital TV has come on to the market predominantly as pay-TV or pay-per-view. Though this is attractive from the point of view of suppliers, since the digital technology facilitates the running and promotion of this kind of offer, it fails to make full use of digital TV's possibilities in terms of either content or technology and in this sense is certainly impeding the introduction of digital TV into a broader market.

24 Screen Digest, Jan. 1998, p. 11
25 See Wired, July 1998, pp. 98-104
26 Originally, an increase by a factor of 5 to 10 has been discussed. However, compression technology seems to have come to a practical limit for the time given.

Fig. 3.9: Market penetration of digital TV in the USA and Western Europe

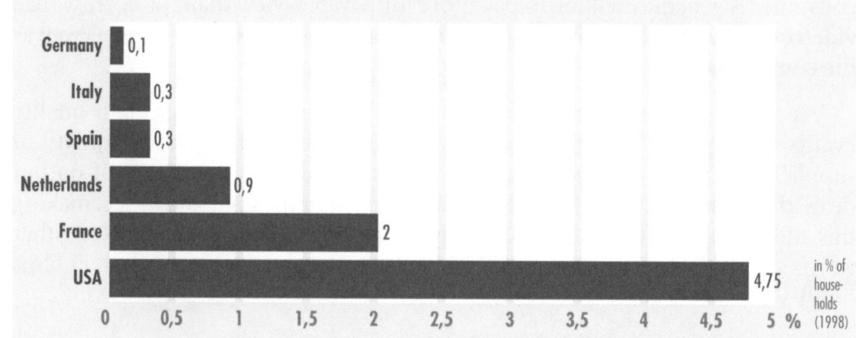

Evolutionary development

The innovation curve described by digital television will thus probably show an evolutionary development. Even this will presuppose a willingness to learn and to cooperate on the part of both the participants in the market and the political authorities. It also remains to be seen whether consumers will show themselves willing to invest in the new technology and buy digital TV receivers in large quantities.[27] While the US regulatory authority FCC has set the year 2006 as the fixed date for the switch from analogue to digital broadcasting, in Europe deadlines of this type are not yet decided on.[28]

For this reason there has not been a common strategy for the implementation of digital TV in Europe, not only as regards deadlines but also the differences in standards. Yet decisive factors for its success also include the distribution infrastructure in the individual countries concerned, the range of analogue TV channels (whether financed by advertising or licences), as well as the extent of analogue pay-TV.

Two stage purchase decision: consumer willingness to pay is more difficult to estimate

The consumer's willingness to pay for digital TV at present is likewise limited. Spending potential tends to be small, while digital TV calls for high conversion costs within the framework of a two stage purchase decision (technology and channel) with a high risk factor as far as the technological development and respective standards are concerned. The strategies so far adopted by suppliers have hardly taken account of the network effects of digital TV's systems technology. They have chosen a high-price strategy inadequate to consumer needs and failed to offer any significant added value in terms of content. In view of the competition from television financed by advertising, this would have been essential for its successful adoption. So far, the chances that digital TV in Europe will function as a motor for growth and innovation are thus best seen as slight.

For a further investigation of the media sector and its financial market mechanisms, the focus will now shift to the possible revenue types and revenue models used to finance the services provided by the media supplier.

27 See Schrape 1997
28 For Germany, in August 1998 an digital broadcast industry initiative has suggested a definite deadline for this switch in 2010.

Source

Fig. 3.9: Schrape 1997; Prognos

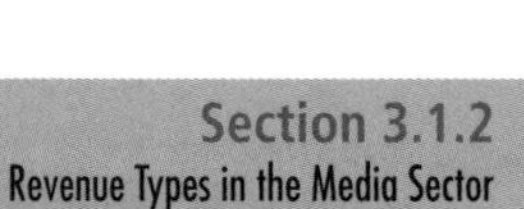

3.1.2 Revenue Types and Revenue Models in the Media Sector

An analysis of the possible revenue types and models used to finance media services reveals a whole range of variants. Their importance and the frequency of their use varies considerably from one European country to another. In the context of this investigation the main emphasis will thus be placed on the financing of the broadcasting sector.

In principle a distinction can be drawn between state revenue sources (divided into sources directly related to broadcasting and those only indirectly related) and revenue from other sources (divided into market-related sources and non-market-related sources).

Fig. 3.10: Revenue types for the financing of broadcasting

Revenue types			
State		Non-state	
Directly related to broadcasting	Indirectly related to broadcasting	Market-related	Non market-related
User fees media based income-based appliance-based	**Electricity surcharge**	**fees** individual fees subscription	**Donations**
Taxation of private suppliers	**Product taxes on acquisition of equipment**	**Advertising** commercials sponsoring teleshopping infomercials Product placements	**Member contributions**
Proceeds from auctioning of transmission licence	**Allocations from public funds**	**Other income** cofinancing licensing merchandising bartering rental of assets interest stock revenues	

Source

Fig. 3.10: ECC 1999; based on Pethig/Blind 1998, p. 28

State sources directly related to broadcasting include all types of user fees, the taxation of private suppliers[29] and – particularly in England – the proceeds from the auctioning of transmission licences. Sources indirectly related to broadcasting on the other hand include electricity surcharges, product taxes on the acquisition of equipment and direct financing from public funds. As the following table makes clear, the range of market-related revenue types is much more diverse and much more extensive, including financing not only from fees and advertising but also from other sources. Non-market-related, non-state revenue sources include donations, member contributions and private resources.

Market-related revenue types more diverse and more extensive

As has already been described, the three most widely used revenue models for the broadcasting sector are financing by advertising, financing by subscription or licence fees, and financing by direct transaction. There now follows an analysis of the revenue determinants in each case.

Revenue Determinants of Media Financed by Advertising

The revenue of media financed by advertising lends itself to analysis by means of a revenue cube. This latter is a product of the interdependent triangular relationship described above between supplier, recipient and advertising customer in the media market financed by advertising. It outlines the variables that the broadcasters themselves can decisively influence. The most important determinant here is the reach achieved, or rather the access to recipients as the true product of a medium financed by advertising. Opposed to the product "attention" is the intermedial and intramedial competition for recipients.

Revenue cube: three dimensions that can be influenced by a broadcaster

Fig. 3.11: Revenue cube: a model of revenue determinants in media financed by advertising

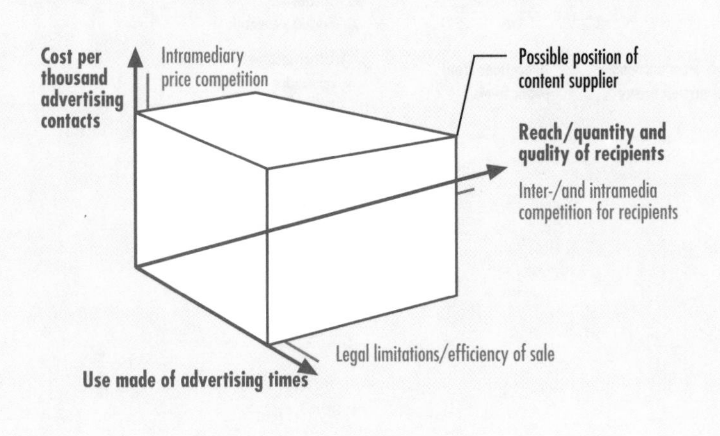

29 As is the case, for example, with radio stations in the German state of Lower Saxony.

Source
Fig. 3.11: Goldhammer 1998, p. 151

A second determinant that is directly dependent upon the others is the "cost per thousand," (CPM) which is a relational price for the attainment of one thousand recipients, a factor that is a crucial criterium in the choice made by the advertising customer. A third determinant, in turn dependent upon the others, is the use made of advertising times. These three determinants allow for a detailed analysis of the competition within a broadcasting market.

Revenue Determinants of Media Financed by Subscription

From the outset a clear distinction should be made between subscriptions that have to be paid in the form of a licence fee (or, as in Greece for example, a surcharge paid on top of the electricity bill) and voluntary subscriptions. In the television sector, voluntary subscriptions are mostly taken out in return for access to a pay-TV broadcaster.

Given the economics of the situation, it can be demonstrated that a single fee-financed supplier with several channels at its disposal should be in a position, ceteris paribus, to offer a programme schedule that is both diverse and in the optimal public interest.[30] Yet while it is certainly the case that the revenue model based on licence fees facilitates the production of programmes with a greater cultural and political emphasis, a concomitant danger of indoctrination or intellectual force-feeding is always present.[31] In the case of nearly all fee-financed broadcasters, their political influence on the programmes is harmfully high. This leads to the problem that the programme planning is thus also oriented towards the interests of the institutions that decide on the fees and not the interests of the recipients.[32]

Licence fee: theoretically a potential for an optimal range of programs, practically too much potential of political influence

With pay-TV offers (in other words, with voluntary subscriptions), by contrast, the relationship between the interests of supplier and customer is optimised: the subscriber wants the most attractive programme schedule possible, while the supplier seeks to maximise revenue. The revenue is thus directly dependent upon the content offered (feature films, sporting events and erotic films), its marketing and availability, and the price of the subscription. Thus, the advantage of pay-TV is in the economically optimal relationship between suppliers on the one hand, and those consumers on the other who are willing and able to pay. However, all other people are excluded from using the formerly public good TV, and this makes pay-TV as a system "Pareto-sub optimal", i.e. it does not lead to maximum social benefit.

Pay TV: direct relationship between supplier and subscriber interest, but "pareto-sub-optimal"

Revenue Determinants of Media Financed by Direct Transaction

In the case of media financed by direct transaction, the revenue level is significantly influenced not only by the pricing and marketing policy but also the availability of appropriate access technology as well as the spectrum of content that is obtainable. While the average price per film with the American PPV-suppliers sunk from around five dollars in 1987 to under four dollars in 1997, the user costs in Europe are often higher. Although the absolute prices in France for

30 See Steiner 1952
31 It needn't even go as far as the credo of the chairman of the German NWDR (a public broadcaster from the fifties), Adolf Grimme, who called for the transmission of "what the people ought to want to see."
32 See Pethig/Blind 1998, p. 78-79. Theoretical economics considerations are, unfortunately, not reflected in the actual situation.

example are twice as high as in Spain, PPV films in both countries are offered at between 30 and 55 per cent more than the respective local average video rental prices.[32]

Fig. 3.12: Prices for PPV-films in Europe

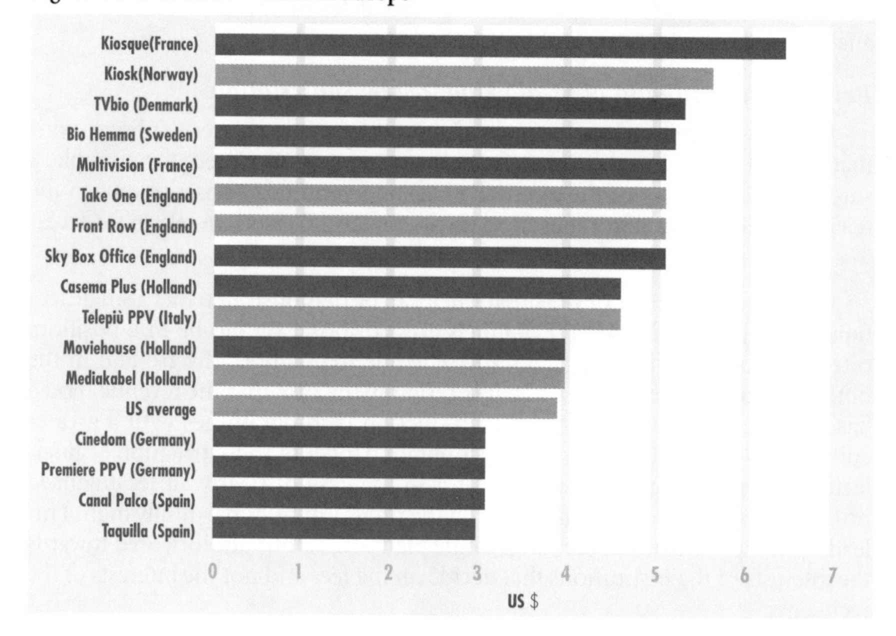

Pay-per-view pricing: one third more than video rentals

The market price for PPV is thus determined by the video rental business. At the same time, however, PPV prices contain an extra charge for the additional advantage of direct access to content without having to make a special trip to the video library for the film and later having to return it.

The pricing itself is likewise one of the decisive revenue determinants, as is made clear by an example from the world of sport. The first three fights involving Mike Tyson following his release from prison each lasted only a matter of a few minutes, yet the prices were between 40 and 50 dollars. Spectators were thus at times paying more than 0.5 dollars per second! For the subsequent fights there was a substantial drop in the number of PPV customers.

32 Screen Digest, Jan. 1998, p. 13

Source

Fig. 3.12: Screen Digest, Jan. 1998, S. 12. Highest prices as of November 1997; darker bars refer to actual prices, lighter bars to prices planned for future offers.

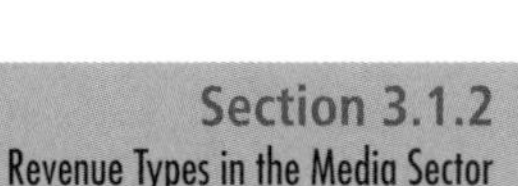

Fig. 3.13: PPV in the USA: Did Tyson box too quickly?

Fight: Tyson vs. ...	Number of PPV-viewers	Length of fight in minutes	Price on US pay - channel	Cost per second	PPV turnover
Peter Mc. Neely	1,7 million	1:29	$ 39,95 – $ 49,95	$ 0,44 – $ 0,56	$ 68 m – $ 85 m
Buster Mathis	1,4 million	8:32	$ 39,95 –$ 49,95**	$ 0,09 – $ 0,12	n.a.
Frank Bruno	n.a.	6:50	Free on Fox		
Bruce Seldon	1 million	1:49	ca. $ 40	$ 0,37	ca. $ 40 m
Evander Holyfield	n.a.	33:37	Up to 49,75**	$ 0,02	n.a.
Evander Holyfield	2 million	9:00	$ 60*	$ 0,15	$ 120 m

A pricing model for boxing

It was not until the fight between Tyson and Evander Holyfield that the supplier Cablevision offered the following pricing model: the first round was to cost 9.95 dollars and each further round another 9.95 dollars, until a final price of 49.75 dollars was reached after five rounds. All further rounds would then go uncharged. Fortunately for all concerned, this fight lasted more than half an hour, so both the viewers (paying 0.02 dollars per minute) and the supplier ended up getting their money's worth.

The turnover level, determined by the buy rates, is also however crucially influenced by the number of channels available, i.e. the choice of content available within a PPV station or a cable network, as well as the choice of call times.

Pay-per-view revenues depend on the number of channels on offer

Fig. 3.14: Buy rates and the number of PPV channels offered

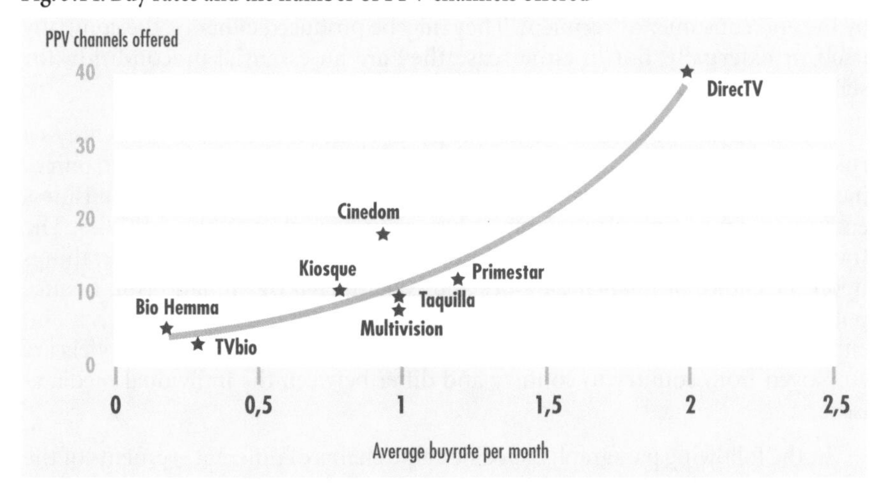

Source

Fig. 3.13: 1997, p. 312
* estimates
** depending on type of ordering (via telephone or by DSS remote control)
Fig. 3.14: Screen Digest, Jan. 1998, p. 12

Buy rates still relatively low

The greater the number of channels, the higher the buy rate, defined as the average number of purchases of a PPV offer per month in all households able to receive a particular PPV service. Analogue PPV with between one and six channels here achieves a buy rate of between 0.2 and 0.5; digital PPV with seven to sixteen channels achieves buy rates between 0.5 and 1.25; and digital PPV (as Near Video on Demand) with forty and more channels obtains buy rates of around 2.0 per month. Even so, there is not a linear relation between the number of channels and the buy rate.

The conclusion from experience with PPV offers up to now is as follows: the buy rate does not rise in proportion with the number of channels. Instead, the buy rate remains rather low even when a choice between a very high number of channels is being offered.

3.1.3 Value Chains in the Media Sector

In the terminology of economics, the media sector comprises the value-adding activities of enterprises concerned first and foremost with the production of (mass-)media content.

Value added in the media heavily depends on packaging and distribution services

From the perspective of a company producing media content, a successful value chain calls for further services or products without which it is impossible either to produce and distribute media content or to develop a demand for them. These services and products permit the manufacture and assembly of content, facilitate their reproduction and distribution, and make possible their utilisation by the end consumer or recipient. They may be produced either by the company itself or externally, but in either case they are an essential precondition for successful communication and thus the generation of value added.[34]

The necessary advance investments, production costs and overheads are as a rule refinanced from a number of diverse sources. Potential financing sources include the available income of the end consumers, the expenditure of business enterprises on communication and advertising, as well as public subsidies. The transformation of improbability into probabilities depends amongst other things upon the choice of financing modes and is hampered by limitations in finance potential. Financing by the end consumer can be either direct or indirect, and either related to media use or independent of it. Different financing models are employed from country to country and differ between the individual media at issue.

In the following paragraphs, specific value chains of different segments of the media market will be discussed.

34 This is an attempt to re-create a concept of production and distribution stages as seen from the perspective of system theory. The improbabilities of successful communication and value added must be overcome regarding the contact with, reception by, comprehension by and willingness to pay of the consumer.

Print Media

The typical value chain for the print media can be represented in functional terms as follows:

Fig. 3.15: Value chain for the print media

(1a) Acquisition of information	(2a) Editing	(3) Layout	(4) Duplication by printing	(5) Sales	(6) Reader service
(1b) Acquisition of advertising customers	(2b) Advertising				

The functions “acquisition of advertising customers,” “advertising” and “reader service” are not applicable in the case of books.

Crucial differences in the value chains of the other print media exist above all at the sales stage and are thus also closely connected to the mode of financing used. While freesheets are delivered free of charge and unsolicited and are financed exclusively by advertising, all other types of newspaper and magazine deploy a combination of various sales systems (individual delivery system or delivery to subscribers by post, retail sale at kiosks or newsagents) and modes of financing.

Print media have developed very different revenue models

In the different European countries there are diverse traditions of direct or indirect state subsidies or support given to the print media.[35] These may operate through price fixing with books, the reduction of VAT rates, subsidised postal rates or direct press sponsorship. In general the aim is the maintenance of diversity and quality in the provision of the press.

The sales revenue (together with state subsidies where applicable) generally only covers a portion of the production costs of press products. Advertising revenue from the sale of space is indispensable for the commercial survival of the vast majority of print media (the extreme example here being freesheets). But the refinancing of production costs through sales revenue alone (retail sales and subscription) would in most cases put too much of a strain on the consumer's willingness to pay.

Combining sales and advertising revenue

In system-theoretical terms, the (partial) financing of a medium through advertising can be interpreted as a successful structural coupling of distinct value chains. To take the example of the print media: success in the reader market, measured in reach or contact probabilities, increases the chances of success in the advertising market.[36] This means that the enterprises doing the advertising, who are interested in the quantity and quality of the media performance (the contact chance), are better able to exploit the convergence of the roles of recipient and consumer for their own value added.

Unlike the electronic media the print media are without an otherwise necessary link in the value chain in product form: the end equipment as a technical interface between the media offer adapted for distribution purposes

35 For example in Austria, France, Norway, Sweden, Germany, Switzerland

36 Contact probabilities are calculated by the monetised "currency" CPM (cost per thousand) and target group or contact qualities.

Source

Fig. 3.15: Sennewald 1997, p. 56

and the media-user. As far as the recipient is physically capable of reading, has the requisite linguistic or reading competence, and is attentive and motivated to read, then the previously bought print medium can be used directly and without any further investment.

Reader aspects

One striking feature of the value chain for the print media as shown above is that the reader – as the purchaser or customer, as the consumer or user of the media content – turns up in the "reader service" function or not at all. This is surprising for two reasons. Firstly, by using the medium's editorial content the reader creates the conditions for success in the advertising market (financing by advertising). Secondly, his acceptance and purchase of the product form the second financing base (sales revenue).

Apart from the reader or customer, there are further links / figures missing from the ideal-typical value chain that are indispensable for a print market to function. Particularly noteworthy are the connection with the staff acquisition market, advance payments in the printing sector, as well as co-operation with advertising agencies and with market and media research (indirect feedback).

Sound Carriers (Music Market)

The main functional elements in the supply-side value chain of the music market are shown in fig. 3.16

Music market: two steps of use and sales

Fig. 3.16: Main elements of the value chain for the music market

Authors and artists	Production management	Sales	First use	Second use
Financial security Remuneration	Combination of creativity financing production	Logistics marketing publicity	Sales through trade and clubs	Broadcasting (radio and TV)

The functional representation of the value chain misses out the end consumers as a target and resource potential in the value added with their various forms of media use (listening, purchasing, private duplication). As well as this, first and second use are as a rule simultaneous.[37] Second use in broadcasting (radio and television) is at the same time also an essential marketing requirement (promotion, publicity) for first use.

An expanded model of the value chain combines three perspectives: the functional, the legal and the institutional.

37 Promotional activities for new releases are today frequently carried out at the level of second-use (through broadcasting) even before first-use on the market.

Source

Fig. 3.16: ECC 1999

"The Internet and "music on demand" potentially dissolve traditional value chain structures."

Fig. 3.17: Expanded value chain for the music market

Creation of a work	Publication of a work Performance	Recording/ production Manufacture of sound-carrier	Duplication	Publication "Release"	Distribution (First use)	Transmission Public presentation (Second use)	Listening Purchase Private duplication
Composer **Songwriter** **Arranger**	**Record-company** **Artist** **Promoter**	Artist Artistic producer Sound technician/ studio Business producer (sound-carrier manufacturer)	Sound-carrier firm technical production		Trade clubs	Broadcasting company discotheques bars shops	**Customer/ listener**

Manufacture and use of sound-carrier

Regulatory implications

The expanded version of the value chain for the music market includes the consumer with his/her use activities but not the reciprocal relationship between first and second use. The classification of functions and legal categories also makes it clear however that the music market is dependent upon political and legal regulations to determine the relations between those involved on the supply side[38] and between these and the demand side (copyright). On top of this the classification of functions and institutional participants indicates the close structural coupling of the value chain for the music market with other value chains: those of the artists involved, the printing press, the broadcasting companies and other second users.

Internet and "music on demand"

To this day the value chain for the music market has been dominated by the sound-carrier firms. The market leaders here are all among the biggest international media concerns. Digital technology and multimedia pose a threat to the dominant position of sound-carrier firms within the music market's value chain. The Internet and Music on Demand (MoD) make it possible for artists to be independent from the music companies, while digital storage devices facilitate high quality copying. The potential is there to dissolve the market's traditional value chain structures.

Film

As regards the value chain for the film industry, it is important to draw a distinction between two closely interrelated aspects. Firstly there is the value chain from the perspective of audiovisual film production, and secondly the profit

38 And thus also the rules for contracts fixing the allocation of revenue and payments for second use.

Source

Fig. 3.17: ECC 1999

window chain for film rights. From the producers' point of view the fundamental structure of the value chain is normally represented as follows:

Fig. 3.18: Value chain for film production

Revenue return

Pre-production	Production	Post-production	Rights Management Library/Asset Management	Packaging Marketing Distribution	Delivery screening
Producer	Producer	Producer/ Service-provider	Owner of rights	Distributor	Cinemas/Broad-casting stations
Research activities Archive acquisition Preproduction Financing	Scriptwriting Direction Production Filming Controlling	Editing Effects Title Prints	Sale of rights Stock management	Advertising/ promotion PR Sale and distribution Synchronisation /copies Bundling	Cinema screening TV screening Video release etc.

Crucial role of the producer

One disadvantage with this representation is that it fails to do justice to the role of the film's prefinancing, which appears under preproduction but in fact has an essential "gatekeeper" function necessary for the value chain to materialize at all. A further fault is the absence of the end consumer, the cinemagoer, whose willingness to pay for a visit to the cinema[39] is decisive for the film's success or failure and thus also for its future value in the profit windows to come.

Asymmetric revenue returns

The advantage with the above representation, however, is that for the first time it not only shows the various stages that have to be combined in order to get a value chain successfully off the ground, it also contains the revenue returns that flow back. This opens up an additional dimension of analysis, concerning whether the (necessary) work performed at the individual stages of the value chain is also rewarded with a corresponding ("fair") allocation of the revenue. Long-term performance/revenue imbalances at the level of the individual categories of participants within the value chain (such as exist, for example, for European film producers) can prove to be an essential factor causing individual or branch specific value chains to fail to come off.

After successful production and marketing or promotion, the profit window chain for feature films begins with the cinema premiere and then typically takes the following course:

Fig. 3.19: Profit windows for film rights [40]

Cinema screening	Video sales/ rental	Pay-TV	Free-to-air television	Terrestr. vs. Cable Syndication	Re-licensing archive exploitation

39 Which depends on the script, cast, director, and PR and marketing strategies
40 In order for a value chain to be successfully established.

Source

Fig. 3.18: ECC 1999
Fig. 3.19: Financial Times, Charles Brown, p. 37 (Example: USA)

In Europe the syndication profit window does not apply. The order in which the windows follow one another is based on the principle of the gradual absorption of the exclusivity yields through degressive pricing for the end-consumer. For whoever purchases the film rights for use in a different value chain (for example TV or video market), the prices are fixed by means of the success indicator "gross box office takings" and the size of the sales area (see 3.1.4).

Value chains and profit windows

The value chain and profit window chain for film production are extremely closely interrelated. This applies equally at the level of performance and of financing. Attractive feature films are an essential and indispensable programming resource and revenue source for the profit windows to come. By the same token, film utilisation in the following stages of the chain both in the home market and abroad is becoming progressively more important as a basis for refinancing increasingly expensive film productions, which can nowadays hardly be financed by "gross box office takings" alone. In the case of so-called "output deals" the acquired utilisation rights are fundamental to costing for the joint financing of film productions by those involved in the later profit windows and/or banks and film-financing funds.

A high-risk setup

This is an example of the ways in which performance flows and financing flows in the value chain cycle or network of emerging markets can be, or indeed must be, linked up with one another in reverse or asynchronous mode. In such cases, the expectations of performance and success as well as the prefinancing precede the performance itself and the proof of its success (or not), creating a highly risky and implausible configuration.

Broadcasting (focus on television)

The value chain in classical broadcasting is characterised by the lack of any direct customer relation between content-provider and user, there being only an indirect and anonymous relation mediated by audience research. The representation of the value chain for traditional television thus generally consists of just four functional components:

Fig. 3.20: Simple value chain for traditional television

Content/ programme production → Programming/ broadcasting → Distribution → Receivers → **Recipient/ consumer**

Yet this fails to provide an adequate reflection of the complex relationships within the old television markets. It leaves out of account the structural links with the markets that come before, during and afterwards and are indispensable for the "functioning" of the television market. It also ignores the financing flows crucial to the economic workability of the market.

Source

Fig. 3.20: ECC 1999

To illustrate the structural couplings with the markets and services that come before, during and afterwards, it is helpful to split the production function up into home productions and outside productions and also to take the acquisition market into consideration.

Fig. 3.21: Value chain for television and structural couplings

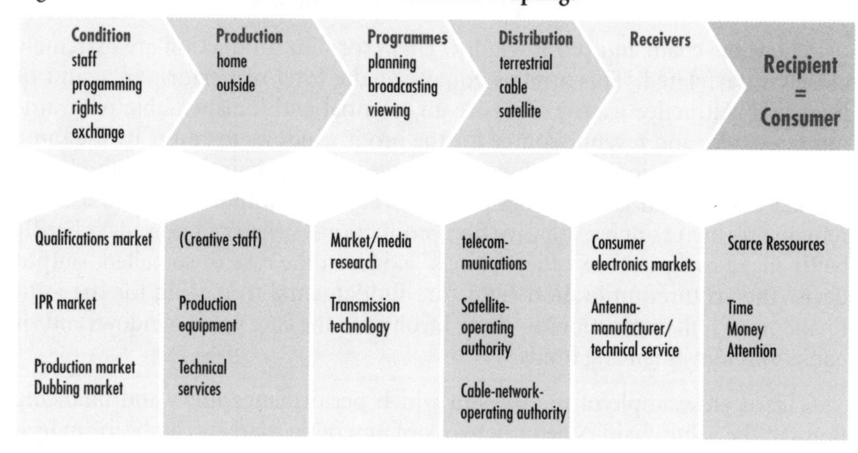

Mutually dependent elements

The introduction of private television in the eighties made it clear throughout Europe that individual links or functions of the value chain are mutually dependent upon one another or, to put it more precisely, presuppose and limit one another.[41] The market success of a television station, for example, depends directly on the technical reach of channels of distribution as well as on the availability of appropriate receivers in private households. Conversely, even when the technical reach and market share are small, programme planners will opt to buy competitively priced programmes from foreign acquisition markets rather than expensive home productions.

Political intervention and planning security

The economic viability of a TV market (in its initial stages) thus hinges decisively on whether it proves possible to establish "just in time" effective structural couplings with the markets that come before and afterwards. The TV market alone is often not capable of achieving this. In order to guarantee those active in these markets adequate planning security (and conversely to keep the investment risk within limits), political intervention (target setting, regulations, midwife-function, subsidies) thus becomes essential.

If the television system is considered from the perspective of financing, the value chain is shown to require further additions. Along with the functional performance flows, the financing flows must also be represented:

41 These constraints can only be dissolved through concerted developments for which time is necessary.

Source

Fig. 3.21: ECC 1999

Fig. 3.22: Value chain functions and payment flows in the (German) television market

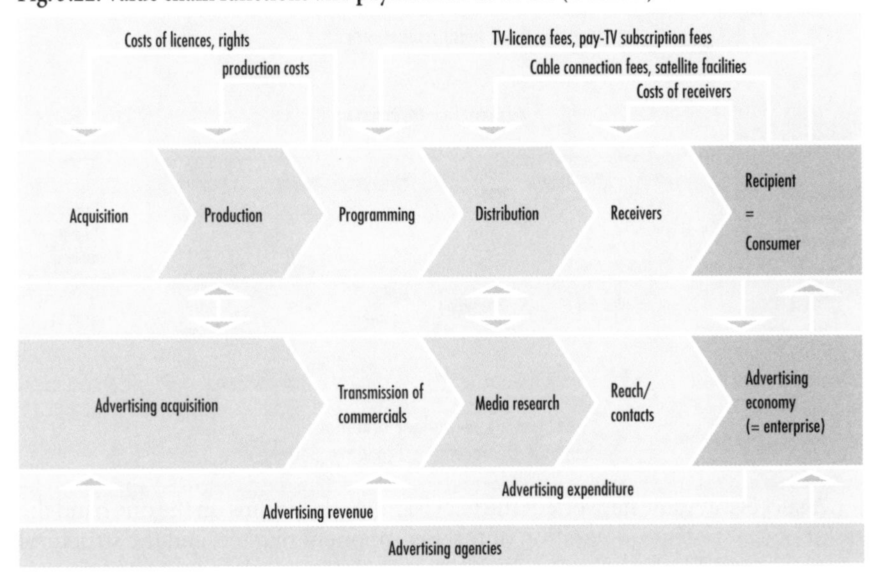

German tv market: insufficient structural coupling

This diagram provides a realistic representation of the relations between the various parties involved in the television market. Television broadcasting companies (both private and public) operate simultaneously in three markets: the viewer market, the advertising market and the acquisition and production market, as well as in a structural coupling with the afore-mentioned markets located before and afterwards in the chain.[42] Both a successful value added on the part of the individual TV companies and the economic viability of the television market as a whole depend on the entire network of factors being taken into account, with considerations of profitability being weighted as appropriate according to participant or function. In the German television market for example, this requirement is at present met only in the case of a small number of the programme suppliers. For the system as a whole it clearly is not.[43]

The market participants have as a rule tended to attribute this sort of failing to the basic legal and political framework, or on occasion also to unfavourable economic conditions (the economies of scale enjoyed by the US television industry, the weakness of the economy or a shortfall of demand). Explanations such as these certainly hit part of the truth, but at the same time they also conceal the many weaknesses in the self-organisation of the market by those participating in it.[44] Even so, what they do bring to light is the dependence of the television market's value chain on its structural coupling with the political system (regulatory policies and fair trading laws) as well as the social background. The overall context of the value network can thus be visualised as follows:

42 See section 3.1.1
43 This results in negative value added, for example in the case of the closing down of the German channels Nickelodeon and Wetter-kanal, or the innovation deficiencies shown by cable operating authorities
44 For example, a failure to support the acquisition and production market sufficiently, poor terms of delivery and payment, handling of advertising customers or agencies, excessive advertising.

Source

Fig. 3.22: ECC 1999

"Political failings and market failings both are faulty structural coupling."

Fig. 3.23: Context and limits of the television market

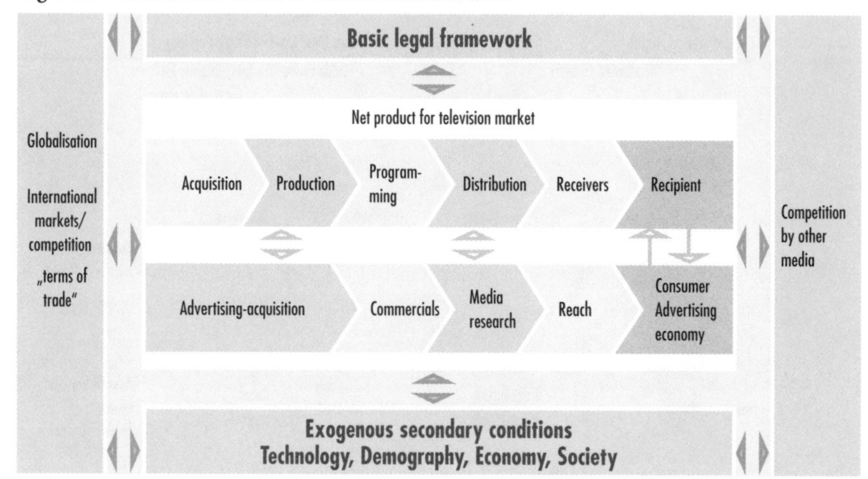

Operational freedom limited by external conditions

A successful value network in the television market is thus on the one hand the result of the strategic integration of three component markets and the structural coupling with the branches located before and afterwards, while on the other hand it depends on a low-friction interaction with exogenous secondary conditions and the legal framework in force under the pressure of international and intermediary competition. In other words, the participants are operating in a highly complex market. Their operational freedom is limited by external conditions that can be influenced only to a small degree, if at all.

Conversely, as far as the legal framework is concerned, direct intervention in the market mechanisms or the value chain mechanisms through de- or reregulation or other measures seems problematic on account of the high complexity of the market and the diverse adaptation possibilities[45] that have to be generated for integration. Over-regulation in the sense of regulation that is too elaborate or concrete brings the danger of misregulation (with negative value added as a consequence) or leads to a permanent need for changes to the regulations.[46] The result can be regulation gaps or uncertainties and a delayed value added. Political failings and market failings can thus be regarded as two sides of the same coin, faulty structural coupling.

3.1.4 Using the Revenue Potential of Profit Windows

The nonrivalry in consumption characteristics of audiovisual content is used by programme-producers and distributors within the framework of what is known as windowing in order to maximise profits. This means that a particular content or programme is distributed through diverse sales channels at different times, so a feature film can be used successively through cinema, pay-per-view,

45 For example the competition (international and intermediary), acceptance, and willingness to pay.
46 The frequent amendments to the broadcasting treaties between the German states (regulating common issues in broadcasting policies) are a good example for this problem.

Source

Fig. 3.23: ECC 1999

video, pay-TV and TV financed by advertising, as well as then additionally through syndication and programme archives. The film is only produced once but – along with merchandising and other licence revenues – is sold at least five times. Utilisation through the various distribution channels or profit windows thus allows for multiple revenues.[47]

The profit window priciple: producing once, selling five times

The individual profit windows are in principle classified according to their revenue potential. The introduction of home videos in the early eighties, for example, led to the profit window of TV financed by advertising being pushed down to a lower spot because the home video profit window yields more revenue. A further factor is the profit window's strategic significance. The cinema profit window thus has greater relevance for revenue potential, since it is by this means that a film's attractiveness is determined and hence also the prices for the further profit windows. The more successful and well-known a film has been in the cinema market, the easier (and more lucrative) is its marketing in the profit windows to come.

Sequence of profit windows

Films generally attain the highest level of profitability in the video profit window, the greater part of the turnover being achieved through videos sold (sell-through) rather than rented out (rentals). After this comes the relatively new pay-per-view window. In the United States the pay-per-view window begins roughly 40 to 90 days after the film's introduction onto the video market; in the various countries of Europe (with the exception of France) it is normally not until six months afterwards. Following the pay-TV window comes TV financed by advertising, which is placed after the fee-financed profit windows because consumers are only rarely willing to pay for a film when it has already been shown for free on television.[48]

Fig. 3.24: Chronological sequence of profit windows in the United States

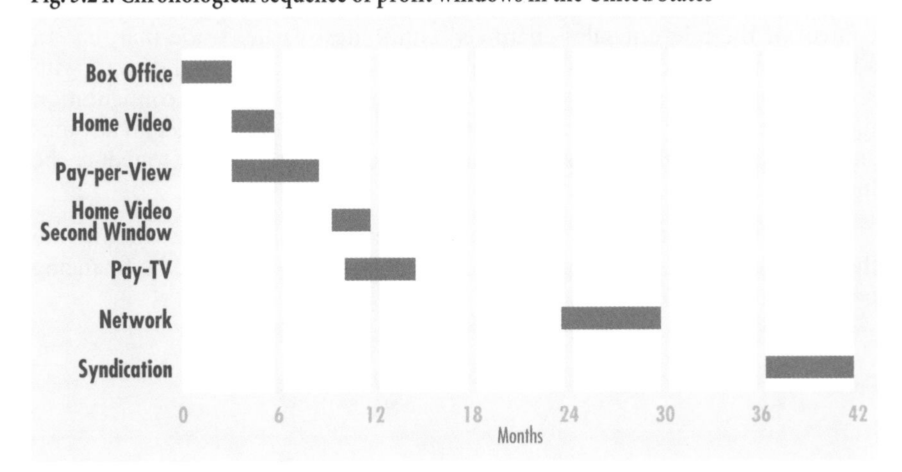

47 See Artopé/Zerdick 1995 for a discussion of these mechanisms.
48 See Owen/Wildman 1992, p. 31

Source

Fig. 3.24: ECC 1997

The market significance of potential new profit windows, such as the digital video discs (DVD) introduced from the end of 1997 onwards and in future also the Internet, remains to be seen.

Fig. 3.25: New profit windows in film and TV markets

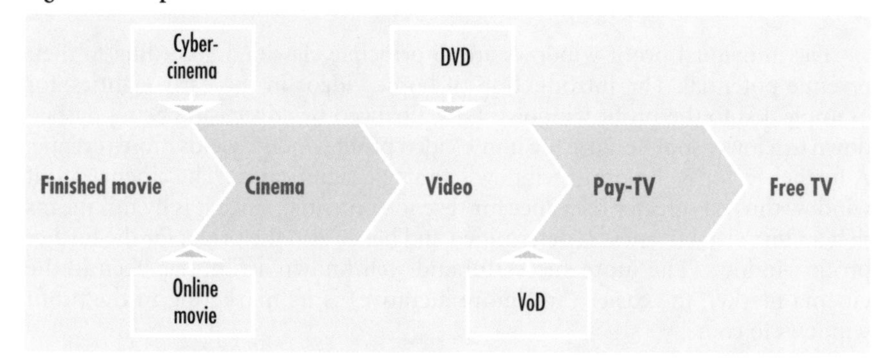

Time lag reduces cannibalisation

The importance of each individual profit window comes to light not only in how close it is to the first showing in the cinema, but also in how long the film remains available in a particular window. As the turnover flows in the later stages are not supposed to be "cannibalised," care is taken to ensure that films have already finished being shown in the cinema before they can be obtained in the video trade. The multiple utilisation of film and TV programmes is not only restricted to the home market, however, but also encompasses foreign markets.

Advantages of vertical integration

In this context, the big media concerns display a high degree of vertical integration at all stages of distribution and on an international level as well. The concept of profit windows thus unleashes considerable synergy effects. From the perspective of business management, the combination of multiple sales and the control of the relevant sales channels "eliminates" earlier trade margins. In addition, prices (and thus also profit margins) can be determined directly with respect to end consumers. With regard to the television market, this configuration means that through their control of the sales channels suppliers get an advance "turnover guarantee" to cover TV production costs and so are able to reduce the financial risk to which they are exposed.

For vertically integrated media concerns in particular, profit windows are therefore one of the most significant concepts for ensuring that the financing potentials of media content are actualised with optimal efficiency.

Source

Fig. 3.25: ECC 1999

3.2 Rulers of the Waves – An Analysis of the Telecommunications Sector

The following section turns to a market that for over a hundred years was dominated worldwide by state-protected monopolies and has only recently – through lasting deregulation – come to experience competition. The story begins with the invention of the telephone in 1876. It was at first unclear what the use of the new medium was. Conversations between people were not initially viewed as its most important function, and the telephone was at this stage used for broadcasting operas and concerts.[49]

"Death of distance"

No one could then have predicted that in the distant future optical fibre networks would be using light signals to transmit data at the speed of light or that satellites orbiting the earth would make it possible for people to communicate with one another from the remotest corners of the planet. As the "network of networks," the Internet in particular has permitted a whole range of new forms of voice and data communication. The consequence of these developments is the "death of distance,"[50] or in other words the end of spatial separation as a factor in communication. And so the historical circle closes once again: in July 1997 Intel put on a music festival in New York that was transmitted live through the Internet and could thus be received through the telephone network all over the world.

Before proceeding any further, an explanation is required of the concept of telecommunications on which the following analysis is based. In its broadest sense telecommunications denotes the transmission of data of any sort (textual, graphic, picture, audio, video, vocal[51] and all possible combinations) in analogue or digitalised form and via the most diverse forms of network. The different networks and the usual applications for the associated data transmission are shown in the following table. It should be noted here that the applications given in the table only refer to the original use to which the network was put.[52]

Fig. 3.26: A systematic overview of the telecommunications networks and their original applications

Type of network	Wireline Networks			Wireless networks	
Means of transmission	Twisted copper	(Copper) co-axial	Optical fibres	Terrestrial	Satellite
Applications	Telephon, ISDN	Cable television	Broadband data transmission	Mobile, telephony, broadcast	Satellite television

49 See Picot/Reichwald/Wigand 1998, p. 117
50 See Cairncross 1997
51 A distinction is being made here between "voice" and "audio data" in order to draw attention to the different applications based on them. The transmission of "voice" is particularly important in telephony, while the term "audio data" refers to the transmission of sound documents.
52 Telephone, television and data transmission are now all possible through all these networks. This has been described relating to the concept of convergence in section 3.2.3.

Source

Fig. 3.26: ECC 1999

The applications and their associated networks can further be subdivided into two groups according to their fundamental purpose. Co-axial, satellite and terrestrial radio networks are broadcasting networks, used first and foremost for the unilateral transmission of content from a broadcaster to a large number of receivers. The transmission of content is available without exception to anyone connected to the network. The other category consists of the twisted copper, optical fibre and terrestrial mobile telephone networks, which constitute telecommunications networks in the narrower sense of the word. Their core purpose resides in reciprocal communication between a (relatively small) number of participants. The data transmission in these networks targets a specific, limited portion of the potential receivers connected to the network.

Broadcasting and telecommunication

The concept of telecommunications in its broader sense can accordingly be subdivided into broadcasting on the one hand and telecommunications in the narrower sense on the other. For the purposes of this report, which is concerned to investigate the fields of media, telecommunications and information technology as well as the existent and emergent interrelations between them, the narrower concept of telecommunications is more appropriate. Broadcasting belongs to the field of media. This is intended to avoid unwanted overlaps and reduce the risk of inaccuracy.

Only in the course of the convergence of the most diverse networks, leading – as will be shown in section 3.4 – to the convergence of the media and communications sectors, will broadcasting networks be considered in the analyses of this section.

3.2.1 Size and Structure of European Telecommunications Markets

A commonly applied definition of the market is provided by the European Information Technology Observatory (EITO). This market definition is based on a more extensive conception of telecommunications, which will here be drawn on simply for the purpose of providing a numerical description of the telecommunications market in order to give an idea of its size and structure.

Infrastructure, equipment, services

On the basis of this conception of telecommunications, the telecommunications market comprises the two fields of network infrastructure and equipment on the one hand and telecommunications services on the other.[53] The first field contains whatever relates to the equipping of the telecommunications infrastructure. Public networks should here be distinguished from private ones. While public networks are available to the general public for telephone and data communication, private networks refer to ones in business enterprises or other organizations. Equipment for public networks includes above all lines and switches, as well as components necessary for operating mobile telephone networks. Equipment for private networks incorporates not only private switches

53 See EITO 1998, p. 386

(such as the PABX), but also stationary devices (telephones and telephone connections), mobile devices (mobile phones, pagers), as well as peripheral devices (such as fax machines, answering machines, video conference equipment).

According to EITO, the second field, telecommunications services,[54] includes alongside the classic telephone service also the mobile telephone, data communication and cable television services. These four services can be distinguished as follows:

Four types of services

* telephone services in the classic sense cover those services necessary for the establishment of communication via fixed networks;
* a distinct field is represented by mobile telephony, which constitutes a separate segment offering mobile communication services via terrestrial networks. These services perform a different function from fixed telephone networks, taking into account the customer's demand for mobility;
* the field of data communications relates to the transmission of data (textual, graphic, audio, video) via a variety of data networks. It specifically excludes voice communication,[55] instead covering the realm of the Internet, as well as data transmission within and between firms;
* the field of cable television is limited solely to the transmission services of the network operators and accordingly excludes both the production of content and the packaging, which have already been dealt with in section 3.1.

Services dominate the European telecommunications market

The structure of the telecommunications market as thus classified is shown in the following table. In 1998 the Western European telecommunications market had a volume of 199.1 billion ECU. With a market share just touching 17 per cent, the field of telecom equipment here plays a relatively minor role. It is the telecom services with 83 per cent of the market value that dominate the market. This field is in turn dominated by the classic telephone services with a share of 62 per cent of the telecom services and 51 per cent of the market as a whole.

54 These services incorporate both the simple telephone service provided by the network operators as well as integrated telecommunication services (see also section 3.2.5).

55 Although voice communication can in general be seen as an aspect of data communication, for the sake of clarity of classification it is excluded here. Instead it is included into the sectors of telephone and mobile telephone services.

Fig. 3.27: The structure of the telecommunications sector in Western Europe (1998)

Markets	Market segments	Market value in billion ECU	Market value in %
Telecom equipment		**33.5**	**16.8**
	Public network equipment	**33.5**	**16.8**
	Transmission	15.0	7.5
	Switching	4.0	2.0
	Mobile communications infrastructure	4.6	2.3
	Private network equipment	**18.5**	**9.3**
	PABX and key systems	3.2	1.6
	Telephone sets	5.3	2.6
	Mobile terminal equiment	4.0	2.0
	Other terminal equipment	6.0	3.0
Telecom service		**165.6**	**83.2**
	Telephone services	102.5	51.5
	Mobile telephone service	32.3	16.2
	Switched data and leased line service	24.3	12.2
	CableTV service	6.5	3.3
Total telecom		**199.1**	**100.0**

Two essential developments: liberalisation and data communications

The following analysis of the telecommunications market concentrates on the two essential developments within the field of telecom services.

In the first part (3.2.2), the spotlight falls on the liberalisation of the telephone markets, a trend currently dominating the market development in virtually all European states. From this starting point the analysis will then turn to strategic aspects of the competition behaviour shown by the old and new players in the market.

The second part (3.2.3) is concerned with the connection between voice and data communication. After an outline of the functionality and significance of the Internet, there follows a survey of the general differences between data communication and voice communication, before moving on to the convergence

Source

Fig. 3.27: EITO 1999, p. 376
* Western Europe includes the 15 EU countries as well as Norway and Switzerland

of the two fields and an overview of the relevant technological trends and economic ramifications.

3.2.2 Liberalisation of Telephone Markets in Europe

Telephone monopolies in Europe are over ...

The market for telephone services in Europe is at present under the sway of EU deregulation guidelines. While the competition for public telephone services in Finland, Sweden, Great Britain, Denmark and the Netherlands has been open for a number of years now, in almost all the other European countries the telephone monopoly came to an end on 1 January 1998. Further extensions to the deadline were given only to Greece, Ireland, Spain, Portugal and Luxembourg.[56]

The European telephone companies play a significant role in the worldwide market. This comes to light in the following table, which juxtaposes the five biggest of the former telephone monopolists in the European markets with the world's biggest enterprises in telecommunications services.

... but the initial strength of incumbent companies is impressive

Fig. 3.28: The world's 15 biggest telecommunications servicesuppliers

	Company	Country	Turnover in 1997 (in billion US-$)
1	NTT	Japan	71.1
2	AT&T	USA	51.3
3	Deutsche Telekom	Germany	40.6
4	BCE	Canada	33.2
5	Bell Atlantic	USA	30.2
6	France Télécom	France	29.4
7	SBC Communications	USA	24.9
8	Telecom Italia	Italy	24.2
9	British Telecommunications	Great Britain	24.1
10	GTE	USA	23.3
11	Bellsouth	USA	20.6
12	MCI Communications	USA	19.7
13	Ameritech	USA	15.9
14	Telefonica de España	Spain	15.3
15	Sprint	USA	14.9

56 See EITO 1998, p. 36

Source

Fig. 3.28: Elstrom et. al. 1998, p. 53

Before analysing the features of the new competition in the liberalised markets, there follows a discussion of the general advantages and disadvantages of liberalisation. This discussion is based on the experiences gleaned in markets where liberalisation has already taken place, such as the United States, Great Britain and Scandinavia.

Advantages and Disadvantages of Liberalisation

Liberalisation means more than lower rates

The economic advantages of liberalising telephone markets are compelling, applying as they do to all areas of any economy. In general the resulting competition forces the suppliers of telephone services to make price cuts, develop and apply new technologies, as well as provide innovative services.

The provision of new services was evident in the form of the call extensions, brokers and three-way systems already being offered in anticipation of the competition to come by the state telephone companies even before their monopolies were broken up. But in addition to this, competing network-operators have been pursuing the development of other innovative services. These projects range from the integration of mobile and fixed-network-based telecommunications services through to personal telephone numbers. By means of this number the telephone customer can be reached both in the office and at home, in the car or even through a network-integrated "mailbox." "Unified messaging" in turn permits this "mailbox" to store telephone calls, faxes, e-mails and other data which the customers can then retrieve and use wherever they may be.

Business customers have most advantages

It is above all business customers who profit from the development of such innovative services, which enable them to increase efficiency in the structuring of company operations. On top of this, drastic falls in charges can also be observed. Companies with high telecommunications expenditures in particular enjoy considerable savings.

Private customers react to the improved range of offers and the falling charges above all through increased voice traffic. More frequent and longer phone calls have meant that in liberalised markets the voice traffic has risen almost twice as quickly as in monopolistic markets.[57]

Even so, this range of positive effects brought about by the breaking up of monopolies is offset by disadvantages too.

Regulation necessary for liberalisation

The liberalisation of markets is marked at its outset, for example, by a great need for regulation. Questions of detail crop up that remain unaccounted for by the law or leave excessive scope for interpretation. Examples of this include the question of switching fees to be paid by customers if they wish to change their supplier, or the fixing of network access costs (interconnection fees) to be paid by new suppliers making use of the infrastructure of the former monopolist. What is at issue is the smoothing over of conflicts of interest. The regulatory measures should as far as possible serve the consumer's interests in a well-functioning

57 See anon 1998a, p. 59

competitive situation, without thereby putting any of the parties concerned (either ex-monopolist or new competitors) at a fundamental disadvantage. As a rule, the interests of the competition in general stand above the interests of the individual parties. At the same time, however, competition-oriented decisions should not force the ex-monopolist to subsidise its competitors. Deregulation pure and simple is consequently not enough.

Risk of highly unstable markets

Such unresolved difficulties lead to the risk of instability for all market participants. Suppliers often find themselves deprived of a reliable costing base for assessing and planning their future activities. On the customer side too, uncertainties arise that make it difficult to work out the financial repercussions of a change of supplier.

Customer adjustments

The early stages of newly existing competition, moreover, tend to be characterised by a lack of transparency. A large number of suppliers enter the market with complicated price structures that hamper comparability, in some cases intentionally. Discrepancies in time units and the existence of minimum or standing charges as well as complicated discount systems make it virtually impossible to compare charges exactly. The sudden increase in supply within a market that has up to now denied the customer any room for consumer choice may quickly come to seem surplus to needs, with the result that the new competition is at first not accepted by many customers.

As has been seen, the liberalisation of telephone markets produces a number of problems in the initial phase of competition that make it difficult for new suppliers to establish themselves and that delay the customers' adjustment to the new circumstances. Provided that these drawbacks associated with liberalisation are only temporary, however, the beneficial effects of the competition for telephone services will soon lead to long-term improvements in market conditions.

Analysis of Competition in the Case of Germany

Biggest European market, most successful liberalisation

The following analysis of the development of competition in the European telephone markets will focus on Germany as a paradigm case. The basic tendencies that emerge can be applied equally well to the situations in other European countries. Several factors speak for Germany as the choice of paradigm. Firstly, the German market for telephone services is the most important in Europe. With a market volume of 24 billion ECU, it is significantly bigger than the French market, which is the second largest in Europe with a volume of 16 billion ECU.[58] This makes it very attractive for new national and international suppliers, who have as a result thronged into the German market in large numbers. Even if some of the new competitors have not so far seen their turnover expectations fulfilled (as is the case with o.tel.o), with others the surge of sales has been overwhelming (for example Arcor, Mobilcom). For long-distance domestic calls, the new suppliers have already managed to total a market share of 12 per cent after just half a year,[59] an indication that the new competition in Germany is meeting

58 See EITO 1998, p. 328ff.
59 See anon 1998o, p. 1

with widespread acceptance among customers. Germany is thus already being seen as "World Liberalisation Champion" or "Telekom Valley."[60]

Segments of the telephone market

For a more in-depth analysis of the conditions of competition in the deregulated market for telephone services, this market can be further subdivided into two distinct fields that differentiate local calls from long-distance or international calls according to distance or destination respectively. Local calls take up above all "the first mile,"[61] in other words the line from the customer's connection to the networks' local distribution points. The requirements for local competition thus diverge markedly from long-distance competition.

Distinction between local and long-distance blurring for business customers

This distinction between the regional field of local calls and the trans-regional field of long-distance and international calls can additionally be juxtaposed with the division into the private customer segment and the business customer segment. In the field of business customers, however, the regional distinction becomes less important, as big companies in particular tend to secure contracts for integrated communication services incorporating both local calls and long-distance or international calls. The two fields thus merge into a single business field in the business customer sector.

Fig. 3.29: The competition structure of the telephone market in the paradigm case of Germany

	Local calls	Long-distance international calls
Private customer	Deutsche Telekom City network operators (e.g. Netcologne)	Deutsche Telekom National and international network operators (e.g. Arcor, o.tel.o, Worldcom) Service providers (e.g. Mobilcom)
Business customer	Deutsche Telekom National and international network operators (e.g. Arcor, o.tel.o, Worldcom)	

Before analysing these three segments in detail, a preliminary look at the competitors in the telephone services market as a whole is in order. Deutsche Telekom's situation report itself contains an analysis of its strategic strengths and weaknesses, as the former monopolist is the enterprise defending market shares in all three business sectors. The new suppliers of telephone services, by contrast, are introduced purely schematically.

The competition situation faced by Deutsche Telekom is marked by a number of serious problems stemming from the past. Lack of achievement-orientation in the company allowed a huge mountain of debt to build up which can only be reduced gradually. In spite of the considerable efforts made by the ex-monopolist to establish a customer-friendly business culture, the service mentality still leaves a great deal to be desired. The fact that many of the employees are civil servants

60 See anon 1998j, p. 13
61 The literature on the subject tends to use the concept of the "last mile" instead of "first mile." This concept is not followed here, however, as it relegates the customer to a marginal role. By contrast, the emphasis in our thinking is on the customer as the starting and focal point of the activity. See also Steinberg 1998, p. 79

Source

Fig. 3.29: ECC 1999

is a further impediment to attempts to create a consistently customer-friendly and achievement-oriented enterprise. A great strain is put on the company's cost structures by the disproportionately high number of staff, which is to be steadily reduced with the help of social support programmes. Finally, the company is weighed down by excessive value added depth, meaning that loss-making business sectors such as cable television and end devices are a burden to the business as a whole. This lack of outsourcing goes hand in hand with a lack of clear positioning and a failure to establish core competencies.

Remaining benefits of past monopoly

On the other hand, the years enjoyed as a monopolist also produce benefits that remain unavailable to the new competitors in the short term at least. These include contact with a sizeable customer base, established and distinctive brand products, as well as an already existent sales and customer service structure. A reservoir of experience in the field, financial strength and the possession of an extensive and high-quality network infrastructure can be regarded as further strengths. These will not only help the company cope with the forthcoming challenges in the German telephone services market, but may well also be put to use in the increasingly international telecommunications competition.

Fast initial competition

In mid-1998, half a year after the breaking up of the monopoly, the competition faced by Deutsche Telekom in the field was structured as follows. By 30 June 1998, 110 licences for transmission lines (class 3) and 93 licences for telephone services with an individual network (class 4) had already been issued. This total of over 200 licences was in the hands of 120 companies, made up of the following:[62]

* six per cent of the new competitors are American suppliers such as Worldcom entering the market;
* another six per cent are offering their telecommunications services on the basis of corporate networks opened up for external customers. Included here are Mannesmann Arcor, o.tel.o and Viag-Interkom. These ventures are backed mainly by powerful associations of foreign telephone companies (such as AT&T or British Telecom) and domestic energy concerns (such as RWE, Veba or Viag);
* 10 per cent of the new licence-owners operate cable television networks which they want to use for the provision of telephone services in the future;
* the numerically largest share of 50 per cent is made up of enterprises founded by energy supply companies, municipal works departments or local savings banks. This field of new suppliers includes above all smaller-scale local companies such as Netcologne. Given their size, these seem likely to be relatively minor players in the competition faced by Telekom;
* the remaining 28 per cent are suppliers of telephone services that do not operate their own networks but connect calls through leased lines. The most

62 Regulierungsbehörde für Telekommunikation and Post, 15.7.1998 (press conference)

significant of these service providers, such as Mobilcom or Talkline, have their origins in the mobile telephone sector.

Proceeding from this structural representation of the telephone market in Germany, the following section will turn to the new competition situation in the three fields of business customers, private local calls and private long-distance calls.

Business Customers

Main area of initial competition

On account of the revenue they bring in, it is above all big companies with high levels of telecommunications expenditure that represent the most attractive customers for the suppliers of telephone services. As the primary scene of competition, the lucrative segment of business custom has thus been a fiercely disputed one. New suppliers are here faced with good prospects. As telecommunications expenses constitute a major chunk of total costs for many companies, special discounts arising from competition become particularly attractive. As well as this, there is a huge demand for specialised services, technological consultation, customised offers and telecommunications solutions all from a single supplier.

Service and consulting

The business customers' demand for services and consultation is focused above all on achieving greater transparency in their own telecommunications expenditure. Although the total level of telephone costs is usually known, the structure of the services made use of frequently remains a mystery. Yet a knowledge of this structure is indispensable for coming to grips with the complex system of charge rates on offer and ascertaining the best offer for the individual company in question.

Deutsche Telekom has been struggling here mainly on account of the high customer requirements in terms of flexibility and service orientation. To prevent the loss of this key customer segment, it has thus been forced to resort to giving generous special discount rates. Even so, a number of the most attractive German business customers, such as BMW, Deutsche Bank, BASF, Lufthansa and Karstadt, have already turned their back on Telekom, turning instead to Westcom, Arcor, o.tel.o or others for the running of their telecommunications operations.[63]

Local Calls

In the private customer segment, the competition for local telephone services has not yet fully flared up. Only a few German cities have seen municipal works companies offer any competition to Deutsche Telekom in the provision of telephone calls over their own local networks. Well-known examples are Netcologne, which has been offering cheap rates in Cologne since the beginning of January 1998, or Hansenet, which serves the Hamburg area. In half of the German cities with over 100,000 inhabitants there are further "city carriers" planning entry into the market. Yet it can be taken as given that the limited size of these local telephone companies will in the long term mean they find it difficult

63 See anon 1997a, p. 95

competing with the performance level and the diversity of services offered by the big companies. For this reason some of the bigger competitors are planning to set up as one-stop suppliers in the near future. Arcor for example is intending to take up business in the "first mile" within the second half of 1998.

At present there are still problems concerning the complete takeover of private telephone connections, as the new suppliers are still without a basis for costing. What is necessary here is the fixing of the so-called interconnection fee. At issue is the technical access to the private households' telephone connections. As excessively high investment costs make it unviable for new suppliers to install a network of their own for the 37 million or so German households, they are instead dependent on Deutsche Telekom's facilities. Telekom is legally obliged to relinquish use in these cases.

Unbundled access and interconnection charges

For a long time there was uncertainty regarding what form the network access was to take. At stake here is the question whether the suppliers can only use the final section of the Telekom line (simple network access) or whether they should also lay claim to switching services which they could then provide themselves (bundled network access).[64] The decision in favour of simple, non-bundled network access has far-reaching consequences for the second problem, the fixing of the use-price or interconnection fee. This decision too involves a weighing up of interests between Deutsche Telekom's rights of ownership and the public interest in the unimpeded market participation of competitors. The problem is that obstacles placed in the way of competition in the field of local calls could also prove harmful to the other spheres of competition. In Sweden for example, the ex-monopolist Telia was thus seen to finance cheap rates in the hotly disputed long-distance and international sector with the help of price rises in the unthreatened local sector.

Long-distance and International Calls

While local calls have so far continued to be the virtually unchallenged territory of Deutsche Telekom, private long-distance and international calls have already been the object of fierce competition. In general customers are faced with two possibilities as regards the choice of telephone company. The greater flexibility is offered by the so-called "call-by-call" system, where customers can choose their supplier from case to case for each individual call. The other option is known as "pre-selection," which involves a fixed choice of telephone company for all long-distance and international calls. Once the customer has opted for a company's pre-selection offer, then all his trans-regional calls will automatically go via this company's network.

"Call-by-call" and "preselection"

In the early stages of the new competition, it is the call-by-call option that is meeting with greater resonance among private customers. Few customers have been willing to commit themselves to one new supplier from the outset, most preferring to start by trying out the range of choices available. The call-by-call

64 See Weishaupt 1997, p. 2

system thus appears the more attractive alternative and a precondition for market success in the initial stages of liberalisation. This also means telephone companies doing without registration when a customer is connected up to the call-by-call offer for the first time. This painful lesson is being learnt by o.tel.o for example, whose registration-based offer has so far fallen considerably short of expectations, while Arcor, whose offer has from the outset excluded compulsory registration, is already enjoying widespread use.

Further factors conducive to success that have emerged in the early days of competition include not only the persuasive power of cheap rates but also a simple tariff structure. The most successful example is Mobilcom, whose offer of long-distance calls seven days a week round the clock at 19 pfennigs a minute soon won over numerous customers. As well as this, new standards seem to be gaining long-term acceptance, such as charges made by the second (as with Arcor, Tele2, First Telecom, Telepassport or Worldcom) that dispense with charge calculations based on units of time (various units in the case of Telekom or fixed minute-units with Mobilcom).

The new suppliers can be clearly distinguished in terms of business objectives and accordingly split into two separate groups, simple service providers on the one hand and suppliers that operate their own network on the other.

Service providers

The simple service providers act as resellers of line capacities. They do not own networks of their own but switch their customers' calls through leased lines in networks belonging to Deutsche Telekom or other network operators. As the cost of leasing line capacities is fixed at a very low rate of 2.7 pfennigs, the incentive to build up an infrastructure of their own is small. New suppliers can thus avoid the huge initial investments amounting to billions of marks. The strategic advantage they have over Deutsche Telekom lies in the lower number of staff required and their more flexible structures. With a good customer service, reasonable tariffs and the intelligent use of modern computer and communications technology, the newcomers to the market will provide it with an attractive alternative to existing services.

Integrated companies: network and service

Along with these service providers, the second category of new suppliers consists of those that operate their own network in the market. These tend to have strategic objectives that go far beyond that of the service providers. With a view to establishing themselves in the long term in the international telecommunications market, these ventures are squaring up to be one-stop suppliers offering the most diverse and high-quality services in voice and data transmission. The operation of an individually owned network is the underlying prerequisite both for the more immediate aim of becoming a fully-fledged rival to Telekom and in particular for the long-term aim of participating in the worldwide telecommunications market. Yet any such capital intensive venture remains the prerogative of companies with sufficient financial weight behind them. There are thus only a few new telephone companies in Germany in this category of network-operators. The table below

provides an overview of the four competitors running nationwide networks of their own.

Fig. 3.30: Structure of the national operators of telecommunications networks in Germany

	Deutsche Telekom	Mannesmann Arcor	o.tel.o	Viag-Interkom
Owner	Federal Government of Germany shareholders	Mannesmann, *(AT&T, Unisource)*[65], Deutsche Bahn, Deutsche Bank	RWE, Veba	Viag, British Telecom Telenor
Planned investment	not known	4 billion DM by 2001	7 billion DM by 2005	8.5 billion DM by 2006
Infrastructure	Nationwide blanket telephone, mobile phone and data network	40,000 km of digital telephone and data network	Network cooperations	Construction of an integrated fixed and mobil telephone network
	World`s biggest optical fibre network	5,000 km of optical fibre network	11,000 km of optical fibre network	4,000 km of optical fibre network
International partners	France Télécom, Sprint (both through Global One)	*(AT&T, Unisource)*[65]	sought after	BT, Concert, Telenor

Comparison with the Process of Liberalisation in the United States

US liberalisation: 14 years earlier, but not ahead

By way of contrast with the process of liberalisation in Europe, illustrated above with the paradigm case of Germany, the following section will sketch the two-phase deregulation process as it took place in the United States. Although this actually began as early as 1984, the competition situation in the States is not noticeably more advanced than the new situation in Europe.

The first phase of deregulation was initiated in 1984 by the US Department of Justice with its decision that AT&T ("Mother Bell") should be divided up on account of its excessive dominance in the telephone market. The upshot of this was the creation of seven regional telephone companies known as "Baby Bells." This splitting up of AT&T was based on a distinction drawn between local telephone services and transregional (or long-distance) services: AT&T was no longer allowed to operate as a supplier in the field of local calls, while the long-distance sector was for the first time opened up to new competitors. The Baby Bells however, who now dominated the local sector as monopolists, were to be excluded from this competition for long-distance calls. As a result of this first phase of deregulation there was intense competition in the long-distance sector. This success was reflected in a statement made by the FCC (Federal Communications Commission) in 1995 to the effect that AT&T's market position was no longer dominant.

65 Ownership structure is undergoing change.

Source

Fig. 3.30: anon 1998c, p. 18; anon 1997b, p. 108 f.

The local telephone companies were thus paradoxically forced to look for new business fields outside their own original core market of voice telephony. When an increasing convergence of the different telecommunications services began to become apparent in 1992, the possibility of expansion into the sphere of data communication presented itself. The Baby Bells attempted to compensate for their deficiencies in this sphere by means of alliances with cable companies. Noteworthy is the attempted merger, announced in 1993 and called off a few months later, between Bell Atlantic and the biggest American cable company TCI. John Malone, TCI's chief executive officer, in this context coined the concept of the "500 channel universe," and at the same time revealed the strategic objective of both concerns to be the provision of multimedia content.

Baby Bell strategies

Along with alliances, the market for voice and data communication in the United States was dominated by the predicted struggle between telephone and cable companies. It was this period that saw many of the now-discarded test projects in interactive television. The realisation of this vision that sought to combine broadcasting and interaction instead became the reserve of one network that entered the household through the classic telephone network, the Internet.

1996 Telecommunications Act

A second fundamental change to the situation ensued in 1996 with the Telecommunications Act. AT&T was by now faced with considerable competition in the long-distance market, in particular from big network operators such as MCI, Sprint and Worldcom. In theory, the Telecommunications Act made it now possible for there to be competition for the local call market between these long-distance suppliers and the seven Baby Bells, while also permitting the entry of the Baby Bells into the market for long-distance calls.

This opening of the market has curiously led to the Baby Bells withdrawing from the data communication sector for the time being and instead concentrating on telephony again. By way of preparation for the long-distance market the Baby Bells have since been involved in a number of mergers. Following the amalgamations of SBC Communications and Pacific Bell as well as Bell Atlantic and Nynex there are now just five companies remaining. Yet SBC is already planning its next takeover and has made a bid for Ameritech. If this takeover bid is given the go-ahead, current developments would seem to be running directly counter to the objectives of the 1996 act as regards competition.[66]

The situation in the United States since 1996 has been further characterised by the attempts of participants in the long-distance market to penetrate the market for local calls. The Telecommunications Act here envisages the possibility of leasing unbundled network elements from the Bells. These lease prices have so far only been fixed in a few states, however, and so up till now there have not been any real possibilities for suppliers of long-distance calls to enter the local market. The situation is reminiscent of the state of interconnection regulation in Germany.

66 See anon 1998l, p. 73f.

Cable network operators have thus again come to assume particular significance for the telephone companies. As the big telephone concerns from the long-distance sector such as AT&T, MCI and Sprint have up to now lacked direct access to customers in the first mile, cooperations with or takeovers of cable network operators present an interesting alternative. These could facilitate direct access to the individual connections of private or small business customers without having to rely on the networks belonging to the Baby Bells. It was in this context that AT&T in June 1998 announced their merger with TCI, envisioning it as a means of attacking the monopoly of the Baby Bells in the first mile. By means of TCI's cable connections, AT&T now reaches a third of all American households, offering them local calls, cable television and access to the Internet.[67] On balance, in spite of a two-phase deregulation process dating from 14 years back, competition in the USA is not much more advanced than in the European countries. The reason for this lies above all in the initially tentative deregulation of the first phase, which led to the paradoxical situation of the Baby Bells turning away from their core operations in order to expand in other fields. The second phase of deregulation in 1996 countered this tendency, bringing the telephone companies to revert to their core operation, telephony. As a consequence, the American telephone market is marked on the one hand by tendencies to concentration. On the other hand, decisions necessary for the development of competition in the sector of local telephone services have been delayed and impeded. For this reason, the deregulatory impulse behind the Telecommunications Act has not yet come to full fruition.

Competition in data services: broadcast cable and the Internet

3.2.3 Netheads versus Bellheads: The Fusion of Data and Voice Communication

The following section is concerned with the convergence of voice and data communication as the second essential trend characterising the telecommunications market. The developments associated with this trend imply lasting changes to future economic conditions that will extend well beyond the telecommunications market. The trend's starting point is the spread of the Internet as a universal network being joined by more and more other networks and incorporating more and more telecommunications applications within itself.

Internet as a universal network

A description of these developments and an analysis of their consequences follows in four steps:

* first comes a description of data communication in the Internet, including the development and structure of the Internet. A sketch will also be given of the technological functioning of Internet data communication;
* attention then turns to the specific developments in data communication that led to the convergence of voice and data communication. The ensuing fusion of the markets entails a whole new range of conditions of competition;
* the third section looks at these conditions of competition from a technological perspective. The possibilities for development of the various networks on the level of infrastructure are here presented and evaluated;
* the fourth part is concerned with the momentous strategic ramifications of the new conditions of competition. The focus is on telephone companies and Internet Service Providers, as well as the starting situation for their forthcoming mutual competition.

The Structure of the Net

Combine of decentralised networks

Telecommunication on the Internet refers to the transfer of data within and between various networks. The term "Internet" itself contains the implication that it is not a single unit of network that is at stake. Instead the Internet is a worldwide combine of distinct subnetworks and computers operated largely autonomously. The operators of these networks are business enterprises, universities or other state or private organizations. Key aspects such as connection, administration and transmission are subject to a largely decentralised organisation.

TCP/IP: protocols as linking elements

The link element between subnetworks is formed by the unified Transmission Control Protocol (TCP) and the Internet Protocol (IP). These follow and steer the transmission of data through the various networks and so ensure that the data reach the correct destination address. The fundamental architecture of the protocols was fixed in 1974 by Vinton Cerf and Robert Kahn.[68] The TCP here guarantees the construction of a logical connection for the data transmission and monitors this transmission, while the IP is responsible for addressing the data.[69] The use of these protocols is independent of the physical network used for data transmission and so allows the connection of diverse types of network.

Network hierachy

The fundamental construction of the Internet can be described as a hierarchical network structure. The lowest level consists of numerous smaller networks with a local extension (Local Area Networks, or LANs), as they exist in firms, universities and other organisations. These are limited mainly to single buildings or properties. The LANs are linked to one another through regional networks (Metropolitan Area Networks, or MANs). In a third stage, these MANs are in turn then connected to data networks with a nationwide or intracontinental extension (Wide Area Networks, or WANs), which manage data communication within a particular country or even continent. Intercontinental data transmission is

68 Cerf/Kahn 1974
69 See Rupp 1996, p. 7

carried out by means of so-called backbone networks, which regulate (for example) the transatlantic data transfer between Europe and the United States via a small number of junctions.

Deviating from the Internet's hierarchical structure are the big, internationally operating companies that run networks to manage their own data communication activities. These networks, some of them global, are only linked to the Internet at a few points.[70] Similar data networks are also operated by commercial on-line services, connecting private customers and smaller organisations to the Internet.

Internet access at home and the multitude of options

A private household can thus participate in worldwide data communication by getting connected to the Net through a nearby local connection point. Such points may be university networks or the servers of Internet Service Providers (ISPs). The data transfer from a household to the connection point takes place at present mainly through the telephone network. Yet the same function could also be fulfilled by numerous other networks that private households are either already connected to (such as cable networks or electricity mains) or that they could get connected to without any difficulties (such as satellite or terrestrial networks). The technical feasibility of interactive data transmission is in many respects still at the stage of test runs and trial phases.

The significance of the telephone network for Internet traffic comes to light in the claim of the American local carrier Pacific Bell that it is responsible for the construction and completion of 40 per cent of Internet connections. The implications of this for what would happen if there really were an opening up of competition in the field of local calls in Germany are clear. The resulting fall in local charge rates or even the introduction of a "flat fee" containing local calls in the standing charge could prove an important stimulus to the further development of on-line use in Germany. This is a point emphasized by Nicholas Negroponte, founder member of the Media Labs at MIT in Cambridge:

> *"Make local calls a flat fee and you would change Germany's economy more than any president could. It's a cultural demand."* [71]

How Data Transmission in the Internet Works

Packet switching

Data transmission in the Internet is based on a process of "packet switching" whereby a quantity of data is divided up into small packets. The type of data is here without significance for the transmission of the packets, as packets with voice data are not fundamentally different from packets with text or graphic data. Packets of distinct data flows can thus be transported through the networks simultaneously, using the same lines. By contrast with networks with "circuit switching" (such as telephone networks), where data transmission between two communication partners means the line is occupied, packet switching does not require any permanent physical connection.[72]

70 See Rupp 1996, p. 4
71 Negroponte 1998
72 See Rupp 1996, p. 5

The principle underlying how packet switching works can be clarified by means of the analogy with motorway traffic. The data packets are each sent individually onto the motorway, which is also used for transporting other packets. At various junctions they are sent on through the motorway network until they come to the "exit" – i.e. to their receiver. Using the same analogy, circuit switching by contrast signifies that individual sections of the motorway are closed off for the connection of a particular data transmission.

Efficient use of networks

Data transmission through packet switching can be applied on all types of network and is independent of the network's physical composition. A great advantage of this transmission technology lies in the efficient use of a network's transmission capacities that results from the lines being used for several data flows at the same time. A further special feature is the greater flexibility it imparts to transmission. If individual network lines break down or are overloaded, the data packets can be conducted through alternative routes, leading to an increased resilience in the face of network disturbances. This flexibility in packet transmission leads to a phenomenon known as "hot potato routing."[73] This refers to the principle whereby network operators try to conduct a given packet through their own network and into the network of the next network operator using the shortest possible route. It is a way of keeping to a minimum the costs that arise through the utilisation of their own line capacities.

"Hot potato routing"

Also related to this, however, is a problem of packet switching based on the "best effort" method, where the attempt is made to transmit the packets as quickly and as reliably as possible from sender to receiver. Owing to the flexible transmission, it can happen that heavy use of the network leads to considerable delays or even individual packets containing a quantity of data being dropped. Yet this only proves a disadvantage for time-sensitive applications that require a continuous flow of data and do not tolerate delays in transmission. Examples are telephone connections or the transmission of moving pictures.[74]

Best effort service

Many applications on the other hand, such as e-mail or fax, react very flexibly to bandwidth fluctuations, needing neither a real time transmission nor a continuous flow of data. For applications of this sort, packet switching technology has great advantages as regards the efficient use of line capacities.

All in all, a shortage of transmission capacities thus constitutes a bottleneck factor for certain applications in the Internet. The ensuing problems are made even worse by the continuous increase in data communication. This tendency is countered, however, by the steady extension of transmission capacities, which should be able to rectify the existing problems within the next few years. Time-sensitive applications should then be able to be comfortably carried out on the Internet.

73 Hal R. Varian, Dean of the School of Information Management and Systems of UC Berkeley, during the ECC Berkeley Workshop in April 1998.
74 See Rupp 1996, p. 9

The Convergence of the Communication Services

WWW – a European invention brings multimedia to the Internet

In its beginnings the Internet served primarily for core applications of classical computer networks. These include in particular text communication, the transfer of files and terminal connections. The development of the foundations of the World Wide Web (WWW) by the CERN[75] researchers Tim Berners-Lee and Robert Cailliou starting in 1989[76] and the increasing power of the computers connected to the Internet, however, has led to the Net being used increasingly for multimedia applications, with the transmission of graphics, pictures and video or audio files. The Internet thus combined a number of services that had previously only been provided by separate networks. The traditional computer network consequently developed increasingly into an integrated universal network. Along with electronic mail, this "network of networks" now also unites fax and telephone services as well as the transmission of moving pictures for a steadily rising number of users.

The far-reaching integration of services through the Internet leads to a convergence of telecommunications services that had previously existed independently of one another in different markets. The result is a significant convergence of the markets for telephone services, TV and data transmission. The convergence of data and voice communication in particular deserves special attention.

Shift towards data communications

What can be expected in general are profound changes in the relative importance of data and voice communication. Data traffic, which by comparison with voice telephony currently plays a minor role in the telecommunications market, will come to assume much greater worldwide significance in the next few years. Though on a markedly higher level, the classical telephone business is developing only slowly in comparison. While data communication spending is thus growing at an annual rate of 100 per cent, telephone expenditure is only increasing at a rate of some five to ten per cent. And it looks as though these trends will continue more or less unchanged into the future. Even if telephone services according to EITO at present still represent 65 per cent of total telecommunications services in West Europe, the emphasis is sure to have shifted within a few years time.

75 CERN (Conseil Européenne pour la Recherche Nucléaire) is a European nuclear research institution in Geneva.
76 See anon 1998h, p. 222; see ECC 1997, p. 333

Fig. 3.31: The changing relative importance of telecommunications services, as illustrated by data traffic in the United States

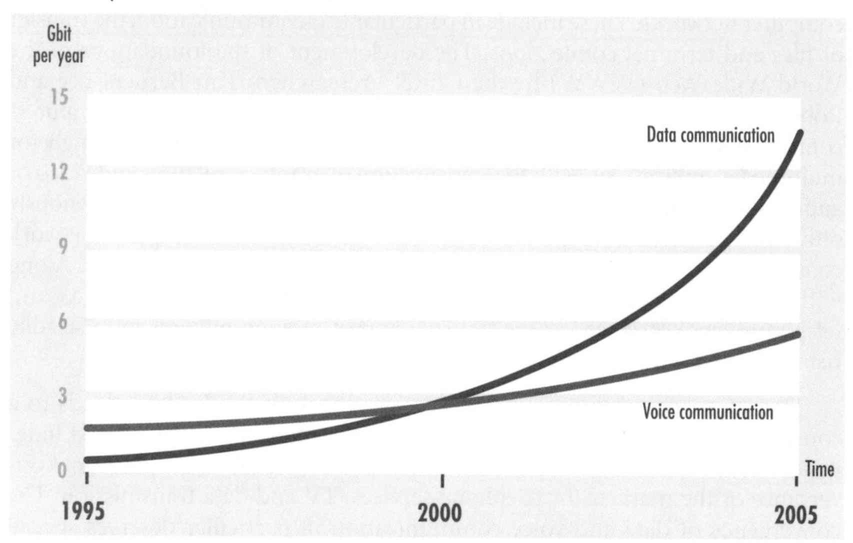

Internet telephony

This changing relative significance among the telecommunications services is further supported by the increasing attractiveness of the Internet for voice communication. The main advantage of Internet telephony lies in the more efficient use of network capacities and the concomitant cost savings for data transmission. It is especially with transmissions over long distances (such as long-distance or international calls) that this advantage makes itself felt. An additional factor is the continuous improvement in transmission quality. Along with the expansion of transmission capacities through broadband optical fibre networks spanning the whole earth, this also includes the continued advances in transmission software. One important step in the direction of intelligent data transmission is the process of prioritising that allows time-sensitive applications to be dealt with first. Once data transmission is ensured that is qualitatively the equal of the traditional telephone network, then customers will be able to enjoy the economic superiority of Internet telephony without any qualitative shortfall.

According to the market research institute Forrester Research, almost 70 per cent of the "Fortune" top thousand companies[77] announced that from the year 2000 onwards they were intending to carry out telephone calls through the Internet.[78] Extrapolating from this trend, it is calculated that within the next four years the data network will be responsible for as much as 13 per cent of worldwide telephone communication.[79] In the long term voice communication will thus become a part of integrated data communication.

Triggered off by the convergence of services, there is also a fusion of the telecommunications markets. Services for the transmission of the diverse data

77 The business magazine "Fortune" annually publishes a list of the thousand biggest companies in the world.
78 See Laube et al. 1998, p. 206
79 See Elstrom et al. 1998, p. 49

Source

Fig. 3.31: Madden 1997, p. 2

forms are no longer offered solely through separate infrastructures but also jointly through the steadily expanding universal network. On the basis of this convergent market, the different infrastructures (telephone networks, mobile phone networks, cable networks, computer networks and so on) and their operators and service-suppliers are thus in competition with one another for the demand for integrated communications services.

Improvement of existing networks

The new competition in this united telecommunications market makes technological and strategic activity necessary on the part of the market participants. The strategic aspects of the new competition situation will be discussed at the end of this section. On the technological side, it is essential that the necessary technical improvements be made to the networks already in existence to allow for the integrated provision of communications services. The following section describes and evaluates the possibilities in this area.

Technological Developments for Speeding up Data Transmission

As has been seen, speed and reliability of data transmission are the factors crucial to the success of the Internet's expansion into other fields of telecommunications. This applies not only to the time-sensitive applications of voice communication but also to the transmission of media content. The use of multimedia services is thus at present still marked by long waiting periods and delays in the loading of large quantities of data that can greatly limit multimedia enjoyment. Even the so-called "streaming method," where the data are fed through for transmission simultaneously, still involves troublesome delays and impairments to transmission quality. If data communication expands as expected, these problems will become even more acute, and it can be presumed that the future will see a huge demand for transmission capacity.

Expanding bandwidth necessary

In the attempt to overcome these difficulties, two kinds of technological problems can be differentiated. On the one hand, the great demand for transmission capacities leads to bottlenecks in the backbone networks that have to deal with worldwide data communication. On the other hand, the process of data communication in general is slowed down by low transmission speeds in the field of local network access, up to now still carried out through the telephone network. These two problem areas are represented in figure 3.33.

George Gilder: "Bandwidths will treble every 12 months." 81

Fig. 3.32: Network levels of data communication

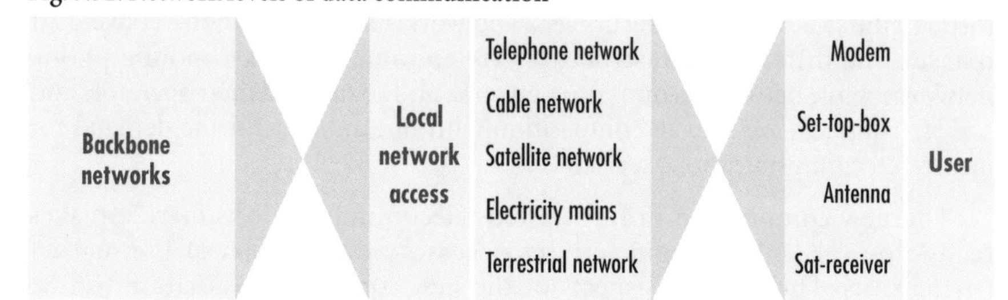

In order to clear the bottlenecks in the backbone networks, capacities worldwide are currently being built up. The growth in transmission capacities seems to comply with Gilder's Law. This law,[80] formulated by the writer and journalist George Gilder, predicts that telecommunications bandwidths will in future treble every 12 months.[81] If the actual expansion of the networks obeys this law, then even a rapid growth in worldwide data communication should not present a serious problem.

Local storage

A further principle for dealing with the current capacity bottlenecks is "local storage," which means relieving global backbone networks of data traffic by storing frequently demanded content and data on nearby servers (known as "proxy" or "mirror" servers), for example Internet Service Providers. When a user returns to this content, the data do not have to be brought back from the original server through the backbone network, but can be fetched directly from the proxy or mirror server. For the user the principle of local storage thus brings the benefit of speedier access to the desired information.

First mile bandwidth

The transmission problems in the local network traffic of the first mile require technological improvements in the existing networks. Adequate transmission speed can here be attained either through greater bandwidths or intelligent transmission technology. There now follows a sketch of the possible technologies in this sphere.

Extending to the end customer the optical fibre networks dealing with worldwide data communication is technologically the optimal means of achieving maximum transmission capacities and speeds. However, this alternative involves higher than average investment and is unlikely to be realised for the time being.

Yet improvements in capacity are also possible on the basis of the existing narrowband telephone networks. What are called for here are transmission technologies which permit rapid data transmission without physically altering the copper networks. ADSL (Asymmetrical Digital Subscriber Line) is the most promising of the numerous xDSL technologies. This makes more efficient use of the copper networks' capacities, allowing broadband communication for downstream data transmission (bound for the end user), while upstream the data

80 The formulation of the law is based on Moore's Law, whose message regarding the development of IT performance forms one of the key principles underlying the network economy (see section 3.3.1).
81 See Kelly 1998, p. 49

Source

Fig. 3.32: ECC 1999, based on Adstead / McGarvey 1997, p. 33

transfer remains narrowband.[82] This technology is specially suited for on-demand services such as Video-on-Demand.

Digital subscriber lines

The advance of ADSL is in full swing. On 11 May 1998 in New York some of the world's biggest telecommunications concerns (Deutsche Telekom, British Telecom, France Telecom, NTT, PLC and Singapore Telecommunications) signed a declaration of membership to the Universal ADSL Working Group, that up to then had comprised Compaq, Intel and Microsoft. The objective of this influential circle is to give an impetus to the development of this technology and establish ADSL as the world standard.[83]

One technology that can be used both in copper networks (such as telephone and TV cable networks) and in optical fibre networks is ATM (Asynchronous Transfer Mode). This transmission technology allows for a flexible use of bandwidths of up to 622 Mbit/s. ATM was established by the International Telecommunications Union (ITU) as the technical basis for data transmission on future broadband-ISDN-networks.[84] The particular advantage of this technology lies in its use of packet-switching data transmission in the telephone networks of the first mile, while analogue transmission technologies and the digital ISDN and ADSL depend on circuit switching. This makes it possible to obtain the gains in efficiency that come from packet switching in the first mile too. The disadvantage of this broadband transmission technology lies in the high adaptation costs it entails.

Interactive cable

Another network capable of transmitting multimedia content as far as the end user is the broadband cable network. Up till now this has been used for the unilateral transmission (from broadcaster to end user) of broadcasting content and as such it already offers sufficient transmission capacities. Even so, the cable network too will require investment for interactive services to be possible, because as yet it is not possible for the user to send control signals upstream.[85] This will either involve adapting the head-ends from which content is sent to the network participants or the user having to switch to the telephone line for interactive purposes.

Further investments will also be needed because the cable network has up to now transmitted its content indiscriminately to all participants within reach. For interactive use however it is necessary to make sure the content desired by the user is transmitted selectively to the specific customer who requested the data in question (for example a particular film). Yet the cable network is associated with the joint use of a transmission line by all the households within a neighbourhood. This has two consequences. As line capacities have to be shared, heavy use may lead to a slowing down in transmission speeds. In addition to this, there is a resulting security problem, and potential risks inhere especially in applications such as Internet banking. Nevertheless, the cable network in general offers an attractive opportunity for providing the user with the information, entertainment and communication services of the Internet through the first mile.

82 For bandwidth specifications see table 3.34
83 See anon 1998k, p. 10
84 See anon 1998f, p. 69
85 See Booz, Allen & Hamilton 1997, p. 105ff

High speed mobile

Data communication in the wireless mobile telephone network is at present limited by the European GSM Standard to very low transmission speeds of 9.6 kbit/s. With mobile phone networks, moreover, the reception quality is greatly dependent upon local conditions. Yet GSM is currently being developed into the Universal Mobile Telecommunications System (UMTS), and with the help of this system it will be possible for data transmission in mobile phone networks to reach speeds of up to 2 Mbit/s. UMTS should also permit the unrestricted combination of the communication services provided by fixed and mobile networks.[86]

Satellite

Broadband data transmission through wireless networks also underlies the plans for a number of future satellite projects. These projects are aimed in particular at the development of smaller reception devices and at improvements in transmission speeds. The need for development here shows parallels to the situation for cable networks. For large-scale data transmissions bandwidths in the range of up to several Mbit/s are required, as all the users of a satellite network in a particular country have to share the same channel and thus also its transmission capacities.[87] Used up to now for the one-way transmission of television pictures, broadband satellite systems have likewise not had a channel available for upstream transmission. Even so, Astra-Net and Eutelsat are now using their transmission capacities for communication services as well, for example offering on-demand services in catalogue pages or videos.[88] The interaction as yet still takes place the roundabout way through the telephone network. The new systems by contrast will be equipped with upstream transmission facilities. The following table provides an overview of the most significant projects in 1998:

Fig. 3.33: Construction projects for satellite networks for broadband data communication

Project	Cyberstar	Spaceway	Astrolink	Skybridge	Celestri	Teledesic
Company	Loral	GM-Hughes	Lockheed	Alcatel, Loral	Motorola	Craig McCaw, Boeing, Bill Gates
Availability (planned)	1998	2000	End of 2000	2001	2002	2002
Orbit	GEO	GEO	GEO	LEO	GEO, LEO	LEO
Speed (Mbit/s)	30	6	9,6	Downstream: 60 Upstream: 2	155	64
Number of satellites	n.a.	8+	9	64	63 LEO, 9 GEO	288
Antenna size (inch)	16	26	33-47	n.a.	24	10

86 See EITO 1998, p. 124f.
87 See EITO 1998, p. 125f.
88 See anon 1997c, p. 120

Source

Fig. 3.33: EITO 1998, p. 126

Of particular topical interest is the satellite project Iridium due to enter the market in autumn 1998.[89] Supported by such companies as Motorola, Siemens, Vebacom and Sprint amongst others, the Iridium network with its 66 satellites offers communication possibilities reaching the remotest corners of the earth. However, the bandwidths are not enough for large-scale data transmissions, so the network will be limited to mobile telephone services and pagers.

Using electric power lines

Finally there is also the possibility of private households using their connection to the electricity supply system for entering the Internet. Projects for developing the use of electricity mains for data transmission have not yet progressed beyond the test phase, but notable efforts are being made in particular by telecommunications companies with a background in the energy industry (e.g. o.tel.o), associations of international telecommunications and energy supply companies, as well as regional energy suppliers. The British firm United Utilities and the Canadian Northern Telecom for example have carried out a large-scale experiment testing the Digital Power Line transmission method (DPL). The results were bandwidths in the region of 1 Mbit/s.[90] The conditions thus exist for broadband data transmission. Decisive for the success of the Power Lines, however, are questions relating to the development of the decoders, the as yet uncertain economic viability of the electricity supply system compared with its main rival the telephone network, and last but not least the resolution of security problems such as data-protection or interference.

Different prospects for different technologies

The above developments in the various networks will in the near future bring considerable improvements in data transmission in the first mile. But the prospects of them superseding the telephone network vary from case to case.

Wireless transmission technologies can expect success in the first mile above all insofar as they meet the users' demand for mobility. Mobile networks will come in especially handy for people often on the move wanting access to the Internet's information and communication services at all times or for those who frequently spend time in remote regions where the construction and maintenance of any sort of cable network is economically impracticable. The demand for mobile communication services is demonstrated by the triumphal advance of the mobile telephone, which in West Europe enjoyed a growth rate of between 26 and 53 per cent from 1994 to 1997.[91]

Among the wireline networks, the cable network offers those households already connected an attractive opportunity to receive broadband access to the Internet. Along with the technological requirements described above, however, this will depend upon the further spread of cable connections. After all, once improved by xDSL technologies, the telephone network also represents a good interim solution for the near future. Within a few years, however, even the bandwidths achieved in this way will no longer be sufficient to meet the increased demands of users.

---

89 See anon 1997c, p. 113f
90 See Schürmann 1997, p. 47
91 See EITO 1998, p. 350

The following table compares the bandwidths technologically attainable by the various competing networks from a present-day point of view, together with the maximal transmission speed for each applied transmission technology.

Fig. 3.34: Speeds of the different data transmission networks

Network	Technology	Downstream	Upstream
Telephone network	Analogue transmission	14.4 to 56 kbit/s	
	ISDN	56 to 128 kbit/s	
	ADSL (Asymmetric DSL)	1.5 to 8 Mbit/s	640 kbit/s
	VDSL (Very high DSL)	12 to 52 Mbit/s	192 to 640 kbit/s
Cable network	Cable modem	1 to 10 Mbit/s	768 kbit/s
Optical fibre network	T1	1.54 Mbit/s	
	T3	45 Mbit/s	
Terrestrial	GSM	9.6 kbit/s	
	UMTS	2 Mbit/s	
Satellite network	Various projects (see above)	up to 155 Mbit/s	
Electricity supply system	DPL (Digital Power Line)	1 Mbit/s	

Strategic Aspects of Competition

Strategic importance of integrated services

The convergence of voice and data communication described above has led to the formation of a new market of integrated communications services which will steadily supplant the two separate markets.[92] Telecommunications companies are at present arming themselves for this new competition for global data transfer.

The competitors can be divided into two camps according to their origin. Powerful national telecommunications companies, stemming from the field of voice communications and with their roots in the previously monopolistic structures of the telephone markets, compete with firms from the dynamic data communications markets. The following analysis treats the starting positions of the rivals in the forthcoming competition as well as the strategic alternatives they have at their disposal.[93] Yet before turning to the big players with their own networks and the potential to provide integrated communications services for the most demanding of customer segments (such as business customers), our attention will focus on the competitive situation and the prospects for smaller Internet Service Providers (ISPs).

The services offered by small ISPs is limited mainly to access to the Internet and a handful of more or less distinctive service components. The current successes of these ISPs are supported by the extremely positive market development impelling the Internet forward. Even in areas of high competition

92 See section 4.2.4 on the processs of market erosion.
93 See also the remarks by Mario Martinoli following on pp. 92-93.

Source

Fig. 3.34: U.S. Department of Commerce 1998, p. A2-14; Mertz 1998, p. 94 ff.; Wayner 1998, p. 2 ff.; Dehn 1998; p. 30

density, most ISPs are thus still registering turnover growth in the region of three-figure percentages.[94] The present market seems to have room for an unlimited number of suppliers of Internet access. Yet it can be taken as given that these market conditions will soon yield to a much tougher concentration of competition.

Future demands in security, reliability and speed

With the increasing requirements of Internet customers resulting from the adoption of multimedia possibilities and the use of the Internet for business activities, there will be greater demands put on the security, reliability and speed of data transmission. As well as this, the costs that accrue and the value added services offered will become crucial criteria in the process of sounding out the competition. To gain a footing in the lucrative business customer segment in particular, it will be necessary to make substantial investments in transmission capacities out of the range of most small and middle-sized ISPs. Strategic options then boil down to concentrating on small, specialised segments or cooperating with powerful partners in order to be able to use their basic infrastructure. This also entails the possibility of being taken over by a financially strong partner.

Competition in the telecommunications market of the future will thus be determined by big companies with huge capital resources that are thus in a position to use their own networks to offer attractive integrated communications services from a single supplier.

"Next-generation Telcos"

This is indeed the objective currently being pursued by a number of firms from the field of data communications, also known as "next-generation Telcos." PSINet, for example, is at present investing in the construction of a global optical fibre network for the purpose of offering high-speed transmission with the highest degree of security primarily for business customers. Qwest is likewise laying a 16,000 mile modern optical fibre network that will ensure it high transmission capacities, and has further strengthened its position through the takeover of the smaller US long-distance call-supplier LCI. The satellite system Teledesic too is building up capacities, supported among others by Bill Gates.[95] The strategy followed by Worldcom is not to build up capacities from scratch, but instead create an Internet empire through the new acquisition of firms such as UUNet and MCI. Even though MCI had to split off its Internet operations so the monopolies authorities would give the deal the green light, Worldcom now controls a large part of American Internet traffic. To reinforce this position on a global scale, Worldcom together with Cable&Wireless have invested half a billion US dollars in a sea cable network.

The competition situation faced by these next-generation Telcos is characterised by a number of strategic advantages. As has been seen, they certainly have the financial muscle necessary to consolidate and extend their own networks. By contrast with the telephone companies, moreover, they also enjoy advantages in terms of customer orientation and flexibility of business culture. In particular, they are used to the pace of innovation in the Internet, where innovation cycles are measured in months.

94 See Froitzheim 1997, p. 40f.
95 See Fig. 3.33

"A tightrope walk between self-cannibalisation and investment in the future."

The quality of the networks being operated is a further decisive factor. As described above, almost all the main rivals are notable for their investments in the most modern optical fibre networks guaranteeing rapid data transmission. These networks are markedly superior to the telephone networks, which are optimised for connections lasting three to four minutes, and this superiority manifests itself in higher efficiency and lower costs. The new networks satisfy the highest demands in security and reliability too. While on the one hand allowing access to what is offered by the Internet, as Extranets they are on the other hand protected by firewalls from the anarchy of Internet structures. For the lucrative business customer segment in particular, the offers here can thus prove highly attractive.

Strategic dilemma of incumbents

The starting situation of the big national telephone companies by contrast is characterised first of all by a strategic dilemma. On the one hand they have to prepare for the vast market in integrated data communication, while on the other their efforts are focused on protecting their core activities in telephone services. The profits from this area are needed for investment in the new technologies, yet these investments may paradoxically in turn threaten their core activities, for example through speeding up the progress made by Internet telephony. The support given by Deutsche Telekom to VocalTec, one of the pioneers in Internet telephony, thus constitutes a sort of tightrope walk between self-cannibalisation and investment in the future.

Challenges

In this context, the big telephone companies are faced with a number of challenges that will necessitate drastic breaks with the past. The priority has to be to introduce greater flexibility into their business culture in order to be able to respond to the quickening innovation cycles. In particular this means a shift from an engineering culture to a market culture where it is the customer who moves to the centre of the enterprise. On the technological side, existing networks have to be replaced by new, more competitive ones. The resulting financial task is thus to cope simultaneously with the rapid depreciation of the existing infrastructure and with huge investments in new technologies.

Competitive advantages

In spite of all this, the telephone companies do enjoy certain competitive advantages that for the time being at least secure them a clear lead over their rivals above all in the original national markets. Their greatest potential lies in their direct access to customers in the first mile. Another factor is that customers are used to dealing with their own national telephone company. A degree of familiarity and confidence has thus been built up which other companies still have to attain. Any definitive switch to new suppliers is as a result always bound up with uncertainties. Finally, the market position of the telephone companies in their own country is bolstered by long-standing business connections, political clout and extensive networks with high standards in security and reliability.

These strengths only carry weight for defending their own position in the domestic markets, however, and are hardly applicable at all in the competition for global data communication. To conquer new markets, it will thus be necessary to

fall back on many years of telecommunications experience and sheer financial might. The capital resources enjoyed by the big telephone companies are as yet still unattainable for the new network operators, giving AT&T, Deutsche Telekom and others considerable scope for action as regards the takeover of interesting smaller companies or awkward competitors.

International alliances

The starting position of the telephone companies is supported by the big international alliances they have already formed. These have particular control over the network connections between Europe and America and include Global One (Deutsche Telekom, France Télécom and Sprint), AT&T-Unisource (AT&T and Unisource, a European alliance between PTT Telecom, Telia and Swiss Telecom) and Concert (British Telecom, MCI and Viag-Interkom). Yet Concert also provides an example of the tough competition that has flared up at the starting posts for the future telecommunications market and of the possibilities already available to the next-generation Telcos. The partnership between British Telecom (BT) and MCI has been broken up following the 37 billion dollar takeover of MCI by Worldcom, who then prevented the further alliance of MCI and BT. Yet in July 1998 BT found a new partner in the American market: the second biggest telecommunications concern in the world, AT&T.

Players and losers

In conclusion it can be said that the competition for integrated communications services is being determined by big, financially powerful businesses from two fields: the next-generation Telcos from the field of data communication and the big national telephone companies from the field of voice communication. Companies such as the smaller Internet Service Providers will tend to lose this competition owing to their lack of size.

Global integrated telecommunications market

The new ranks of next-generation Telcos stand out especially owing to their construction of modern transmission capacities and their flexible business culture. The threat they pose to the telephone companies becomes clear, for example, with the triumph of Worldcom over BT in the struggle to take over MCI. For their part, the telephone companies have to cope with serious problems connected in particular with their rigid business culture and their dual objectives of protecting their core activities while also investing in the future. If they prove able to meet these challenges, however, then the strengths enjoyed by the telephone concerns in terms of capital resources, experience and long-standing connections will come to full fruition. At present they are faced with strategic opportunities that can best be realised with the aid of the appropriate international partners or by taking over interesting smaller firms. Only in this way will they be able to assume a strong position within the global integrated telecommunications market of the future.

Telcos vs ISPs: who will win the race?

By Mario Martinoli

One of the hottest topics in today's debate on the future and the evolution of the Internet market structure is the role played by Dominant Telecom Operators. Their entry in the Internet market opened wide development perspectives, due to their clear advantage in terms of existing infrastructures.

Independent Internet Service Providers (ISPs) looked at their entry in the Internet market as a big threat for their business. Undoubtedly a global redefinition of the Internet market and a redistribution of forces will lead in a short time to a re-assessment of the existing balance between the players, with very high risks for independent ISPs to lose their market shares.

A bandwidth play

The Internet game is indeed a bandwidth play: the telcos have it and the ISPs need it. Right now, with mostly text mail and Web, ISPs survive. But when demand for voice and video on the net increase, the winners will be those who have the wires and fibres in the ground. This will be the moment when the real shake-out will happen.

Internet Access Providers (IAPs) and small ISPs focused only on access services will not be able to survive, unless they move towards the supply of value-added services such as electronic publishing and technical consultancy for the establishment and management of websites. It will be difficult for them, however, to get into the more profitable content aggregation and specialised applications business since at present they seem to lack the core competencies. Therefore their number is likely to decrease dramatically.

Strategies for ISPs

To survive the shake-out, medium-sized ISPs will have to rely on one or more of the following strategies: To look for strong alliances with telcos. This may represent the only chance for independent medium-sized ISPs to stay in the Internet service provision market. This solution is by far the most logical development for them. Oleane has developed in France a network of access partners with about 70 ISPs, becoming their exclusive backbone supplier and distributing back, in exchange, a share of the income collected from the final customers.

To exit the Internet access market and to enter the growing content development and value-added services provision market, thus competing with On-line Service Providers (OSPs) such as Compuserve. Curiously enough, the dramatic explosion of the Web in late 1996 led to a partial outpositioning of OSPs in the Internet market. These organisations then quickly converted their operation to Internet Service Provision, occupying in this way the ISPs business sector. In other words, OSPs became ISPs, retaining however some advantages on them, as they were already well positioned in Internet commercial content provision business and in customised products

delivery. As a logical reaction, ISPs looked at them as a threat for their business, but now the on-going shake-out seems to reverse the game, and ISPs could become major competitors for OSPs.

To focus strongly on market niches, in terms of access, service provision and geographic coverage. Independent medium-sized ISPs are aware of the on-going shake-out, but most of them do not seem engaged in investing into value-added services development and are reluctant to make alliances with telcos. Do they still believe to have real competitive chances in the Internet service provision market?

Concerning the penetration of international operators into the European market (such as WorldCom and AT&T), they will probably not be able to gain more than 5%-7% of customers in large European markets such as France and Germany within the next five years, according to a series of studies made by investment banks.

Dominant position of national European Telcos likely to last – for some years

The dominant position of national European telcos, which is likely to last still for some years notwithstanding the process of liberalisation of the telecom markets, and their penetration capacity into both the consumer and business market will create some problems for international telcos wishing to enter the European markets. To overcome these barriers and to increase their profitability, they have a series of options, mostly consisting in establishing joint ventures with selected local operators or positioning as niche players offering services to a carefully selected base of customers.

This penetration scenario will also impact the Internet service provision market. In a medium-term perspective, with an increasing demand for on-line multimedia services, this market is likely to be largely dominated by large national operators and their ISP businesses, such as TIN in Italy or T-OnLine in Germany, keeping new entrants knocking at the door of the future Internet business.

3.2.4 Revenue Types and Revenue Models in the Telecommunications Sector

Basic model

The individual segments of the telecommunications market exhibit a relatively homogeneous structure in their revenue models. In the monopolistic markets for telephone services the most frequent revenue model was a mixture of monthly standing charges and usage-related payments based on the length of call times measured in specified units. This revenue model oriented itself on the costs that accrued. The monthly standing charge here corresponds to the service of connecting the household in question directly to the telecommunications network. The costs for the maintenance and servicing of the first mile are charged directly to the customer. The usage-related payments by contrast represent the costs arising through the use actually made of the network for telephone calls. In addition to this come the one-off connection fees charged for example for a new telephone connection.

Changes and extensions

The liberalisation of the European telecommunications markets has led to a number of changes or deviations from this basic model:

- The monthly standing charge revenue form can no longer be sustained by new suppliers, as the monthly payments are as a rule not matched by any corresponding service. Only when the new supplier takes over a household's connection complete from the former monopolist is a monthly standing charge still imposed. The suppliers of long-distance and international calls on the other hand have to rely on the usage-related revenue form, regardless of whether the customer has preselected them or chooses them call-by-call. In Germany, Mannesmann Arcor has established another mixed revenue form. Their revenue model comprises two different usage-related revenue types: charges are based on the one hand on service time and on the other hand on service quantity as well. This means a customer paying a fixed price of six pfennigs for each call made (service quantity) and on top of this a further charge calculated according to the length of the calls made (service time).
- There is another new development induced by the liberalisation of the European telecommunications markets that also seems to be catching on. A number of suppliers are now replacing the rather inexact measurement of call time (in longer units) by a more meticulous measurement made in seconds. This appears to be establishing itself as the standard.
- In Sweden the telephone business has adopted its own particular form of media financing. Just as TV advertising means that the media content on offer is not directly paid for by the user as it is the advertising economy that bears the costs, customers of Gratistelefon Svenska AB in Sweden can make long-distance calls free of charge.[96] In return, the conversation is interrupted by occasional 30-second commercials. The two interlocutors each hear different

96 See anon 1997e, p. 8.

commercials that vary according to the information that is known about them. For the person registered as making the call this includes not only the usual data but also personal data relating to profession, hobbies and age. For the person being called the differentiation can only be in terms of regional location.

US flat rates

In the United States there is another model being put into practice, whereby the local calls are contained in the monthly standing charge and no usage-related payments are incurred. The basic model here amounts to a subscription for local calls.

Mobile telephony

The mobile telephone segment uses the same basic model as the suppliers of telephone services. Here too there is a mixture of monthly standing charges and revenue dependent on call time, although the suppliers vary in their combinations of these two revenue forms. The customer is thus faced with the choice between high standing charges and low costs-per-minute (attractive for customers with a high call time) or low standing charges and high costs-per-minute (attractive for customers for whom the most important thing is to be easily reached).

Data communication

In the field of data communication the revenue types for Internet access are of interest. The revenue models do not differ fundamentally from those in the field of telephone services. Noteworthy is not only the mixture of monthly standing charges and usage-related payments, but also the financing by advertising (for example Germany Net) or the subscription system, where a monthly standing charge alone is incurred (a flat fee). Yet the significance attached to the various revenue models varies. While the usage-related revenue form dominates the telephone services, for Internet access it is mainly the monthly standing charge that prevails.[97]

97 See EITO 1998, p. 75

3.2.5 Value Chains in Telecommunications

In the simplest case, the value chain in the telecommunications sector consists of four functional elements, which have to be integrated by the market participants with the end consumer in mind.

Fig. 3.35: Simple telecommunications value chain

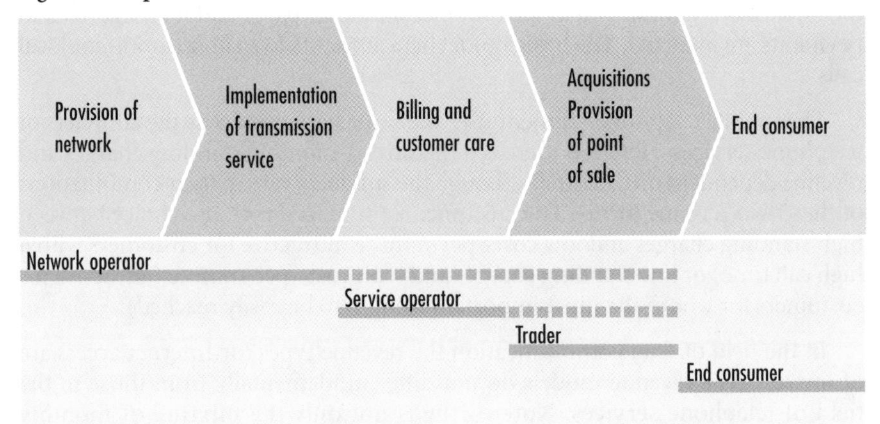

Four basic functions in the value chain

The example of mobile telephony provides a clear illustration of the individual activities belonging to each basic function (investment, production, billing and sales). [98]

* The investment function corresponds to the provision of the network, which refers to all those activities that serve to make the allocated frequency spectrum usable through the physical construction of the network components and their maintenance in working order. These include the acquisition of frequency usage rights, the network planning, the installation and maintenance of the basis stations, the feeding and updating of transmission in the switching equipment, among other similar activities.
* The production function corresponds to the implementation of transmission services. This incorporates in the first place simply the technical process of converting and transmitting the user's information content. Yet once switching is logically separated from transmission, there arise further differentiations within the production function in the value chain. Distinctions can then be drawn between the provision of transmission lines, the provision of switching equipment, the implementation of transmission services and the implementation of switching services.

The implementation of switching services includes the recording of usage data. Usage data are above all a prerequisite for billing, but at the same time

98 See Tewes 1997 for the following remarks.

Source

Fig. 3.35: Tewes 1997, p. 18

provide marketing-strategic insights into customer behaviour that can be applied in a variety of ways (as with data mining or data warehousing).[99]

The activities included in the billing function on the one hand encompass the price-assessment for the services demanded by the customer and on the other the sending of the claim for payment. Collection also refers to the right to impose the claim and demand the sum due in payment.

* The other party to the end customer's contract is the company imposing the charge. Linked to this are further responsibilities and duties, such as customer care in particular, which can also be seen as a marketing instrument.
* The sales function covers all those activities that serve the acquisition of new customers (preparation and conclusion of contracts). These include the provision of a physical or virtual "point of sale."

Functions and their suppliers

The above functions within the value chain can be assigned to the various suppliers of telecommunications services as follows:

* The network operators are responsible for the provision of the network and the implementation of transmission services (the investment and production functions).
* Service Providers fulfil the functions of billing and customer care.
* Traders assume the sales function. These can also include businesses and institutions whose core activities lie in other markets, such as credit card companies, banks or automobile associations.

Integration of functions

Vertical forward-integration of the functions is possible for network operators over the value chain in its entirety, but only without leaving out any individual value chain stages. For Service Providers it is only the sales stage that can be integrated. Backward-integration by contrast is not possible. Integration strategies can further be arranged according to customer segments. As a rule there is a considerable difference in the integration of the value chain for the business sector from the consumer sector.

Mobile telephony service categories

In the simple value chain model, the various categories of services or bundled offers at the individual value chain stages can be easily differentiated. In the market for mobile telephony for example, four service categories can be distinguished, which can then be classified according to three criteria.[100]

99 Data warehousing refers to the systematic storage of large quantities of user-related data over long periods of time. Data mining denotes the processing of the stored user data for specific purposes.
100 See Tewes 1997, p. 19ff.

Fig. 3.36: Classification of service categories in mobile telephony

	Licence content	Network basis	Network-use contract
Basic service	yes	yes	yes
Network-dependent integrated service	no	yes	yes
Network-independent integrated service	no	no	yes
Value-added service	no	no	no

Regulatory shaping of the value chain

Basic services are regulated through licence mechanisms, and thus can be attributed to the production function (implementation of transmission services) of network operators. This is an example of direct regulatory intervention in the structuring of the value chain.

Network-dependent integrated services, such as voice mail or SMS (short message service), are not demanded by licence. As the provision of such services requires an alteration in the network basis, however, these services must likewise be assigned to the investment or production function and so to the network operators.

Which of these two categories a specific service is assigned to varies according to the regulations in force in the country in question as regards the scope of the licence. For this reason they are often brought together under the concept of network-based services.

Network-independent services

With network-independent integrated services, the close connection with the network basis is no longer applicable. These are services provided outside the network system and either made available via a man-machine interface as an information good through the network or delivered as a material good outside the network. They include for example box office or flower services, as well as information and office services. In spite of generally being network-independent, such services can only be used by customers with a network-use contract. Charges come with the telephone bill, so this category of service is assigned to the billing function and thus to the Service Providers. With the increasing substitution of the man-machine interface by a machine-machine interface thanks to software-based network intelligence, the boundaries with network-based services are becoming progressively vaguer (as with computer-aided traffic control systems).

The category of value-added services encompasses all those services not included in the preceding ones. As the utilisation of these services does not even require a network-use contract, they are best assigned to the sales function and so

Source

Fig. 3.36: Tewes 1997, p. 20

to the various traders. Examples of such services are consultation services, the supply of accessories at the point of sale, and telebanking.

Telebanking an example of coupled value chains

The example of telebanking illustrates problems of demarcation that apply analogously to all forms of e-commerce. Considered from the perspective of the mobile telephony value chain, telebanking services are in the first place value-added services provided by the bank. Once the bank grants its newly-won telebanking customers discounts on bank transactions that are dependent on the network-use contract, however, then there is a structural coupling between the distinct value chains of the mobile telephony and the financial services markets, in that utilisation of the discounted bank services is linked to a network-use contract.

Lock-in, bundling and coupled services

In the attempt to achieve the benefits of "lock-in," today's marketing and sales activities are laying increasing emphasis on this sort of "bundling" of distinct use-dimensions with the end consumer. It thus makes sense to rename this value chain stage as coupled services. This opens up broader perspectives for the analysis of all the value chain stages in the telecommunications sector, not just the sales stage of the value chain.

Before taking into account the structural couplings with before, during and aftermarkets, the expanded value chain for the telecommunications sector looks as follows:

Fig. 3.37: Expanded value chain in the telecommunications sector (for mobile telephony)

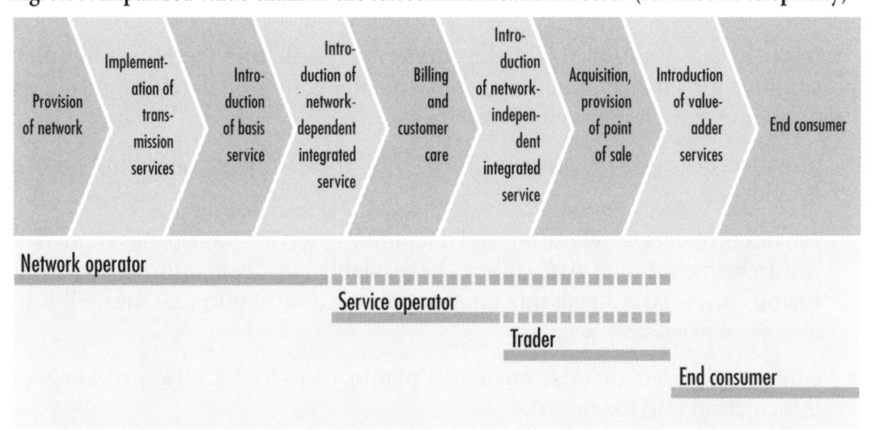

This expanded model makes it possible to identify the relevant players in the telecommunications markets, assign the functions they each perform and also differentiate the important service categories.

The value chain can be further expanded by introducing intelligent network structures into the picture, as shown by the following diagram:

Source

Fig. 3.37: Tewes 1997, p. 23

Fig. 3.38: Expanded value chain in intelligent networks

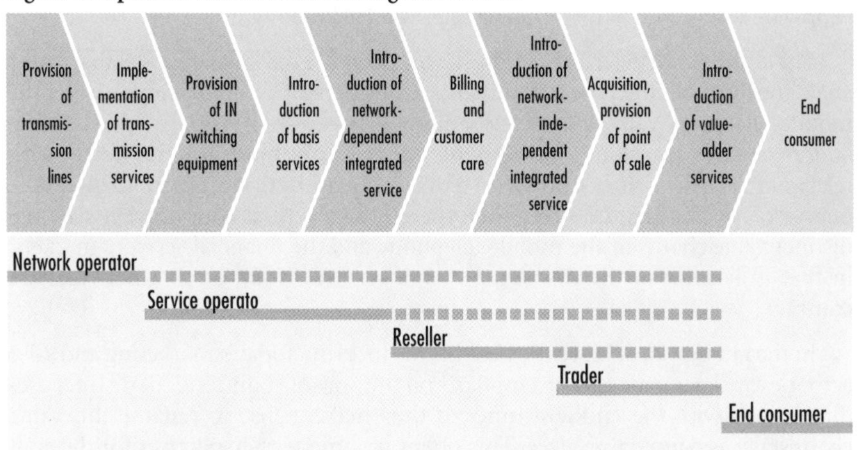

What is decisive here is the functional separation of transmission lines and switching equipment. This entails a widening of scope for network-independent Service Providers in terms of structuring and classifying the services they offer. Traditional Service Providers, limiting themselves to the billing function alone, are from this point of view regarded as resellers. In this sense they are to be distinguished from Service Providers that no longer restrict themselves to the billing function but also run intelligent switching equipment.

Intelligent networks expand the value chain

In open, decentralized and intelligent network structures there are additional possibilities for further provider-groups to enter into the telecommunications value chain. Examples of such groups include:

* providers of network management services for network operators and Service Providers;
* providers of data warehousing and data mining services (analogous to market research institutions) that collect the available customer information from various markets, evaluate this information for diverse purposes and sell it to interested parties, such as
* content providers that use intelligent platforms to feed a variety of kinds of information into the network.

The last case is a typical example of structural coupling. Participants in other value chains – such as software-suppliers, music firms, on-line publishers, radio and TV programme-suppliers, or car companies – here use the "technical reach" of telecommunications networks of any sort (fixed networks, mobile telephone networks, cable networks) as a potential market, sales channel and marketing instrument for offering services outside the field of telecommunications. The strategic integration of previously separate value chains takes place through server

Source

Fig. 3.38: Tewes 1997, p. 66

computers or subscriber management systems coupled to the network with machine-machine interfaces.

A consequence of this kind of coupling for the telecommunications value chain is that the significance and the portion of network operators in the expanded value chain is tending to fall. To the extent that network operators want to participate in the expanded value chain, their only remaining strategic option is vertical forward integration in the direction of the Service Provider function. The objective of such integration is to use their own network to deal with all the traffic generated by the customer. The network operator can here concentrate on taking over the billing function and/or its own marketing of the services provided by the content providers. As these are then being treated as mere suppliers, such strategies clearly run against contrary interests among the content providers.

Falling share of telcos, additional chances for outsiders

Such is the starting position in the competitive struggle to dominate the multimedia value chain.

3.3 Digital Data Crunchers – An Analysis of the Information Technology Sector

Information technology (IT) is the newest of the three Media & Communication sectors. The concept "computer or IT industry" is also used analogously to the concept of the IT sector. In what follows, the concept IT will be used to denote all technologies whose technological base involves digital circuitry and microelectronics.

The beginning of the computer age in the technological sense dates back to the year 1941 in Berlin, when Konrad Zuse built the first working program-run calculating machine. Parallel to this but without knowing about Zuse's work, the American Aiken together with the firm International Business Machines (IBM) was developing the calculating machine MARK I in the early forties.[101] Nonetheless, it was another twenty years before the launching of the IBM S/360 mainframe computer really got the modern computer age off the ground.

Computer origins in Europe and in the USA

The ensuing market development of the computer industry is remarkable in two respects. Firstly, market growth is above average in comparison to other industries. An industry that did not even exist in the fifties had by the early eighties already reached a worldwide market volume of some 50 billion ECU. And this dynamic growth is not letting up at all, so that in 1994 the turnover had climbed to some 453 billion ECU worldwide and in 1997 the figure was around 621 billion ECU. From a global perspective the United States market has absolute supremacy in the IT industry. In the USA an IT turnover of over 259 billion ECU was thus achieved in 1997, while Europe with 175 billion ECU and Japan with 95 billion ECU attained a significantly lower market volume.[102]

Worldwide dominance of US industry in the 90s

101 See Gründler 1997, p. 29
102 See EITO 1998, p. 21; p. 320

Innovation fed by intellectual and venture capital

The second special feature of the IT industry is the speed of its technological progress, which is extremely high in comparison to other industries. This also explains the central importance of the United States, where conditions exist that are exceptionally favourable to technological innovation. Particularly in such areas as Silicon Valley in California or Boston in Massachusetts, there is a concentration of a whole range of factors that exert a positive influence in this respect: a critical mass of intellectual capital on account of being located close to excellent universities and research institutions and above all good access to investment capital owing to the presence of large numbers of venture capitalists. In addition to this, the USA is characterised in general by a cultural acceptance of ventures that fail.[103]

Product development at lightning speed

As a result, the principle of "trial and error" means that technological innovations are brought onto the market in very short development cycles and removed from the market after even shorter product cycles.[104] Control of this speed is a prerequisite for enterprises wishing to play a leading role within the IT industry, as the example of the microprocessor-manufacturer Intel has clearly proved. Even allowing for the fact that in 1997 one of Intel's main revenue-spinners, the Pentium Processor, was brought onto the market in a more powerful version (Pentium II), the change in turnover structure within a year was extreme: in 1997 Intel achieved about 90 per cent of its turnover with products that in 1996 were not yet on the market.[105]

3.3.1 Size and Structure of European IT Markets

Three IT market segments: hardware, software, services

The market structure of the IT industry comprises the market segments hardware, software and services. The hardware sector consists of the personal computer (or PC) market, work stations and servers. Along with classical desktop computers, PCs also include the field of portable computers such as laptops. Target groups are both end consumers and business customers. Work stations and server systems on the other hand are directed exclusively at the business-to-business market. UNIX, Windows NT or other server systems such as the IBM S/390 can be classified according to their operating system.

In addition to this there are three further areas of hardware to be taken into account. The peripherals necessary for the systems in question include printers and keyboards etc. Hardware equipment from the field of data communications facilitates local and geographically dispersed network structures, i.e. so-called local area and wide area networks (LAN, WAN). The field of office communications incorporates typewriters, pocket calculators as well as photocopiers.

The software sector can be subdivided into the two market segments, systems software and user software, both of which enjoy an equally large market volume. The systems software on the one hand includes systems-infrastructure software allowing the operation and control of the most diverse hardware platforms in

103 See Cringely, 4 April 1998, p. 4. This does not mean that failures are accepted without limits in the USA, but refers rather to the conviction that running an enterprise is a learning process of which failure can also be a part: "To make a mistake once is human, to make it twice is a crime."

104 Particularly in the case of software programs, extreme time pressure leads to a situation that would be unthinkable in other industries. Consumers themselves are in many cases a test market for incomplete and faulty early versions. The expression "First ship it, then test it" is a prerequisite from the point of view of the business enterprise, but from the point of view of the consumers it is an annoying fact of life.

105 See anon 1998j, p. 3

systems. Here the range runs from complex systems-solutions for the management of a business enterprise's IT capacities, through server software such as UNIX, to operating systems such as Microsoft Windows. On the other hand systems software also incorporates the various third (e.g. FORTRAN) and fourth generation (4GLs) programming languages.

User software refers to software designed for individual functions such as word processing or spreadsheets, which target both the end consumer and the business customer. It also includes complex applications for business management, which as with the program R/3 from SAP can range from the handling of production supply to accounts and personnel management. These multiple solutions find their application exclusively in the business-to-business market.

Three types of IT services

Owing to the complexity inherent in the IT industry, the services sector has taken on a central role within it. The increasing need to cut IT costs within business concerns has led to a sharp growth in the field of operations management. This refers, for example, to the concept of outsourcing, in other words the farming out of partial tasks or even the complete handing over of IT management to independent firms.

The second biggest market in the services sector is systems-integration, which denotes the technological implementation of individual IT solutions. As a rule this entails all or some of the following steps: the drawing up of operations and functions, the choice of hardware and software, the programming of individual software applications and the implementation of the system and training.

A third area is composed of classical support services such as telephone hotlines. In addition to these, a whole range of electronic support services such as fax-back services, Internet services and CD-Rom based FAQ-applications have been developed in the last few years. Further, the field of management-oriented consultancy analyses the effectiveness and efficiency in the use of IT with regard to the firm's strategic objectives. This branch of services performs mainly in combination with technological services such as operations management and systems integration.

IT industry revenue distribution

In what follows, the significance of the above-mentioned market segments and subsegments within the IT industry will be illustrated by looking at Western Europe. It should be pointed out that the distribution of turnover in the West European market is roughly equivalent to that in the United States market as well as other world markets.

Hardware is by far the largest IT market segment. Owing to its unfavourable cost structure with relatively constant marginal costs, hardware turnover levels are on the whole high. Here it should be stressed that the revenue structure in the hardware sector is characterised by falling margins. The reason for this lies above all in hardware's diminishing value for money from the perspective of the customer.

Fig. 3.39: Market values in the Western European* IT sector in 1998

Markets	Market segments	Market value in billion ECU	Market value as %
IT hardware		86.8	44.9
	Personal computers (desktops, portables, PDAs)	31.6	16.4
	Servers (Unix, NT, other)	25.0	12.9
	Workstations	2.4	1.2
	PC/workstation add-ons (eg. printers)	10.4	5.4
	Data communications hardware (LAN, WAN, other)	7.8	4.0
	Office equiment (eg. copiers)	9.5	4.9
Software products		38.3	19.8
	Systems-software (eg. operating systems)	18.6	9.6
	Application software	19.7	10.2
Services		68.0	35.2
	Operations management	23.1	12.0
	Implementation	22.6	11.7
	Support services	15.3	7.9
	Consulting	7.0	3.6
Total IT market		193.0	100.0

Cost and revenue structures differ between hardware and software sectors

The opposite relation between costs and revenue can be observed in the software sector. On account of marginal costs that are close to zero, turnover levels are all in all lower while profitable margins can be attained throughout the software sector. Because of the rapidity of innovation and the complexity of technological progress the service sector is characterized by strong demand. For this reason the bundling of services such as maintenance, guarantees and hotlines together with hardware products is a common strategy to compensate for the low margins in the hardware segment.

Source

Fig. 3.39: EITO 1999, p. 376
* Western Europe includes the 15 EU countries plus Norway and Switzerland

3.3.2 From Mainframe Computer to Internet-PC

Technological quantum leaps, new industry structures

The market development of the computer industry is marked by the exponential rise in the performance of information technology. As a consequence of continuous innovations, technologies have come into being that can be regarded not just as involving a technological quantum leap but above all as triggering and laying the foundation for a whole new industry structure. Three technological paradigm shifts in particular come to light in this context. Below, these shifts are drawn on to classify distinct market phases: the system-based, the PC-based and the network-based market phases.

In the system-based phase it was the mainframe computer and the minicomputer that occupied centre stage, while in the PC-based phase it was the personal computer that played the key role. The network-based phase, which began around 1994, is characterized above all by the increasing connectivity of computer and other networks, in particular the Internet.

Fig. 3.40: Phases in IT market development

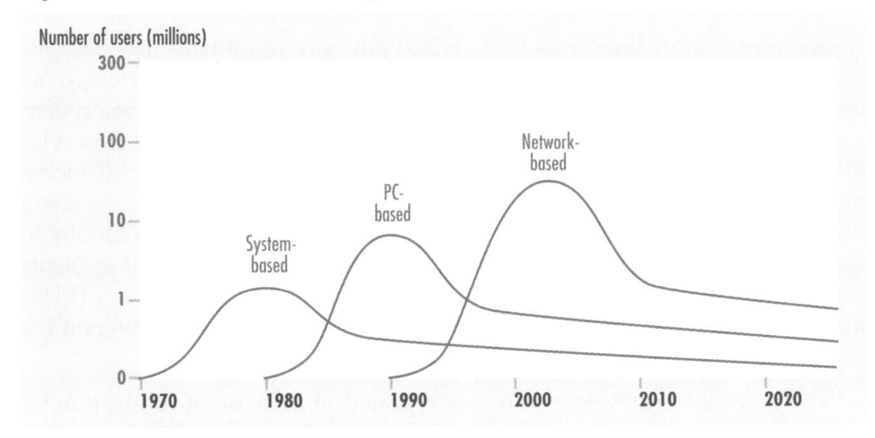

* *System-based Phase (1964 - 1981)*

The modern computer age began in the year 1964 when International Business Machines (IBM) introduced electronic mainframe computers onto the market with the product line S/360. Through its great commercial success IBM succeeded in transferring the already existing monopoly in the field of mechanical calculating machines into the electronic era. IBM's product line and the products in competition represented a technological paradigm shift in that in contrast to the data processing equipment in use since the mid-fifties transistors were for the first time integrated into electronic circuits. The consequences were a drastic reduction in the size and significant improvements in the work that could be performed by data processing equipment.[106]

106 See Gründler 1997, p. 33

Source

Fig. 3.40: ECC 1999, based on Moschella 1997, p. viii

The "minicomputers" available in the early seventies such as the PDP-10 from Digital Equipment Corp. (DEC) were based on circuitry that was highly integrated, i.e. a technological development of the electronic circuits already in existence. Minicomputers were owned mostly by large-scale business enterprises and state-run or university institutions, and like mainframe computers could only be centrally operated. Data-processing was possible through input devices (terminals) that were connected with a mainframe. As it would have been too expensive to make demands on the computer facilities continuously, it was possible to buy computer time (through "time-sharing"), which could then be turned to account for the most varied of tasks. Computer time was thus a scarce commodity requiring goal-oriented use to be made of computer facilities.

Grosch's law

This cost relation can be expressed through Grosch's Law. In the forties, the computer pioneer Herbert Grosch ascertained that computer power increases as the square of cost. This means, for example, that as the cost of a computer doubles, four times the processing power becomes available. This message was one of the chief arguments supporting the acquisition of bigger systems and continued to apply extensively throughout the sixties and seventies.[107]

IBM mainframe: the vertically integrated business model

With its product line S/360 IBM created not only a profitable new business sector but above all a business model for the entire IT industry. IBM produced the hardware, configured the system, programmed the necessary operating system and the relevant applications, structured the sales market and delivered the service to its business customers. This model of a vertically integrated business concern made good sense, as neither in the hardware nor the software field was there a unified industry standard. Accordingly, it was more profitable economically to concentrate on selling proprietary systems than to focus on individual segments within the value chain. Only a few firms like Electronic Data Services (EDS) managed to become key players in the IT industry while concentrating on the service sector alone.

The vertical business structure was repeated in the case of minicomputer companies, albeit with some differences. A large number of the minicomputer-producers such as DEC or Data General were newly founded companies which on account of low levels of company capital were unable to develop all their components themselves and so inevitably had to rely upon other manufacturers. With increasing growth these companies nonetheless came to produce more and more component parts themselves, so that by the end of the seventies most of the minicomputer companies likewise showed a high degree of vertical integration.

In general the supply structure of closed systems[108] had the advantage of stable IT conditions for a customer firm. A disadvantage for business customers was the so-called "lock-in," which refers to the fact that any change to a different supplier would entail high costs. In addition to the acquisition costs for hardware and software, it would involve the employees concerned learning how to deal with a whole new operating system and the corresponding applications. This option

107 See Moschella 1997, p. 15

108 "Closed architecture" refers to a system that does not lay its specifications open, so that other firms are unable to produce complementary products. The antithesis is the open system. Proprietary architectures on the other hand denote the fact that the system is in the control and ownership of one enterprise. The antithesis is the system controlled by a consortium. Microsoft Windows is in these terms a proprietary and open system. UNIX is an operating system controlled by a consortium, which is also open. Apple is a proprietary and closed system. For further details, see Moore 1998, p. 51

normally seemed so impracticable that any decision to go for a particular system architecture was generally long-term in nature.

Thanks to the product line S/360 and its additional services, IBM was the clear market leader in the IT industry throughout the era of mainframe computers. Its superiority was so manifest that IBM and its competitors (RCA, General Electric, Burroughs, Univac, NCR, Control Data and Honeywell) were referred to as "IBM and the seven dwarfs." This market dominance could only be broken by one firm: IBM itself.[109]

* *PC-based phase (1981 - 1994)*

In the early eighties the attention of IBM's business strategists was attracted by the market for microcomputers. The construction of microcomputers had been made possible by the progressive miniaturisation of highly integrated circuitry, leading in 1971 to the development of the microprocessor on a chip. Early microcomputer models were the ALTAIR 8088 in the year 1975 and the first personal computer (PC) from the firm Apple, known simply as Apple I.[110] As the market was characterized by unbroken growth and by 1980 was already turning over around a billion US dollars, IBM decided to enter the market.

PCs: Apple, IBM and MS-DOS

A decisive factor in what was to follow was that contrary to company strategy IBM responded to pressure of time by taking the most important constituents from outside businesses and thus in this case not creating a proprietary system. The firm Intel, founded in 1968, provided the IBM personal computer with microprocessors from the 8088 series, while Microsoft, founded in 1975, supplied the operating system MS-DOS. In autumn 1981 the IBM-PC came onto the market, effectively getting a whole industry off to a start.

Four fundamental trends

The ensuing development of the IT industry is marked essentially by four fundamental trends:

* the central role played by microprocessors
* the change from closed to open computer systems
* the dominance of de facto standards
* the transformation from a vertical to a horizontal industry structure.

Microprocessors are characterized by technological features that have remarkable economic consequences. These can be described by Moore's Law, which is based on a prediction made in 1965 by Gordon Moore, one of the founders of the microprocessor manufacturers Intel. According to this law, if production costs remain constant, the number of transistors – i.e. the complexity – of microprocessors will double every 18 to 24 months.[111] This will make it possible to achieve an exponential increase in the power of computer architectures and the software run on them. The result is a radical drop in the

109 See Moschella 1997, p. 2
110 See Gründler 1997, p. 33
111 This technical development has physical limits. Experts currently reckon that the year 2005 will see the end of this development. Even so, semiconductor manufacturers such as Intel, AMD, Cyrix and Centaur are working feverishly on technologies to permit the law's validity to be extended. See Madden 1998, p. 64

price of computer processor and memory capacities, as illustrated by the following diagram:

Moore's Law

Fig. 3.41: The development in costs of information processing per command

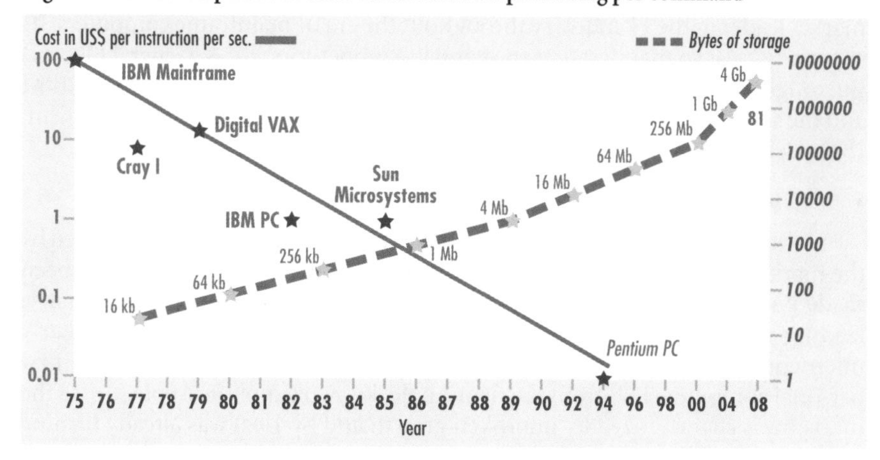

Moore's Law thus became the fundamental principle underlying the dynamics of the entire PC industry.

Open systems

The second central factor influencing the development of the IT industry is the transformation from closed to open computer systems, which will here be analysed taking the example of the development of the PC industry. An account of competition behaviour will bring to light above all the significance of open systems for the success or failure of business enterprises.

In the period of mainframes and minicomputers, closed systems were predominant, meaning that every company had its own hardware-software configuration which was not compatible with other manufacturer platforms. Because of the cost structure of these systems it was the hardware configuration that played the decisive role in gaining competitive advantages, while software was of secondary significance.

Given its experience of these conditions, IBM in 1981 aimed at market leadership first and foremost through a superior hardware configuration. Although the "open" construction of the IBM-PC (i.e. the disclosure of the most important specifications) for the first time made it possible to imitate the IBM platform, the danger of this was judged by IBM to be slight. For a start the cost-effective production of PCs did not appear feasible without the economies of scale enjoyed by IBM. As well as this, other PC manufacturers such as DEC, Tandy, Zenith and Hewlett-Packard were likewise trying to impose their various closed hardware architectures onto the market.

Source

Fig. 3.41: ITU 1995; World Bank

Thanks to their brand name and the sales structures already in existence, IBM-PCs were able to achieve a position of market dominance relatively quickly. A further factor in their favour was the spreadsheet program "1-2-3" from Lotus, written specifically for the IBM-PC. Coming onto the market at the beginning of 1983, this proved to be a "killer application," in other words a huge seller, and as a consequence the sales of IBM-PCs trebled. As the de facto standardisation of the IBM platform became apparent, a large number of firms producing peripherals and software soon began to match their products to this standard. The result was that the IBM standard gained acceptance, so that by the end of 1983 IBM-PCs had a market share of some 70 per cent of the global PC market.

Clones

Initially regarded as unlikely by IBM, a further consequence of this success was the appearance of competition from low-priced, IBM-compatible PC-copies. Though only founded in 1981, the firm Compaq had by the end of 1982 succeeded in becoming the first PC-manufacturer to produce a virtually identical imitation of the IBM-PC.[112] Inspired by the success of Compaq, a large number of imitations, or so-called "PC-clones," came into being.

Rapid product cycles

IBM recognized that the de facto standardisation of the IBM-PC was no necessary guarantee of market leadership. The consequences of its "open" construction proved as serious for IBM as they were pivotal to the development of the PC industry as a whole. Unlike in the mainframe era, when closed platforms had offered protection from aggressive competitors, IBM found itself forced to introduce rapid product cycles with constant improvements in cost-effectiveness in order to defend and consolidate its market shares in the PC industry. Large quantities of firms were now concentrated on each segment of the value chain, and these were thus perpetually driven forward by the impetus of technological innovation. As the business giant's vertically integrated organization structure had difficulties coping with the development speed characteristic of this type of horizontal industry structure, the market share of IBM was slowly but surely eroded away.

The Dominance of Standards and their Consequences

IBM gave away the "key to a kingdom"

The firms best able to profit from the de facto standardisation were Intel and Microsoft. As IBM did not insist on an exclusive licence for the two components concerned, both Intel and Microsoft were in general able to licence their products to any other enterprise, and this they did. Through their control over the microprocessor and operating system respectively, the two companies gained the "key to a kingdom."

The strategic significance they enjoyed can be explained by the fact that the microprocessor and the operating system of a PC form the interface between hardware and software. Only with a high level of coordination between these two components is it possible to achieve optimal efficiency in transforming the technological advances made by microprocessors into an improved performance

112 One particular obstacle had been the production of the BIOS, i.e. the connection of the operating system with the hardware. By means of so-called "reverse engineering" Compaq managed to overcome this hurdle and produce cost-effective PC clones. See Cringely 1996, p. 159ff.

in the systems and user software. The following period consequently saw close cooperation between Intel and Microsoft in microprocessor and software development.

An alliance for success

The spread of IBM-compatible PCs allowed both companies to attain and then steadily strengthen a leading market position. Firstly the close coordination of development activities in comparison with competing PCs produced a superior operating performance. Secondly the imitation of microprocessors and operating systems is considerably more difficult than the simple reconstruction of a hardware configuration such as the IBM-PC's.

In general the development of microprocessors is not only time-intensive but also characterized above all by extremely high investment costs.[113] On account of its continuous revenue growth, Intel was able to re-invest a considerable portion in research and development, for a long time permitting them to keep ahead of the field technologically. In particular the "X86" product line resulted in a near-monopoly situation in the PC field.

The magic of positive feedback

By comparison with the manufacture of microprocessors, the "production" of systems software is appreciably more low-cost, entailing in theory at least that the barriers to market entry for Microsoft's competitors have been and still are relatively low. In practice, however, so-called indirect network effects have proved a substantial advantage for the continued market growth enjoyed by Microsoft.[114] These dictate that the attractiveness of a particular systems software is crucially determined by the number of application programs available for the platform in question.

As a great number of independent software-suppliers manufactured applications based on MS-DOS, the attractiveness of MS-DOS from the perspective of the consumer rose increasingly. Bill Gates has described this as a "positive feedback effect."[115] These network effects thus contributed to the exceedingly strong market position of Microsoft's operating system, for with the rising number of application programs based on MS-DOS not only was the attractiveness of MS-DOS boosted but at the same time it became progressively more difficult for competitors to gain any significant market share.[116]

In the following years, more and more hardware and software manufacturers clustered around what was known as the "Wintel"[117] value network. By means of licence contracts, the "shapers" Intel and Microsoft cleverly bound their partners, so-called "adapters," to the technologies they were marketing.[118] But it was not until the early nineties that all the implications of the "Wintel" partners were to come to light.

For in the mid-eighties it was not Intel or Microsoft who played the leading role along with IBM in the PC industry, but a firm whose technological supremacy is every bit as legendary as its incapacity to transform this supremacy into long-term market dominance: Apple Computers. The Apple II computer, intro-

113 This can be expressed by another rule of thumb recognized by Gordon Moore, which has also come to be known as Moore's Second Law: not only is it the processing power of computers that doubles every 18 months, but likewise the costs for the manufacture of a new chip generation. See Lewis 1997, p. 44.
114 See section 4.2.1
115 Gates 1995, p. 98
116 See Economides 1998, p. 16ff.
117 "Wintel" is a term coined by combining "Windows" (the Microsoft operating system) and "Intel" (the dominant producer of PC processor chips).
118 The "per-processor" licence contracts in particular guaranteed Microsoft a licence fee for every PC sold by the PC-manufacturers regardless of whether the operating system was MS-DOS or otherwise. This practice was later prohibited by the United States monopolies commission.

duced onto the market in 1978, had by 1983 acquired a worldwide market share of some 20 per cent.

The Macintosh

In spite of this, the following years saw Apple come under severe pressure from IBM-PCs and PC-clones. In 1984 Apple reacted to this negative market development with the introduction of the Apple Macintosh, which set new standards in user-friendliness with the first graphic user interface and its mouse. The founder of Apple Steve Jobs called it "the first computer for the rest of us."[119] Despite certain teething troubles the Macintosh meant technological leadership in the PC industry.

Apple's strategic dilemma

What is remarkable is that along with the programming of the operating system Apple also carried out the production of the hardware architecture within the company itself. The advantage of this was that an improved operating performance could be achieved in the software owing to the exact coordination of hardware and software. The drawback lay in it being a closed system, since Apple refused to license its hardware configuration to other PC manufacturers. On top of this, Apple's microprocessor was from Motorola and so not compatible with the IBM-PCs and PC-clones.

Bill Gates' free advice to Apple: excellent (and futile)

On 25 June 1985 Bill Gates addressed a confidential memorandum to Apple's leading managers, John Sculley and Jean-Louis Gassé. Against the background of the horizontal structure of the PC industry, Gates pleaded for the licencing of the hardware platform. Considered to be one of the most important documents in the history of Silicon Valley, the memorandum was not made public until 1997.

Bill Gates: "The Licensing of Mac Technology"

It would all have become one big corporate ecosystem centered around the Mac. Put another way, Apple would have created an industry standard, a playing field that it controlled and everyone else would have to buy into. This standard was envisioned by Bill Gates and outlined in one of the most important documents in Silicon Valley history, a highly confidential three-page memorandum from Gates to Sculley and Gassé dated June 25, 1985. Entitled "Apple Licensing of Mac Technology", the document read:

Background:

Apple's stated position in personal computers is innovative technology leader. This position implies that Apple must create a standard on new, advanced technology. They must establish a "revolutionary" architecture, which necessarily implies new development incompatible with existing architectures.

Apple must make Macintosh a standard. But no personal computer company, not even IBM, can create a standard without independent support.

119 See Carlton 1997, p. 3ff.

Only an open computer architecture ...

... results in combining forces of many companies

Even though Apple realized this, they have not been able to gain the independent support required to be perceived as a standard.

The significant investment (especially independent support) in a "standard personal computer" results in an incredible momentum for its architecture. Specifically, the IBM PC architecture continues to receive huge investment and gains additional momentum ... The investment in the IBM architecture includes development of differentiated compatibles, software and peripherals; user and sales channel education; and most importantly, attitudes and perceptions that are not easily changed.

Any deficiences in the IBM architecture are quickly eliminated by independent support ... The closed architecture prevents similiar independent investment in the Macintosh. The IBM architecture, when compared to the Macintosh, probably has more than 100 times the engineering resources applied to it when investment of compatible manufacturers is included. The ratio becomes even greater when the manufacturers of expansion cards are included.

Conclusion:

As the independent investment in a "standard" architecture grows, so does the momentum for that architecture. The industry has reached the point where it is now impossible for Apple to create a standard out of their innovative technology without support from, and the resulting credibility of, other personal computer manufacturers. Thus, Apple must open the Macintosh architecture to have the independent support required to gain momentum and establish a standard.

The Mac has not become a standard

The Macintosh has failed to attain the critical mass necessary for the technology to be considered a long term contender:

a. Since there is no "competition" to Apple from Mac-compatible manufacturers, corporations consider it risky to be locked into the Mac, for reasons of price and choice.

b. Apple has reinforced the risky perception of the machine by being slow to come out with hardware and software improvements (e.g. hard disk, file server, bigger screen, better keyboard, larger memory ...)

c. Recent negative publicity about Apple hinders the credibility of the Macintosh as a longterm contender in the personal computer market.

d. Independent software and hardware manufacturers reinforced the risky perception of the machines by being slow to come out with new key software and peripheral products.

e. Apple's small corporate account sales force has prevented it from having the presence, training, support, etc. that large companies would recognize and require.

f. Nationalistic pressures in European countries often force foreign consumers to choose local manufacturers. Europeans have local suppliers of the IBM architecture, but not Apple. Apple will lose ground in Europe as was recently exhibited in France.

Recommendation

Apple should license Macintosh technology to 3-5 significant manufacturers for the development of "Mac Compatibles":

Licencing is the ideal mechanism to gain momentum and critical mass

United States manufacturers and contacts: ideal companies – in addition to credibility, they have larger account sales forces that can establish the Mac architecture in larger companies:

- AT&T, James Edwards
- Wang, An Wang
- Digital Equipment Corporation, Ken Olsen
- Texas Instruments, Jerry Junkins
- Hewlett Packard, John Young other companies

(but perhaps more realistic candidates):

- Xerox, Elliott James or Bob Adams
- Motorola, Murray A. Goldman
- Harris/Lanier, Wes Cantrell
- NBI, Thomas S. Kavanagh
- Burroughs, W. Michael Blumenthal and Stephen Weisenfeld
- Kodak European manufacturers:
- Siemens
- Bull
- Olivetti
- Phillips [*sic*]

Bill Gates: how positive feedback works

Apple should license the Macintosh technology to U.S. and European companies in a way that allows them to go to other companies for manufacturing. Sony, Kyocera ... are good candidates for OEM manufacturing of Mac compatibles.

Microsoft is very willing to help Apple implement this strategy. We are familiar with the key manufacturers, their strategies and strengths. We also have a great deal of experience in OEMing system software.

Rationale:

1. The companies that license Mac technology would add credibility to the Macintosh architecture.
2. These companies would broaden the available product offerings through their "Mac-compatible" product lines:

* they would each innovate and add features to the basic system: various memory configurations, video display, and keyboard alternatives, etc.
* Apple would lever the key partners' abilities to produce a wide variety of peripherals, much faster than Apple could develop the peripherals themselves.
* customers would see competition and would have real price/performance choices.

3. Apple will benefit from the distribution channels of these companies.
4. The perception of a significantly increased potential installed base will bring the independent hardware, software, and marketing support that the Macintosh needs.
5. Apple will gain significant, additional marketing support. Everytime a [sic] Mac compatible manufacturer advertises, it is an advertisement for the Apple architecture.
6. Licensing Mac compatibles will enhance Apple's image as a technological innovator. Ironically, IBM is viewed as being a technological innovator. This is because compatible manufactureres are afraid to innovate too much and stray from the standard.

This heretofore unpublished document essentially provided a blueprint for how Apple could save itself from long-term debilitation – a course that, had it been taken, would have put Apple into the driver's seat in the 1990s and possibly beyond".

Quoted from: Carlton, Jim: Apple. The Inside Story of Intrigue, Egomania and Business Blunders. New York, NY 1997: Times Business Books

With permission from Jim Carlton

Background of Gates' advice to Apple

While the advice passed on to a direct competitor may appear incomprehensible from today's standpoint, it is plausible in the context of the time since Bill Gates had nothing to lose and everything to win. The differences in size were manifest: Apple's yearly turnover in 1985 amounted to around 1.5 billion US dollars, while Microsoft's was only just touching 89 million dollars. But two reasons in particular moved Bill Gates to write.

Firstly, his suggestion was aiming primarily at an expansion in the revenue potential of Microsoft. The applications already developed exclusively for the Macintosh such as word-processing (Word) and spreadsheets (Excel) brought in significantly higher revenue by comparison with the licensing of Microsoft systems software to other PC-manufacturers. The spread of the Macintosh operating system to other hardware platforms would thus also multiply the demand for Microsoft applications. The second reason was that on account of its Motorola microprocessor Apple was not IBM-compatible and so moved in a different market segment. For this reason the danger of Macintosh operating systems eroding the turnovers of IBM-compatible PCs with MS-DOS operating systems was relatively slight.

Dangers for Microsoft

The greatest danger for Gates lay in a different scenario: the licensing of the hardware in connection with a simultaneous conversion of the operating system to the Intel processor. With this strategic move Apple would have jeopardised the entire sales market of MS-DOS at one fell swoop.

After intensive discussions (which were to be taken up again at later dates), Apple turned down the proposal made by Bill Gates. The reason lay above all in Apple's own hubris. The company was convinced that the technological supremacy of the Macintosh operating system in conjunction with the hardware configuration would inevitably make it the standard of the PC industry. The licensing of the Macintosh operating system to other hardware manufacturers was thus considered unnecessary.

Apple's grave misjudgement and its consequences

The consequences of this grave misjudgement are well-known: as Apple sank into insignificance, the Intel and Microsoft value network – urged on by Bill Gates's battle cry "a computer on every desk and in every home running Microsoft software" – set the de facto standard for the whole PC industry.[120] What this demonstrates is that in a phase of open computer systems it is not just the technological quality of hardware or software but above all the establishing of standards that has the decisive influence upon the market position of the various supplier-firms.

The shift in emphasis from mainframe computers to PCs not only led to a bitter battle for standards, but also to a paradigm shift as regards the centralised structure of networks. Technologies such as Ethernet or Token Ring made it possible for PCs whose principle field of application was in the business customer sector to be connected to local networks (Local Area Networks, or LANs). The

120 In 1997 Apple had a 2.6 per cent share of the worldwide PC market, while in the mid-nineties Microsoft and Intel in their respective market segments had a market share of around 90 per cent. See Moschella 1997, p. 34. They following comeback of Apple products has not been a result of a fundamental change in strategy. It may still allow for either success in a niche market or total failure.

PC's in local area networks overwhelm mainframe computers

centres of these LANs were individual PC servers, able to cope with increasingly complex applications thanks to the improved cost-effectiveness of microprocessors. As a consequence, many of the existing mainframes and minicomputer-systems became redundant, since the decentralised structure of the PC networks could fulfil equivalent functionalities at a more competitive price. In this segment of the PC market the software company Novell with its program NetWare had a near-monopoly position which in the early nineties manifested itself in a market share of around 70 per cent.[121]

Dis-Integration of the IT Industry's Value Chain

Fragmentation of the value chain

On the whole, what can be ascertained is that the introduction of open computer systems entailed a fragmentation of the value chain. As a consequence of these developments, mainframe and minicomputer-firms found themselves forced to contend with conditions that imposed a whole new set of demands. The PC industry was characterised by the use of completely new technologies, cost structures transformed by constantly improving cost-effectiveness, ever-quicker development cycles, innovative market strategies and above all an extended target group structure comprising both business and private customers.

Horizontal industry structure

At the same time, the rapid establishment of de jure[122] and de facto standards in the individual segments of the value chain was a fundamental precondition for the development of a horizontal industry structure. Particularly noteworthy is that in markets with so-called de facto standards such as microprocessors and operating systems network effects proved conducive to market barriers, making higher margins possible. By comparison, market segments with de jure standards such as SCSI for disk drives or VGA for monitors were exposed to more intensive price competition. In some markets, the resulting predominance of certain standards engendered a quasi-monopoly situation. The situation in the early nineties can be sketched as follows:

Fig. 3.42: Consequences of the paradigm shift: specialists dominate the PC industry

Market segment	Technology Leader	Competitors
Services	Electronic Data Services - EDS	Andersen Consulting, CSC
Database management systems	Oracle	Informix, Sybase
Systems software	Microsoft	Novell
Data communications	Cisco	3Com, Bay Networks (Nortel)
PCs	Compaq	IBM, Apple, Dell
Printer	Hewlett-Packard	Epson, Canon
Hard disks	Seagate	Quantum, Conner
Microprocessors	Intel	Motorola, AMD, Cyrix

121 See anon 1998m, p. 81
122 These refer to standards formally established by industrial consortiums or official institutions.

Source

Fig. 3.42: ECC 1999, based on Moschella 1997, p. 30

By and large, the focusing on individual stages of the value chain can be seen to have been conducive to high learning curve effects within a short time in the specialists concerned, in turn producing a further strengthening of core competencies. The advantage of this horizontal industry structure with de facto standards is the availability of an increased pool of resources, once again benefiting the entire IT industry. The days when a single company could cover the whole IT value chain are gone. The IT industry's fragmented value chain led to a substantial speeding up in market development, while vertically integrated firms like IBM or DEC showed different (and differently successful) modes of adaptation.

US industry dominance enhanced

From a global point of view, this transformation in the industry structure has meant a huge increase in the importance of businesses from the USA. The majority of players to emerge from the changes in structure were companies of American origin. Just as it had in the United States, the speed of the technological development and product cycles that accompanied the new industry structure caused severe problems of adaptation for the older IT firms in Europe and Asia as well.

The vertically integrated system-suppliers in the individual countries were slow to react to the changeover to open computer systems. Traditional suppliers such as Fujitsu, Hitachi and NEC in Japan, Siemens Nixdorf in Germany, Bull in France, Olivetti in Italy and ICL in England lost considerable ground in the PC industry.

Marginalisation of European companies

On the whole, what emerges is that from 1981 to the early nineties it was American companies that succeeded in attaining clear market leadership in decisive markets such as microprocessors, operating systems, network equipment and software. Only in relatively minor niche markets such as DRAMs or CD-ROMs have Japanese enterprises achieved clear predominance. With a few exceptions such as SAP, almost all European IT companies have been restricted to their respective home markets and marginalised in the world market.

* *Network centered phase (1994-2005)*

According to estimates from the International Data Corp. (IDC), the current industry structure will continue to characterise the greater part of the IT industry's market volume for about five more years. But as with the transformation from the mainframe industry to the PC industry, significant structural alterations can already be discerned.

The most palpable indication of a paradigm shift is the new weighting of key technologies. With the spread of the Internet, networks and related technologies move centre stage. Communication is thus of central importance for the future development of the IT industry, and technological innovations boosting transmission speed or bandwidth have a greater effect on the IT industry than a further improvement in processing power. Not surprisingly, the focus of the industry has switched mainly to the improvement of network technologies.

Metcalfe's Law fundamental to Internet industry

Along with Moore's Law, a further axiom will prove crucial to the further development of the IT industry. It was first formulated by Bob Metcalfe, the inventor of Ethernet technology and founder of 3Com. Metcalfe's Law dictates that the value of a network (W) rises exponentially as the square of the number of users (n). $W = n^2 - n$.[123]

The law manifests itself especially through the classical example of telephone. The exponential rise in value of the network arises from direct network effects. Accordingly, a telephone is without value for one user as long as there are no other telephone users to communicate with. But the value of the telephone rises with each additional telephone connection and thus so too does the value of the network.

For the IT industry this means that the increasing networking of computers of any sort through an universal network – the Internet – produces an exponential rise in the value of this network. In other words, connection to this infrastructure is not only necessary but a prerequisite for future success in the IT industry. Equally significant is that the economic features intrinsic to the networks are appreciably different from "normal" markets (see also section 4.2.1). An understanding of these features along with the ability to cope with the speed involved is consequently for IT firms one of the most important preconditions for market success in the network phase.

The Structural Transformation of the PC Industry

PC's for less than 1,000 US-$ make for a paradigm change

In the private customer segment of the PC industry, the dimensions of the changes in the market first became apparent in the spring of 1997 when the PC-manufacturer Compaq broke through the "magical" price limit with its offer of a multimedia PC ("Presario") for less than 1000 US dollars. Up to then in the PC industry, the increasing performance of new generations of processors had been observed to be offset by a rise in the complexity of the corresponding software. While there had been a continuous increase in the performance of PCs, the price level had been stagnating since the early nineties at a level of around 2000 US dollars.

The rapid fall in prices can be traced back to two developments. Firstly, any further increase in PC performance is seen as relatively unimportant by a great number of end consumers. The PC models present on the market and the corresponding software applications (word processing, spreadsheets) have by now reached a performance level that adequately covers the needs of most consumers. The willingness to pay for new PC models with increased processor and memory performance or more complex software is accordingly low.

Secondly, the increasing significance of networks, and in particular the Internet, has induced a shift in strategic perspective on the part of PC-manufacturers. This shift in perspective goes back to the idea of a network computer (NC), introduced in late 1995 by Larry Ellison, the chief executive officer of Oracle. The core of the concept consists of low-priced terminals costing some 500

123 See Shapiro/Varian 1998, p. 143

US dollars, which – equipped with a simple operating system (Browser / Java) and functioning as pure input and output terminals – use the software applications and integrated services offered on networks. The principle is that instead of processing and storing data on local hard disks the desired services (information, entertainment, communication) can be taken from the network as needed.

The PC-manufacturers' offer of low-priced PCs is regarded as a strategic measure designed to anticipate the threat to their regular business posed by the NC. The strategy has so far proved successful, since the cheap multimedia PCs have resulted in new customer groups being opened up which by the end of 1997 already made up more than 30 per cent of the total of PC-buyers. PC-penetration in the USA thus rose to 43 per cent by early 1998, having stagnated for three years at 40 per cent.[124]

Powerful hardware is taken for granted ...

As these new customer groups are interested primarily in simple applications such as word processing and (Internet) services such as games or shopping, the value for money of the operating system and hardware components is low from the perspective of the consumer. The demand for simple Internet terminals at low prices will thus continue to rise. In general the IT industry in the USA is assuming that the sound barrier of 500 US dollars will be reached within the next year or two. All in all, hardware will be sold more and more in conjunction with attractive content, i.e. as "bundling." It is even possible that the terminals will end up being given away, a development that has already come to pass with mobile phones.[125]

... and the share of application revenue will rise

Consequently, the share of total costs of value added taken up by software will continue to rise, while the hardware-share will fall visibly. This is leading to considerable changes in the business patterns of PC-manufacturers such as Compaq, IBM or Hewlett-Packard. While the decrease in prices leads to a rise in numbers sold, the effect of this on turnover and profit margins is relatively slight.

The number of PCs sold in the USA thus rose by over 54 per cent from the beginning of 1997 to the beginning of 1998, while the turnover only went up by 10 per cent.[126] What this means is that for most manufacturers the present strategy must lie in a pronounced increase in quantity in order to be able to defend market shares and current revenues through economies of scale.

"Webifying your company"

In the business customer segment too, the spread of the Internet and the shift to network activities are giving rise to structural changes that are putting the companies' classical business models under mounting pressure. Particularly noticeable is the trend towards "webifying," or the fact that businesses are increasingly aligning their hardware and software services for use in networks. Interestingly, this means a partial return to the centralized structure of the mainframe phase, as highly powerful servers move to the heart of the (network) activity.

124 See Burrows 1998, p. 28
125 See Kanellos 1998. At the time of writing, this was the future – now it has happened, and it may be a lasting phenomenon.
126 See Burrows 1998, p. 29

Network computer architecture attractive to companies

It is for this reason that NC-architecture too has sparked off considerable resonance. For not only can the NC be used in large-scale business-networks but as well as this it has economic advantages that help companies lower the costs of IT use. According to the IDC, the central administration of data and the decreased expenditure on maintenance allow some 8000 US dollars per PC workplace (so-called "costs of ownership") to be saved per year.[127] In this way using NCs becomes a cheap alternative to replacing mainframe terminals of the old sort.

Even so, by mid-1998 there has still not been a single product to come onto the market that exactly corresponds to the idea of the NC. Nonetheless, the concept of NC architecture has triggered off a whole sequence of product developments that have followed it in placing themselves in a markedly lower price segment with reduced functionality. These will be described below.[128]

As the NC architecture was conceived as using the Netscape browser in connection with Java as its operating system, the strategy of Wintel in particular found itself under threat. Since the mid-nineties both companies have thus been making a more concerted effort to extend their value network in the business customer segment beyond the PC and into other spheres as well. One example of this is the work station and server operating system Windows NT introduced by Microsoft in 1993, which in West Europe achieved a growth in its market share from under one per cent in 1994 to some 12 per cent in 1997.

To the unanticipated threat from the low-price segment for business customers Intel and Microsoft responded with the NetPC in mid-1997. This computer, designed in collaboration with 12 PC-manufacturers, offers a distinctly reduced functionality by comparison with classical PCs and is equipped with an Intel processor and an operating system from Microsoft. In addition, Microsoft is offering so-called Windows terminals, representing a further variation on the NC. Worldwide the NetPC managed sales of some 44,000 in 1997, while the Windows terminals had sales of over 150,000. The same year saw sales of around 150,000 for further variations on the NC. The majority of these were achieved by NetStation, IBM's version of the NC.

The Java concept

The attempts of Sun Microsystems's "JavaStation" or the alliance of Oracle and Netscape, "Network Computer Incorporated" (NCI), proved less successful. It is the virtual operating system Java – propagated as a "killer application" – that has proved to be the biggest headache of the original NC architecture. Java was introduced onto the market in 1996 by Sun Microsystems (a manufacturer of powerful servers and work stations) as a free licensable programming language. It was originally intended simply as an aid for improving the representation of Webpages. The underlying thought was to transmit small Java programs via the Internet from a host to a client, and these could then be started up as so-called Applets independently of the operating system in question (Macintosh, UNIX or Windows).

127 See Booz, Allen & Hamilton 1997, p. 125ff.
128 See Madden 1998b

The fundamental precondition for this universal usability is provided by a "Java Virtual Machine" or "VM" integrated into almost every WWW-browser. This program simulates an operating system of its own within the operating systems it is located in. It was quickly recognized that owing to its complete independence from the operating systems in question the potential of the programming language lay above all in its cross-platform usability. Java thus had the potential to be a "lingua franca" of the World Wide Web: "write once, run everywhere."[129]

By 1998 Java has become one of the most important programming languages, and worldwide over half a million programmers work with it. Yet its most frequent use is not as systems software for terminals like the NC, as was originally conceived. Instead it is used principally for cross-platform communication between different computer systems. The advantage of this is that the corresponding systems and user softwares do not have to be individually adapted to their respective hardware platforms but are programmable through Java.

Unintended consequences

The original idea for its application, however, as a virtual operating system on the browser platform Netscape Navigator in NCs, is difficult to put into practice. Owing to Java's unexpectedly high complexity, the terminals in most cases require their own processing and memory capacity, so a cheap PC can in fact fulfil the same function.[130] NCs using Java have thus met with hardly any resonance among businesses. Larry Ellison consequently announced in 1998 that Oracle will not be continuing any further with the NC for the business customer sector.

The facts of the matter are that the number of NC variants sold in 1997 amounted to no more than 350,000. The comparison with the 43 million PCs sold in the business customer sector in the same period shows that NCs have so far not even reached a market share of one per cent.[131] The reason for this is that the advantages of NC architecture in comparison to PCs are relatively slight. On the one hand the drop in PC prices in the end consumer segment in early 1998 has meant that the business customer segment too has seen price cuts of up to 20 per cent.[132] On the other hand these low-priced PCs offer a substantially more extensive functionality than most NCs, making them all in all more cost-effective.

For these reasons PCs will in all probability continue to enjoy the greatest market share in terminals for the business customer segment for the next five years, while NCs can expect moderate growth. Along with these developments in the lower performance sector of the PC market, it should also be borne in mind that the upper market segment for business customers is seeing a development of quite a different sort.

On the whole, the significance of the Internet for most companies has changed from a simple "publish" model (in other words a Webpage) to a more extensive "transact" model (in other words the implementation of an Intra- or Extranet). This demands the integration and connection of internal and external business

129 It was in practice not always possible to uphold this claim, as cross-platform programming predictably enough entails a lot of problems. Cynical programmers thus counter: "write once, debug everywhere."
130 See anon 1998f, p. 82
131 See Madden 1998b
132 See Burrows 1998, p. 29

processes in network structures that are characterised by an increasing level of complexity. Along with a heightened demand for powerful mainframes and Web servers this constellation is also making itself felt in the service sector. In particular the demand for "solution providing" has led to the service sector in West Europe having the strongest growth of all IT segments with an annual level of some 12 per cent. The acquisition of DEC by Compaq in January 1998 should also be seen in this context. DEC offers a large number of regular customers in the business customer sector, making it possible for Compaq to become a "solution provider."[133]

Company web presence shifts from publishing to transaction model

Remarkably, the Internet and the accompanying shift to network structures has indirectly led to a revitalisation of the vertically integrated computer giant IBM. Thanks to its core competence in the field of powerful servers and mainframes and to Chairman Lou Gerstner's relatively early decision at the beginning of 1995 to focus on the Internet, "New Blue" with special "e-business" solutions can make up considerable lost ground. [134]

Revival of IBM strenghts

The Triumphal March of Internet Terminals

The paradigm shift currently taking place shows itself above all in the changes undergone by the terminals. Analogous to the transition from mainframes to PCs, a large number of terminals will come onto the market whose primary function is the connection with the Internet. Such Internet appliances or subminiature computers can be Internet terminals for TV (NetTVs and set-top-boxes), for games (advanced game consoles), Personal Digital Assistants (PDAs), handhelds or Internet smart Phones. The largest part of this expansion of the market will come from new types of appliances, characterised by ease of use and user-friendliness and thus significantly different from today's IT terminals.

Diversity will rule net access terminals

According to IDC estimates, this market will within just five years reach a market volume almost as large as PCs.[135]

Fig. 3.43: Predicted US-market growth for PCs and Internet terminals

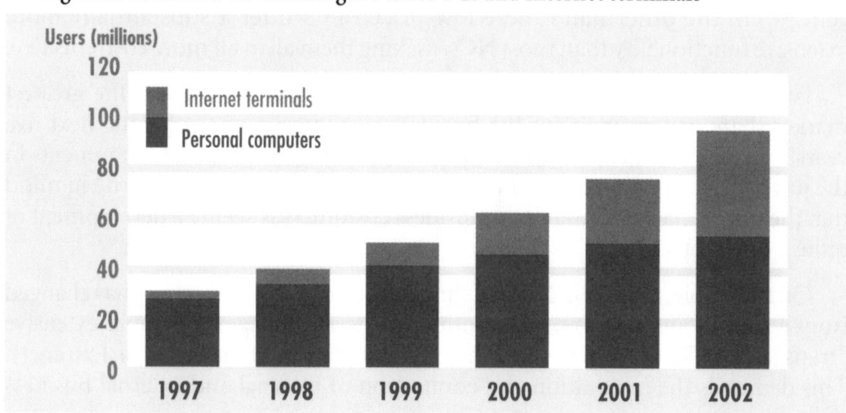

133 See anon 1998e, p. 14
134 See EITO 1998, p. 322; see anon 1998n, p. 80
135 See EITO 1998, p.177

Source

Fig. 3.43: International Data Corporation (www.idc.com) 1998

These Internet terminals here include not only terminals in the end consumer segment, e.g. for television (NetTVs, set-top boxes), games (games consoles), personal digital assistants (PDAs), handheld-PCs and Internet smart phones, but also NCs for the business customer market. Most of the market expansion will come from Internet terminals that differ significantly from the concept of the NC or PC.

Challenges to PC leaders

Companies in the PC industry will have to react promptly to these changes in market conditions, for otherwise they will be threatened with the same loss in significance as large numbers of mainframe and minicomputer firms in the early eighties were.

Intel and Microsoft strike back

The microprocessor-manufacturer Intel has already expanded its product line with the emerging market changes in view. The lower performance sector, which also incorporates Internet terminals, is covered by the acquisition of the chip-manufacturer StrongARM in July 1998 and by the new product Intel Celeron, while the Pentium II is targeted at the middle market segment. With the micro-processors Pentium II Xeon, which came onto the market in mid-1998, and the Intel Merced, due to enter the market in mid-2000, following on from the Pentium Pro, the upper business customer segment of servers and work stations is provided for too.[136]

Microsoft is likewise prepared for the changes to come. One reason for this is the threat posed to regular business by the Internet, a threat not recognized until 1995. Within the framework of a strategic reorientation, Bill Gates in mid-1995 realigned the focus of the entire company. The operative adaptation of the whole range of product lines was carried out inside six months. Microsoft was afraid that the Netscape Navigator together with the virtual operating system Java might come to replace the MS operating system Windows. This is verified by an internal memorandum from 1995:

"Make-or-break time" for Windows

> *"This is not about browsers. Our competitors are trying to make an alternative platform to Windows. They are smart, aggressive and have a big lead. ... Netscape/Java is using the browser to create a virtual operating system. Windows will become devaluated, eventually replaceable. This is make-or-break time: the next six months are critical".*[137]

As a result of the measures taken in 1995, the situation for Microsoft in 1998 looked very different. Firstly, in the context of the much-cited "browser war" Microsoft had succeeded in achieving a market share of some 50 per cent with their own browser, the Internet Explorer.[138] There is a good chance that this market position will be further strengthened in relation to the competitor product Netscape Navigator. Secondly, Microsoft has developed its own version of Java, ActiveX, which has already been integrated into the greater part of the Microsoft products such as Internet Explorer and Windows NT.[139]

Like Intel, Microsoft is following a franchising-strategy, allowing variations of the core product to be transferred onto the most diverse of hardware platforms.

136 See anon 1998r, p. 17; anon 1998q
137 Anon1998c, p. 72
138 According to AdKnowledge, Jupiter Communications and IDC, late July 1998.
139 See anon 1998f, p. 81

The identical basis of this systems software ensures the compatibility of the different versions and thus makes possible the exchange of files and programs between platforms. With this "Windows Everywhere" strategy Microsoft is striving to extend its quasi-monopoly in the field of systems software for PCs to other market segments as well.

Windows CE and NT

The systems software Windows CE is aimed at the fast-growing market segment for Internet terminals of all types. The current version Windows CE 2.0 can be found as systems software in so-called handheld-PCs and palm-PCs. As well as this, there are software applications such as Word, Excel, PowerPoint or Outlook, which have a markedly reduced functionality by comparison with the range of PC applications. In addition, Windows CE 2.0 is also used in set-top boxes.

In the case of PCs the core product Windows is directed at the mass market for end consumers and business customers of the lower market segment. The integration of the Internet Explorer browser in Version 98, which appeared on the market in mid-1998, guarantees the expansion of the operating system onto the Internet's WWW-platform. Windows NT was developed in the early nineties as systems software for the hardware platforms work stations and servers and is aimed at the middle and upper business customer segment. As in the browser market, Microsoft was able to make use of its already existent monopoly in the field of systems software for PCs to secure a market share of 50 per cent by 1997.

"Wintel" value network has found a response

On the whole it can be said that the "Wintel" value network has recognized the paradigm shift and already responded to it. By comparison, the rest of the PC industry has so far been slow to prepare itself for the expected expansion of the PC industry in the lower market segment. The challenge consists in adapting the business models in effect up to now to the mass market of Internet terminals. PC-manufacturers can take it as given that the cost structures for high volumes at simultaneously low prices will approach the business model for consumer electronics.

The future market for Internet terminals thus calls for new business models from PC-manufacturers in order not to be marginalised by firms from consumer electronics. The following table provides an overview of the developments in the IT industry described above.

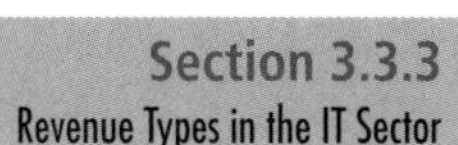

Fig. 3.44: The development of the IT industry

	System (1964-1981)	PC (1981-1994)	Network (1994-2010)
Key technology	Transistoren	Microprocessoren	Communications-bandwidth
Laws applicable	Grosch's Law	Moore's Law	Metcalfe's Law
Main features	Propietary systeme	Standardized products	Integrated services and content
Type of network	Data and computer centres	Internal Local Area Networks (LANs)	Internet (global, regional, local)
Value chain structure	Vertical integration	Horizontal	Unified IT and telecom-munications value web

3.3.3 Revenue Types and Revenue Models in the IT Sector

Business models for hard- and software producers

Various revenue types can be characterised according to market segment.The hardware manufacturers refinance themselves directly in all cases. There are several possibilities for doing this. The hardware can be sold to the customers through appropriate sales channels, which constitutes a direct, usage-related revenue type. This applies equally to the sale of PCs to end consumers, the sale of set-top boxes to TV companies, the sale of servers to business customers etc.

Systems software manufacturers are generally characterized by a mixture of revenues consisting of direct, usage-related licence fees both from PC manufacturers and also from end consumers for software-upgrades. On the one hand a company such as Microsoft licences a copy of the operating system Windows to PC manufacturers such as Compaq or Dell at a flat rate of some 40 to 50 US dollars for a PC in the price segment around 2000 dollars. In general it is the case that the more expensive the PC sales price, the higher the licence fees too, and vice versa.

Software licencing

Thanks in particular to the cost structures of software products (see 4.2.2), these licence fees make for high profit margins. This explains why Microsoft is trying to ensure that the PC is the dominant receiver for multimedia applications. If for example the television becomes the predominant receiver, then Microsoft will have to adapt to the revenue model for consumer electronics. At present manufacturers in this field pay no more than 40 to 50 cents per receiver for systems software licences.[140]

On the other hand, with the same version of the operating system (for example with the introduction of Windows 98), it is not just from licensing to PC-

140 See Cringely 1998

Source

Fig. 3.44: ECC, based on Moschella 1997, p. ix

manufacturers that a software company such as Microsoft makes its money. A second flow of income can be generated through the sale of licences to end consumers at a price of approximately 90 to 100 US dollars (so-called upgrades).[141] The relation of these two income flows cannot be explicitly quantified, as it is heavily dependent upon the underlying price strategy. Up to now it has been refinancing through the PC-manufacturers that has formed the greatest part, while the upgrades for operating systems have played a secondary role. This relation has changed above all on account of the platform presented by the Internet, since browsers for example are now offered free of charge while the upgrades have to be paid for.

Basics and upgrades

Software applications are likewise sold both through indirect channels, i.e. PC-manufacturers wanting to offer attractive software to go with their hardware, and directly to the end consumer. The relation between the two revenue flows, however, is the inverse of what applies for systems software. While indirect sales to PC-manufacturers are secondary in importance, direct sales to the end consumer form the greater part of the revenue mix. Revenue from service charges likewise comes directly from the client in question.

3.3.4 Value Chains in the IT Sector

The value chain in the IT sector can be represented on the one hand from the perspective of the supplier company and on the other hand from the perspective of the customer company that utilises the technology within its business enterprise. Most commonly, it is traditionally represented as a "supply chain."

Fig. 3.45: Value chain for the IT industry

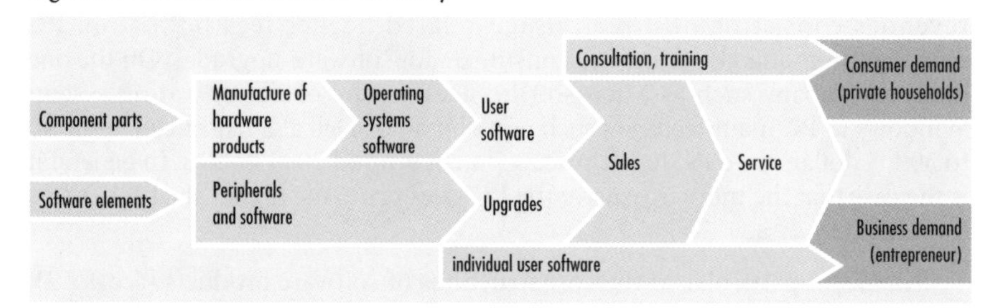

Examples of products include so-called "stand-alone products" such as personal computers, work stations, copiers. On the demand side, in general at least two segments can be differentiated, business customers and private customers.[142] The weight granted to each individual value chain stage varies according to whichever partial market is being considered.

On the supply side, it is possible in retrospect to discern five trends, which will also continue into the future:

141 See Edstrom/Eller 1998, p. 197. It is only in the rarest of cases that consumers who do not already have a PC buy a hardware platform without an operating system. It can thus be assumed that for most consumers the purchase of systems software is a purchase that will be replacing the old operating system.

142 On top of this come public customers (the state).

Source

Fig. 3.45: ECC 1999

Five trends in the IT industry ...

* the progress of microelectronics (component parts) will continue unabated (Moore's Law);
* the portion of software in products and systems is continuing to rise sharply; by the same token the hardware share in total costs and the value added is sinking;
* the innovation and product life cycles are getting shorter and shorter;
* the spread of networking continues inexorably;
* the supply of new multimedia applications (in particular computer games and the Internet/WWW) impels these trends forward, working in the manner of an accelerator catalyst.

These trends have a number of notable consequences for the IT value chain:

... and their consequences for the value chain

* the value added share of component parts and hardware production is declining enormously, while economies of scale are becoming more and more important;[143]
* in the software segment there is a shift in the value added towards (standard) user software. In the case of operating systems, network effects and "lock-in" strategies together with economies of scale are leading to quasi-natural monopolies;[144]
* in the business sector standardised and individually adaptable branch solutions are gaining in significance (the system of "supply chain management" for e-business in the business-to-business segment);
* the sales stage is coming under increasing pressure from shorter and shorter product life cycles and "profit windows." Services such as consultation, training and maintenance and thus also customer relations are falling by the wayside.

Particularly in the consumer sector, part of the consequence is that customers – wanting their problems solved and their consumer expectations fulfilled – are being lost from sight and left to fend for themselves with supposed "plug and play" solutions. For this reason, the acceptance of PCs and other multimedia applications has up to now only been moderately pronounced. Improved service and user-friendliness on the part of the suppliers is called for. Complementary measures should be taken by the state too, particularly in the sphere of school education.[145]

In the business sector there is a somewhat different set of problems. As the original IT customers, big businesses today have at their disposal a more or less complete supply of PCs / work stations and the appropriate internal IT networks. IT investments have been extremely high in recent years, as have the concomitant training and reorganisation costs.

143 For example the discontinuation of PC-production at Siemens-Nixdorf.
143 Money is made above all by means of an increasingly rapid succession of up-dates.
145 Which in the USA applies after all to more than 50 per cent of households.

Product life cycles too short compared to use life cycles

The rapidly accelerating innovation and product life cycles on the supply side of the value chain have come into conflict with the effective market or use life cycles on the demand side. Thanks to on-line sales, Netscape for example was able to reach its critical mass with buyers within a matter of a few months. On the customer side, by contrast, it took much longer, in some cases even years, for new software such as SAP R/3 to be integrated into the existing IT infrastructure and for those concerned to learn how to work with the software efficiently.

The advantages gained by the introduction of IT tools into the optimisation of a firm's business operations are clear. Constraints on the further optimisation of the supply chain include the following:

Bottlenecks

* the high initial investments required;
* the need for organisational changes;
* problems of acceptance and deficiencies in know-how among employees;
* the costs of technical adaptation for the firms involved;
* he building up of confidence necessary between the firms involved.

Market dynamics

For the consumer sector and the business sector alike, the IT value chain can be seen to be exceedingly "supply-heavy" in structure. A strengthening of the value chain stages that form the interface with the customer is indispensable and highly likely. At present demand for IT products is being pushed forward less by IT technology than by context related factors and by the adjacent market segments and adjacent stages of the value chain which are in structural coupling with the IT markets. Favourable context related factors include for example state and EU initiatives,[146] state-sponsored PR and media campaigns for better acceptance of technology by the general public, and professional training in IT.

3.3.5 Information Technology and Productivity – A Paradoxon?

Ever since information technology was first used to support and reproduce company functions and processes, the economic viability of these technologies has been a bone of contention. The discussions here take place on a wide variety of levels. Yet in both an economic and an organisational context it is difficult to come to definitive conclusions on the correlation between investments in information and communications technology on the one hand and productivity on the other. The discussions often come under the heading "the productivity paradox of information technology."[147]

On the whole the assumption is that computers have speeded up the growth of productivity and thus ensured a general increase in the growth of the economy. The levels of expenditure on information technology support this assumption. In

146 Such as the German programme "Schulen ans Netz" (Schools on the Net) and Wissenschaftsnetz (Science Net).
147 See Wigand/Picot/Reichwald 1997, p. 151

1996 for example, some 40 percent of the expenditure of American companies on investment goods was on information technologies. Yet the yearly average growth in productivity in the American business sector has slowed down from a level of 2.6 percent in the period between 1960 and 1973 to roughly 1 percent in recent years. In economic terms this is a reflection of the "productivity paradox."[148]

Theoretical aspects

On the level of the individual firm, the work done by Brynjolfsson and Hitt[149] created a considerable stir, as it was here that (notwithstanding the difficulties of the analysis from the viewpoint of the economy as a whole) first proofs were presented indicating a rise in productivity as a consequence of IT. The study indicates a significant positive contribution made by information technology capital and staff expenditure to output and productivity. All in all, however, the various empirical studies have continued to furnish results that are highly contradictory as far as a causal connection is concerned. Most discussions have thus tended to take place on a heuristic level. The factors that make it particularly difficult to achieve exact results include confusion as regards the object of research, the insufficient data material, and the general structural shift from production and processing to service-based activities.

Along with an understanding of the way figures for productivity are gauged, the insights relevant for practice thus focus on explanations for the existence of the productivity paradox. The diverse approaches do nonetheless show interactions that on occasions are striking.[150]

One explanation for the productivity paradox invokes the intensification of competition and an ensuing redistribution of profits between the various companies in a sector, as a result of which information technology may benefit the individual firm while proving unproductive with respect to the sector as a whole, or at least failing to lead to any rise in productivity.

Problems of measurement

Another explanation looks to the impracticability of measuring input and output. This is especially the case in the service sector, which is where some 75 percent of the technologies are used in the economies of most of the industrialized nations. Here the output is exceptionally awkward to measure, since advantages often emerge not in the form of cost savings or gains in output but as improvements in quality or comfort. Yet it is precisely these categories for measuring the value of information technology that fail to find their way into the statistics of companies or the national account. In this context, the broad shift from centrally automatised data processing to integrated information technology geared to the individual workplace makes measurement even more of a problem.[151]

Lessons learnt from technological developments of the past, where there have frequently been substantial delays between the introduction of changes and any demonstrable increase in productivity, form the basis for a further explanation. The prime example here is the electric dynamo. From its invention in 1881 it took

148 See anon 1997d, pp. 77ff
149 See Brynjolfsson/Hitt 1996
150 See Gründler 1997, p. 75f.; see also Wigand/Picot/Reichwald 1997, p. 152f.
151 See Picot/Reichwald/Behrbohm 1985, p. 4f.

some 40 years before factories began to use it effectively to turn electricity to greater account. The reasons for this can be traced back to manufacturers and users alike. On the one hand what is required is for a branch to develop that makes the resulting benefits available to the market; on the other hand consumers need time to adapt themselves to using these benefits.

What comes to light above all is an inadequate reorganisation of company processes and a concomitant mismanagement of information technology. The explanation for this failure to make full use of the potential inherent in information technology doubtless lies in the fact that up to now little has been known about this potential. The academic field concerned with the influence of information technology on economic processes is broad in scope and diverse in its range of viewpoints and judgements. In this context it should be easy to pinpoint any mismanagement resulting from a lack of knowledge.

As far as concrete insights into the causal connections between information technology and productivity are concerned, what can be ascertained on balance is that we will have to wait and see whether greater ex-post-knowledge and more uniform methodological approaches will soon yield more precise findings.

3.4 Trailblazer for the Internet Economy – The Convergence of the Media and Communications Sectors

Alliances of companies from different industries

Companies from the media and communications sectors are at present in a state of considerable uncertainty. They are asking themselves questions like "What is my market and who are the decisive players in it?" and "Which companies are my allies and which are rivals?" Given the increasing integration of segments that have up until now been separate and the greater complexity in the conditions of competition that this has entailed, it has become progressively more difficult to provide a functional classification of product-market combinations. Is the main competition to the portal Yahoo (www.yahoo.com) to be found in other Internet providers such as Lycos (www.lycos.com), Excite (www.excite.com) and Netcenter (home.netscape.com) or also in the classical media suppliers such as NBC or CNN?

Convergence as a process

The reason for the insecurity currently besetting companies that supply information, entertainment and communications services is a development known as convergence.[152] "Convergence" does not refer to a single event, but rather an evolving process of progressive fusion between three industrial sectors or markets that originally operated more or less independently from one another, i.e. the media, telecommunications and information technology sectors. The concept designates both an overlapping of technologies (in particular transmission technologies) and also the integration of the respective value chains and the general unification of the markets concerned.

152 The underlying meaning of the concept of convergence stems from the realms of mathematics and medicine. Convergence accordingly denotes a meeting, flowing together or intersection, a tendency towards a common goal, a correspondence.

There were points of contact between these three industries even before the actual process of convergence set in. A number of technologies that have been applied in the telecommunications and media sectors for simplifying, speeding up and expanding the production processes there, if not actually making them possible in the first place, have stemmed from the realm of information technology (information technology as an "enabling technology"). Yet recent developments have gone far beyond this sort of mutual complementarity. The driving forces behind the convergence lie in four areas: technology, regulations, demand and competition.

Digitisation technological basis for convergence

The technological basis underlying the fusion of the three sectors is digitisation. The unified use of digital technology in media, telecommunications and information technology has led to the union of the three branches to form an integrated media and communications sector that is of particular interest to consumers.

While information has always been produced, processed and transmitted digitally in the IT sector, in the telecommunications sector the networks have been adapted to digital technology by degrees. Up to recently, transmitting digital information from the IT realm by means of telecommunications networks was a complex process. The digital information first had to be translated by modem into analog information, then transmitted through analog telephone networks, before finally being reconverted by a modem into digital information that could be processed by a computer. This interplay of information technology and telecommunications is now substantially simplified.

The onset of digitisation has hit the media sector too. Along with the print media, which are to a large extent already produced digitally as it is (DTP), digital television is gaining ground as well. The great advantage that digitisation offers TV is the possibility of compressing content. With a compression rate generally in the region of one to ten, the existing channels can be used much more efficiently for the transmission of content. Digitisation is thus not only contributing to the spectacular development of the media and communications market but is also the very precondition for the convergence of the individual sectors.

Yet digitisation constitutes only the basis for this convergence. In the telecommunications sector the process is being further supported by technological advances in the networks. These include in particular improvements in transmission capacity, the development of return channels in some of the networks, as well as the installation of intelligent network structures. In the IT sector, the process of convergence is being underpinned by further increases in processing and memory capacities.

Deregulation and increased competition add momentum

Along with these technological factors, the process has received a further boost from the legal sphere by the deregulation that has led to new levels of competition in many markets. This development has had ramifications in terms of both

competition and custom. With their increased competition, liberalised markets unleash a creative potential that accelerates innovation and improves service quality. At the same time, the demand for integrated services and multimedia products is jacked up by the fall in prices for media and communications services.

The Process of Convergence

The process of convergence was characterised at the outset by the existence of three fundamentally independent markets satisfying distinct customer needs with their value chain activities. Recent years have seen more and more structural couplings between the value chains of the media, telecommunications and IT sectors. It has thus become clear, for example, that the demand for information technology is stimulated largely by the telecommunications applications (services) and media content that can be used through IT terminals and platforms.

Convergence as structural coupling of value chains

The structural coupling between the value chains concerned can be illustrated by the following diagram.[153]

Fig. 3.46: The three separate value chains at the outset of convergence

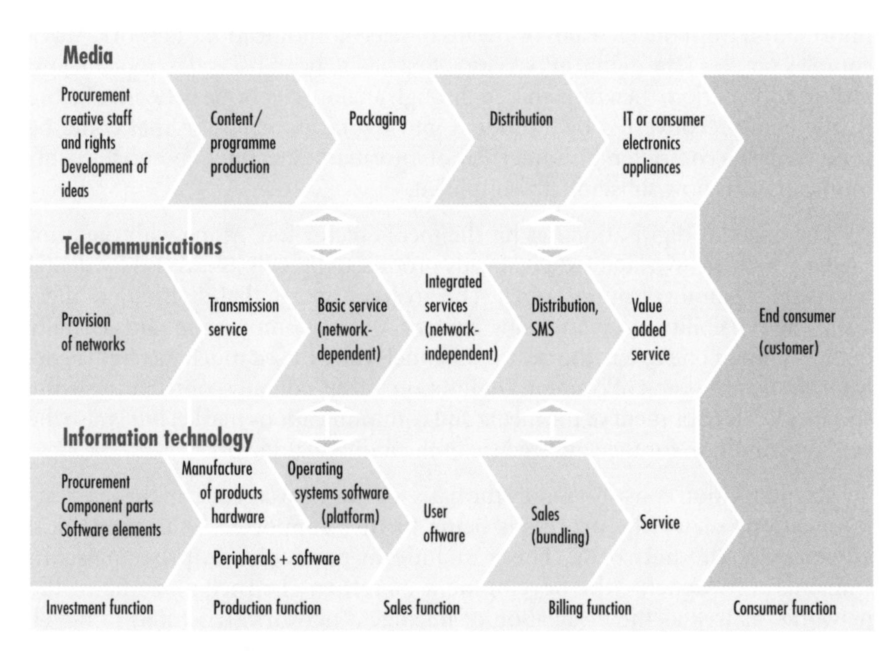

The changes in the underlying conditions of the three markets – and digitisation in particular – have reinforced the structural couplings between the three value chains and given rise to a situation in which all three markets are appealing increasingly to the same consumer needs. An evolving process of convergence has

153 Underlying the three value chains in the following account there is a difference in economic perspective. The IT value chain shows a production-based economic structure, as the most important thing in this market is the manufacture and marketing of products. The value chains of the media and telecommunications sectors by contrast are geared towards the provision of a service.

Source

Fig. 3.46: ECC 1999

thus been set in motion, bringing together the media, telecommunications and IT markets. Two stages can be discerned in this process.

First stage: IT and telecommunications converge

The first stage of convergence consisted in the merging of value chain activities between the telecommunications and the IT sectors. Ever since the mid-sixties there have existed networks of mainframe computers and minicomputers that could be accessed from decentralised terminals located anywhere. In the following period business enterprises in particular used telecommunications connections between individual computers for data transmission and later for Electronic Data Interchange (EDI).

The increasing integration of previously separate computer networks (LANs) has been a product above all of the exponential expansion of the Internet since 1993. More and more firms have connected existing company networks, some of them (such as IBM's) quite sizeable in themselves, to the Internet, in turn thus generating dramatic growth rates. The consequence is the increasing fusion of the value chains of the telecommunications and IT sectors. In Internet markets the two are no longer distinguishable.

Second stage: media get into the process

The second stage of convergence is taking place at present. In this case it is the value chains of the media, telecommunications and IT sectors that are increasingly merging. Two trends especially demonstrate the effects of this process.

* The transmission of media content is no longer the exclusive domain of the broadcasting networks[154] (cable, satellite and terrestrial networks), but classical telecommunications networks too (telephone and computer networks) are becoming more and more important in the dissemination of content. By the same token, broadcasting networks are gaining ground in the provision of communications services. Network operators previously working in different markets thus now find themselves competing with one another.
* New conditions of competition have also emerged in the field of reception appliances. Appliances from all three sectors (television, telephone, computer) can be used for the reception or operation of the distinct information, entertainment and communications services.

This two-stage process of convergence is summarized in the following diagram.

154 Broadcasting networks are networks serving mass communication in the media sector. In principle they permit only unilateral downstream transmission, from the transmitters to the associated participants, although technological advances have made upstream channels possible as well.

Fig. 3.47: The two stages of convergence

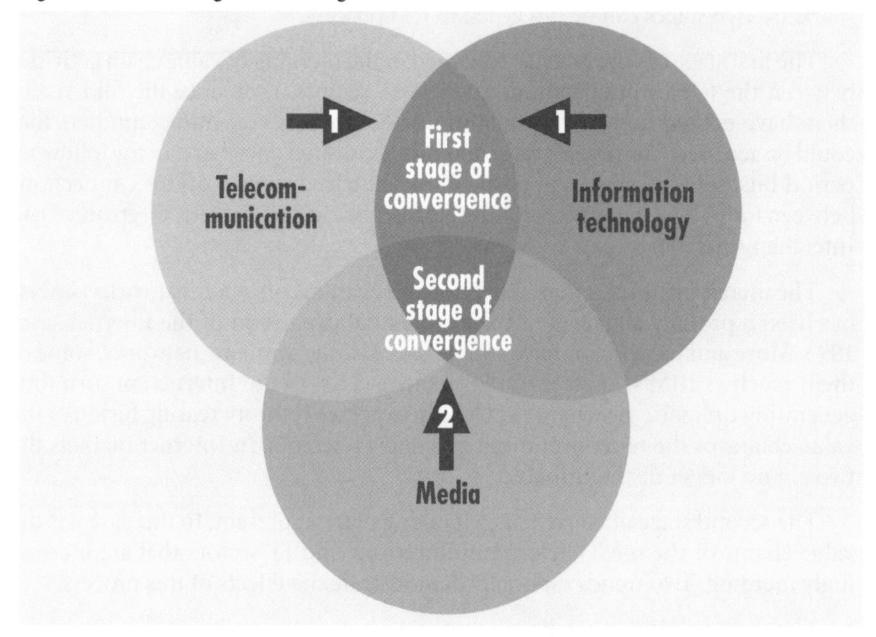

Result: A new multimedia value chain structure

The result of this two-stage process of convergence is the steadily increasing significance of the areas of overlap, in the end leading to the dissolution of existing systems boundaries between the media and communications sectors. New conditions of competition are thus produced between the technologies, suppliers and services of the three sectors concerned. In addition to this a joint market is emerging that combines the services offered by the media, telecommunications and IT sectors and unifies the formerly separate market branches. This joint market, which comprises the integrated provision of media, telecommunications and IT services, contains the value chain fields of content, packaging, transmission, navigation,[155] value-added services and reception appliances.

The diagram below represents this process of unification and shows how three separate vertical value chains combine to form six horizontal market areas. It also gives an idea of the relative significance of the three sectors in the distinct areas of the convergent media and communications market.

155 As a value added field, navigation denotes the manipulation of the infrastructure, i.e. the use of hardware and software components to facilitate and improve orientation within and control of the physical infrastructure. These components include above all operating systems, browsers and intelligent agents.

Source

Fig. 3.47: ECC 1999, based on Adstead / McGarvey 1997, p. 10

Fig. 3.48: The emergence of the multimedia market from the three media and communications sectors

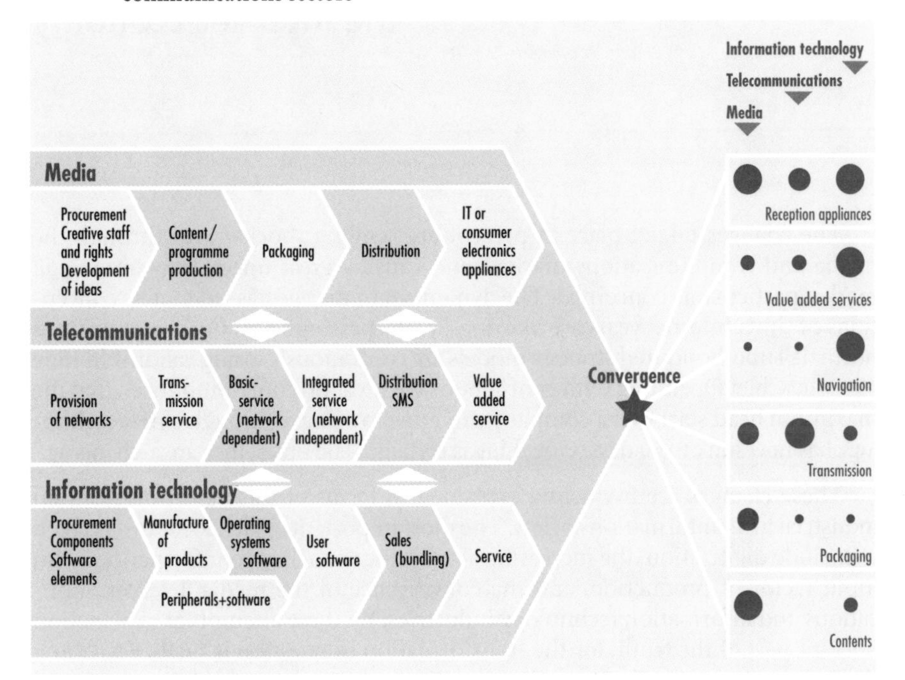

The convergence of the three media and communications sectors entails fundamental changes for the media, telecommunications and IT industries. Yet the impact of these changes goes beyond the individual markets themselves, bringing economic consequences that touch numerous areas of the economy and society in general. Information is now available any time, any place and at very low costs. Such changes harbour new potentials that can be – and indeed already are being – tapped by other areas of the economy such as financial services, trade and tourism.

The economic changes are thus not limited to the new media and communications market, since new communications infrastructures, communications services and the universal availability of information make it imperative for the players in other branches of the economy to adapt to far-reaching structural couplings. The virtual marketplace provides a second, complementary market level.

Network economy: a new marketplace with new economic structures

The convergence of the media, telecommunications and IT industries is the precondition for the formation of a new marketplace bringing new economic structures with it: the network economy.

Source

Fig. 3.48: ECC 1999, based on Collis/Bane/Bradley 1997, p. 3

Chapter 4

The Emergence of the Internet Economy

The changes taking place at present are sending shock waves through the media and communications markets and causing great uncertainty among all market participants concerned. The dynamism of the age has led to new conceptions of time: Internet years are like dogs' years, passing seven times as quickly as normal! Time-honoured strategy models are continuously losing validity. In June 1998 the Chief Operating Officer of Yahoo (www.yahoo.com) announced that the maximum head start that a company can hope to achieve through a new technology has now sunk to 60 days. One thing is certain: "The times, they are a-changing."

Internet years are like dog years

These changes are frequently explained in terms of the transition from an industrial to an information society. The most important evidence for this is taken to include digitization, the increasing significance attached to information as an input factor of production, and the convergence of the media, telecommunications and information technology industries. Yet these attempts at explanation only hit part of the truth, for the transformation in question is further reaching than this. Owing to the explosion in the performance of information and communications technology and the prevalent Internet software standards, what is being produced is a progressive networking of infrastructures both inside and outside the media and communications sectors.

Progressive networking of infrastructures

Even today, for example, it is already possible to use communications technology to link up telephone and office communication networks, voice mail systems, wireless networks such as mobile telephony, Personal Digital Assistants (PDAs) such as the PalmPilot, set-top boxes such as the d-box, radios and television sets, as well as other objects such as cars, watercraft, aircraft, track vehicles or even refrigerators. In addition there are global satellite, radio, electricity and cable networks, as well as logical networks and communication services that make either occasional or permanent contact possible. The result is that the retrieval of information has become something utterly unconstrained by time or space.[156]

The result of these developments is a new network economy.[157] And with increasing networking, moreover, the sway of the laws inherent in the networks is continuously spreading. The accompanying paradigm shift from atoms to bits[158] has so far only been noted sporadically. With the spectacular decline of the centuries-old reference book "Encyclopaedia Britannica," Netscape giving away its browser for free, and CISCO, a provider of network solutions, transferring its

156 See Picot 1998b, p. 3ff.
157 The concept of a new network economy was chosen to maintain a clear conceptual differentiation with respect to the traditional business management network theories. See for example Sydow 1991, 1995
158 See Negroponte 1995, p. 11-20; Kelly 1998, p. 7

entire business activities into the Internet, the signs have been there. But only when seen from the perspective of a network economy do the connections become apparent and accessible to further analysis.

At the end of this process of change there will be a new economic structure that functions differently and has little in common with the reality of today. The transformation already underway has not escaped the notice of industry: in 1995 there were no companies talking about e-commerce, whereas by mid-1998 there were no companies not talking about it.

The proverbial calm before the storm

The following section consists of a detailed analysis of the new network economy. Firstly, the most important determinants that have led to the emergence of the network economy are identified. The second section turns to the economic consequences of these changes. The most remarkable thing about the current situation is the proverbial calm before a storm, for although the consequences are radical, they are sneaking up on us without us realising their full import...

4.1 Decisive Factors in the Emergence of the Internet Economy

Evolution marked by unforeseen contingencies

It is noteworthy that the various factors that have influenced the development of the network economy do not display any systematic cause-effect relationship. The course of the network economy's evolution is marked instead to a great degree by serendipity and unforeseen contingencies.

The starting point for this analysis is a theoretical and empirical description of the role played by information in economic activities. Of particular relevance here is the increased capacity of information and communication technology thanks essentially to digital technology. On the one hand this permits a considerable expansion in the performance achieved by a company or national economy. On the other hand the continuous miniaturisation and standardisation of information and communication components is leading to the opening up of new areas of applicability for microelectronics in virtually every branch of industry. These developments mean an effective and efficient utilisation of the available technological potential and thus also a substantial transformation of the economic activities of almost every industrialised economy.

First commercial browser 1993 triggers explosive growth of the Internet

Nonetheless, it was not until the connection of these technological advances with the explosive growth of the Internet, which up until the early nineties had progressed in relative isolation from the other developments, that the network economy was really triggered off. With the introduction of the first commercial Internet browser, 1993 marks a provisional conclusion to a development that can be regarded as the preparation and origin of the network economy. The result of this development is the formation of a new electronic market place where the most important currency is information. The services offered range from media and telecommunications services to every form of e-commerce imaginable.

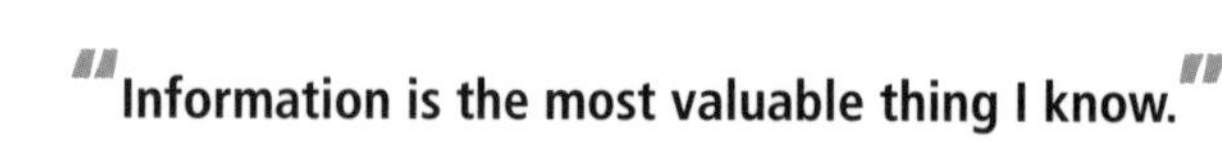

4.1.1 Central Role of Information in Markets

A mere glance at what is happening on the stock exchanges shows the significance attached to information in economic activities.[159] The following section looks in two steps at the role played by information in a network economy. Firstly, a brief theoretical analysis shows to what extent information and knowledge function as integral components of market processes in a country's economy. Secondly, there follows a description of one of the most palpable developments characteristic of the transition to a network economy: the rapid increase in importance attached to information.

Austrian School of economics

Markets are economic places where supply and demand meet one another.[160] By contrast with neoclassical microeconomics, which assumes the free and frictionless coordination of economic activities, this account will conform with the theory of the "Austrian School," where markets are viewed as a process-based phenomenon. Prices here no longer contain all the necessary information but instead cause high levels of uncertainty in the market process and the transactions occurring within it. As in reality transactions do take place, however, there must be forms of reducing these decision making uncertainties. In general this can be achieved by acquiring economically relevant knowledge about the services being sold and the transaction partners in question. Time here becomes a key economic factor, because having a head start in terms of information – through a timely analysis of the latest customer trends or recognition of technological innovations – is the only guarantee of gaining a competitive advantage. Knowledge is thus crucial for all market participants. A market process is therefore not just an economic meeting-place for supply and demand but also and above all a sequence of changes in knowledge.[161]

Markets are also a sequence of change in knowledge

These theoretical considerations are reflected in economic reality too. The unequal distribution of information between market participants calls for a high level of coordination in order for transactions to be carried out. These activities as a rule include the arrangement, implementation, control and adaptation of a given transaction and give rise to information and communication costs, or so-called transaction costs. On account of the highly advanced division of labour and the intensified coordination that this entails, transaction costs form a substantial part of an economy's gross national product. The importance given to information in a country's economy can consequently be expressed as the proportion of national income made up by transaction costs. According to an investigation carried out by Wallis/North, the share of the United States gross national product consisting of transaction costs within and between firms rose from 25 per cent in 1870 to 55 per cent in 1970 and is now estimated to lie somewhere above the 60 per cent mark.[162] It thus becomes clear that a great portion of the national income goes on information and communication.

159 Following Wittman, 1959, information is here understood as goal-oriented knowledge. By way of distinction, knowledge and data can be defined as follows. The starting point is the state of affairs in reality, the reproduction of which generates data. The storage of these data leads to knowledge. Finally, the goal-oriented deployment of this knowledge leads to information. See Jacob 1995, p. 82
160 On the following see also Wigand/Picot/Reichwald 1997, p. 17ff.
161 See Kirzner 1973
162 See Wallis/North 1986, p. 121

A further indicator of the increasing significance attached to information is the continuing spread of media and information goods. Today's rapidly growing markets thus include media, software, consultation, databases and financial services. All these products are characterised by being highly information-intensive and the fact that they come as part of equally information-intensive, non-material services:[163] "Instead of pushing atoms they push bits around."[164]

"Pushing bits": information gains in relative importance in products and in services

The significance of information as an input factor of production has thus steadily risen in recent decades. The dynamism of the present situation, however, as has come to light in the last few years, goes even beyond this. Information has come to play the key part in a development that is often described as a revolution. The reasons why the processing, usage and transmission of information have become quicker, cheaper and more efficient than ever before are to be found in the explosive rise in information and communications technology capacities. These will be analysed below.

4.1.2 Explosive Rise in Information and Communications Technology Capacities

Technologies are an essential impetus to economic development. This emerges particularly clearly when basic innovations such as the steam engine in the eighteenth century or electricity in the nineteenth gave rise to whole new industries. Examples from the realm of information and communications technology are the telegraph and the telephone, which in the nineteenth century made radical changes in value chain structures possible. The consequence was the emergence of completely new trade institutions and large scale enterprise structures such as commodity futures exchanges or distance bridging systems of mass production.[165]

Features of the "tidal wave"

The present transition to a network economy has likewise been impelled to a large degree by advances in information and communications technology. The fundamental difference from the afore-mentioned technologies such as electricity or the steam engine is the speed with which the innovations in information and communications technology are spreading worldwide and thus transforming economic activities. The most notable features of this technological "tidal wave" are above all the following factors:

* digitisation,
* increased cost-effectiveness,
* miniaturisation,
* standardisation.

163 See Picot 1998a, p. 2
164 See Kelly 1998, p. 7
165 See Chandler 1977

The technology underlying all these changes is digitisation.[166] The conversion of information into digital units, or so-called bits (expressed as 0 and 1), has meant that information can be assimilated and applied by processors. It can also be transported in networks, moreover, with the costs remaining unconstrained by the distance covered.

From atoms to bits

The consequences of this change from physical atoms to digital bits are radical: bits have no weight and move at the speed of light. The marginal costs for the production of further bits are more or less zero.

> *"They do not require any storage space. They can be sold and yet at the same time kept. The original and the copy cannot be distinguished. They cannot be stopped at customs borders or any other sort of border. Governments cannot tell where they are currently hanging out. Authorities find it impossible to impose appropriate legal controls on them. The marketplace for bits is global."*[167]

Digitization lies at the heart of these far-reaching changes. On top of this, however, there have been the technological advances forecast by Moore's Law. This law describes an exponential growth in computer processor capacities, which are predicted to double at regular intervals of eighteen months.[168] The resulting increases in computer power with the accompanying fall in prices have produced progressively improving cost-effectiveness in the realm of information and communications technologies.

Cheap computing power

The low-cost availability of computer power has also led to a broadening of the area of application for microelectronics – with remarkable results that have made themselves felt throughout industry as a whole. A mobile telephone from 1998, for example, contains more processing capacity than the NASA computers in operation for the landing on the moon less than 30 years ago. An average PC from 1998 could have controlled the entire Apollo mission. Processing and transmission capacity has thus become a commodity that is cheaply available everywhere.[169] This can be seen for example in the fact that cheap microprocessors are now present in almost all of the more complex consumer durables and increasingly so too in simple consumer goods such as children's toys.

A further determinant is the miniaturisation of almost all components in information and communications technologies. The driving force behind this process of miniaturization is the increasing integration density of micro-processors, i.e. the increasing number of transistors contained on a chip. The effects of this are illustrated by the development of the Intel processor. While the first microprocessor 4004 from Intel in 1971 contained 2,300 transistors, the Pentium II Processor introduced onto the market in 1997 has space for 7.5 million transistors.[170] The number of transistors on a chip has thus risen by a factor of 3260 within just 26 years. The consequence of this miniaturization of information and communications components is a continuous reduction in the material and energy required for use. This gives rise to new possibilities in terms of spatial

166 As already described in section 3.4, it was the increasing use of digital technology that made possible the convergence of the three M&C sectors.
167 See Negroponte in Downes/Mui 1998, p. X
168 See section 3.3.1
169 See Picot 1998b, p. 9
170 See Madden 1998, p. 64

organization and in turn to virtually unlimited flexibility in information and communications services.

One of the most important preconditions necessary for this information and communications potential to be realized is standardisation. As has already been seen in the case of the PC industry,[171] standardisation permits even industries with a highly fragmented value chain to enjoy considerable combination possibilities, something that benefits both the various suppliers and also the consumers.

Compatibility of systems, planning security for investment

For business concerns it is particularly important that the standardisation should ensure the compatibility of sub-systems for all concerned, rationalise the learning necessary, and above all foster planning security for further investments. At issue here are above all the so-called de facto standards, which gain general acceptance either through market based selection processes, as was the case with the PC operating system, or occasionally also through the activities of associations or cooperations, as with the GSM standard in mobile communication. A number of important standards have thus come into being as it were unplanned, catching on globally on account of their practical superiority but without any prior design.

These advances in information and communications technologies constitute one of the fundamental prerequisites for a network economy to develop. Yet this technological potential only wielded its full revolutionary weight with the previously unforeseen take-off of the Internet in 1993.

4.1.3 Causes of the Internet Revolution

The Internet in a shadowed niche

The Internet was the product of a US-government military research project from the year 1969 (ARPA).[172] Until the early eighties it was used mainly for academic purposes and lead a somewhat shadowy existence. It was only with the construction of a backbone network (NSFNET) between 1983 and 1986 that the greater part of the North American universities were able to link up with one another on the basis of the unified transmission protocol TCP/IP, bringing about a substantial rise in the number of users.

1993 breakthrough by WWW

For a long time the worldwide computer network served solely for communication among academic institutions and other research establishments. Not until the development of the graphic user interface of WWW and the software browser Mosaic between 1989 and 1993 did de facto standards emerge, making possible – to the surprise of all concerned – the breakthrough for commercial applications. 1993 can be regarded as the year in which the network economy was born.

As a result the Internet changed from being a computer network with applications such as file transfers or terminal connections to being an increasingly integrated network with a whole wealth of multimedia applications.[173] The

171 See section 3.3.1
172 For the organisation and structure of the Internet, see also 3.2.3
173 See Rupp 1996, p. 9

Multimedia content grows in importance

changes in Internet use make this clear. Up until 1993 the Internet services File Transfer Protocol (FTP), Gopher, Telnet and Usenet were predominant, but after the introduction of the WWW these suffered a steady decline in significance. Considerably more than half of Internet use is now apportioned to the WWW. Electronic communication – e-mail – has from the outset been one of the most frequently used Internet services.[174]

A headache for media companies

The increase in user numbers that resulted from the greater attractiveness of the WWW has in turn brought about an increase in connections with further networks. As the Internet was utilized by private users primarily for access to information, it appeared to be first and foremost a newly emerging form of mass media – and as such it posed a big threat to the established media. This has proved, and is still proving, quite a headache for media companies, especially since in terms of speed the diffusion of the Internet up to now has far exceeded anything experienced by the other mass media. By contrast with television, which needed 13 years, or radio, which needed 38, the Internet has hit the 50-million-user mark within just five years.

Speed of Internet growth exceeds that of previous mass media

Fig. 4.1: Internet growth beats all other mass media

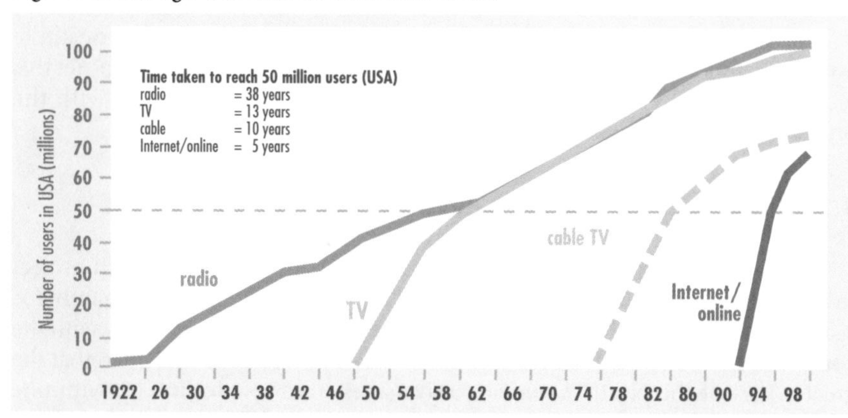

Similar tendencies can be seen in Europe, even if diffusion is not yet quite as advanced as in the United States owing in particular to the lower levels of PC penetration and the higher telephone costs. In Europe the number of Internet users has doubled within two years: first, from 12.3 million in 1995 to 23.9 million in 1997,[175] and then again to 42.7 million in June 1999.[176]

This explosive diffusion on the demand side can be considered a driving force behind the Internet's growth. The reaction on the supply side, i.e. from the businesses concerned, is reflected in the number of hosts. According to the definition given by Network Wizards, hosts are computer systems connected with an IP address as a domain name, like "www.yahoo.com".[177] Hosts contain the most diverse of offers, ranging from information and entertainment through to

174 See Databank Consulting 1998, p. 14
175 See EITO 1998, p. 67
176 See www.nua.ie/surveys/how_many_online/index
177 For this definition by Network Wizards see under www.nw.com.

Source

Fig. 4.1: Morgan Stanley Dean Witter 1997, p. 14

accounts of company performances. The increase in hosts is shown in the following diagram.

Fig. 4.2: The increase in Internet hosts from 1981 to 2000

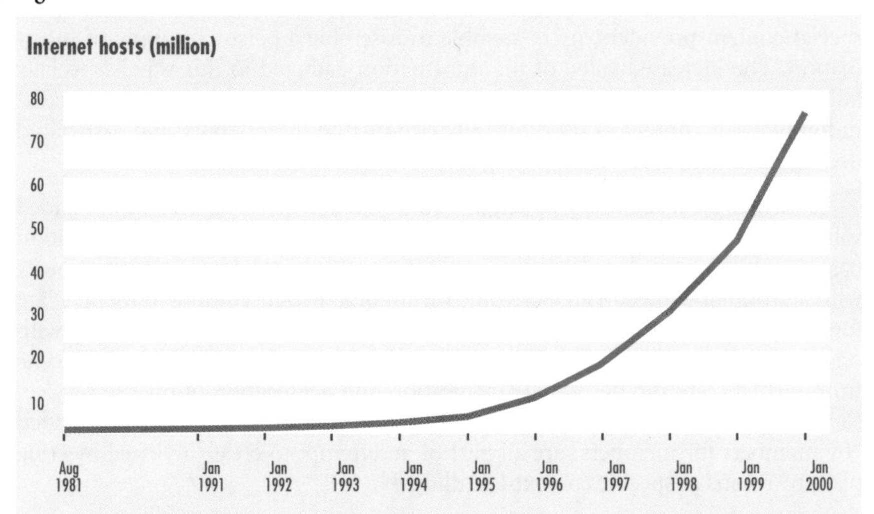

The steadily rising number of users has led to more and more companies offering services in the Internet. This broadening in the supply of offers has served not only to keep existing users but equally to make the Internet significantly more attractive for potential new users. Such positive interaction between suppliers and customers is based on network effects and designated a positive feedback loop.[178]

Added value of Internet offers

The reason for the rapid diffusion of the Internet is the added value of Internet offers by comparison with traditional offers. This added value can be attributed to four factors:

* interactivity / individualisation,
* immediacy of access,
* fall in transaction costs,
* multimedia form of offers.

Interactivity permits the individualisation of content and thus provides the user with great added value. It can range from simple user interaction by means of control signals through to a dialogic component of active communication. With content based offers in particular, interactivity can give rise to the possibility of individualised mass communication. Examples of this include the information pages of portals such as Yahoo (www.my.yahoo.com) or Netcenter (www.my.netscape.com).

178 See section 4.2.1

Source

Fig. 4.2: Internet Software Consortium (http://www.isc.org/). In 1981, there were 213 host, and only 1,961 in 1985. Figures for 1991 through 2000 are for January.

Mass-customisation

So-called "mass customisation" of this order has its advantages for suppliers and customers alike. By means of the modularisation of content suppliers are able to provide relatively low-priced, individualized information with a high added value. From a preconfigured selection of information suppliers such as news agencies or special content providers, users are able to assemble a personal supply of information. The increased value of the information gathered in this way is specially tailored to meet user needs. Compared with traditional forms of mass media, individualised content is on offer but without the interactivity and associated decision making processes becoming too complex for the user.

Extended forms of social communication

A further great advantage springing from interactivity is the creation of added value through extended forms of social communication. Numerous communication based applications such as "chatting" – communications conducted simultaneously with a group of people in an electronic forum organised according to theme – are for many consumers now the main reason for an Internet connection. Along with on-line services such as AOL, these communication oriented offers likewise come from virtual communities.[179] The integration and aggregation of diverse services such as free e-mail, discussion and communication forums, and content provided "by members for members" are all part of an attempt to create a virtual meeting place by means of specific context-bundling.[180]

Immediacy of access

Immediacy of access means that desired information can be retrieved in real time unconstrained by time or space. Internet offers are thus generally characterised by high information value. Examples here include extensive databases on all listed companies in the United States (www.sec.com) or a constantly updated calendar of events for Berlin (www.berlin.de). As WWW information is represented in HTML (Hypertext Markup Language), moreover, there is a decentralised hierarchical structure, which through Hyperlink facilitates instant access to more detailed or extensive information. Furthermore, on account of the digitalised data form, the information can be processed, applied and stored.

Low transaction costs

The fall in transaction costs is a further advantage enjoyed by Internet offers. Whether it is a German student spending half an hour researching literature in the American Congress Library in Washington or complex building plans being sent from India to Hong Kong, the details are of little import: in the real economy, these transactions would entail markedly higher costs. E-mail in particular was the first application to gain widespread acceptance among users owing to the low-cost and high-speed opportunity it presents for global communication.

If it proves possible to improve market transparency, then the current potential for lowering transaction costs will be able to be exploited even more efficiently than up till now. Existing search engines such as Yahoo or Lycos can only partially rectify these problems, as the results found rarely correspond straightaway to the information being sought. For this reason the future will see a further increase in the importance of so-called "intelligent agents." These are software programs with automatic filter or search functions, and they offer consumers high added value. Up

179 See Hagel III/Armstrong 1997, p. 8f. These authors even argue that the great growth enjoyed by AOL in comparison with its former rivals CompuServe, which has since been taken over, and Prodigy, which has since been closed down, is attributable mainly to their early emphasis on user-interaction and communication.

180 Examples of this include "community organizers" such as geocities.com, tripod.com or angelfire.com.

until now, software agents such as Firefly for example have been used mainly by suppliers such as the bookseller Amazon.com. Using previous purchase decisions and customer recommendations, Firefly puts together a user profile based on calculations of statistical probability. For registered customers, the added value thus consists in the fact that each time they pay a visit to Amazon.com they will receive tips on new publications that could prove of particular interest.

Multimedia attraction

The multimedia form of Internet offers refers to the many different forms – audio, video, text, image, graphics – these offers may take.[181] So far the multimedia potential of Internet offers has been limited above all by the low transmission capacities for reproducing moving pictures at a level comparable with television. Even so, there are some radio and TV programmes already being broadcast in the form of so-called webcasting.

All in all, what comes to light is that Internet offers generate high added value for consumers on account of the four factors described above. This attractiveness explains the rapid diffusion and also hints at continued dynamic growth to come. Juxtaposing the above user side features with the trends on the technology side traced in section 4.1.2, then the following overall picture emerges:

Technology push and market pull mutually re-enforced

Fig. 4.3: The interplay of technology push and market pull

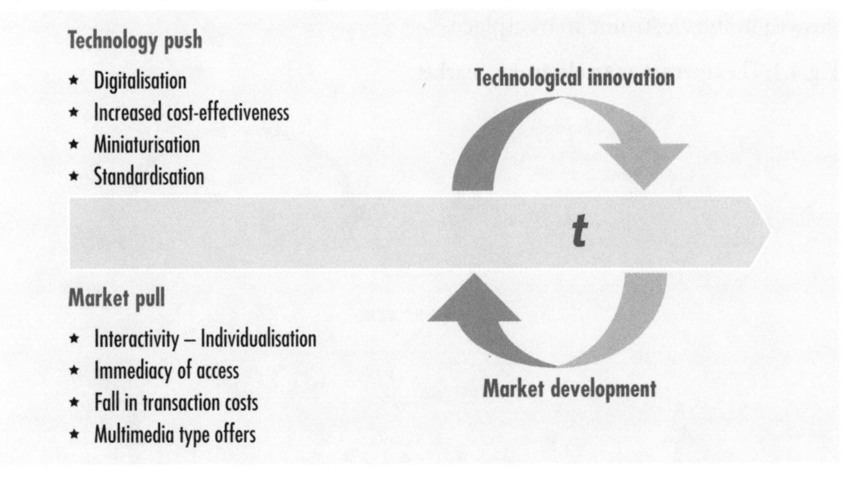

The dramatic increase in capacity on the technology side is thus conjoined with factors conducive to a rapid diffusion and adaptation of the technologies and their new applications. This interplay of technology push and market pull has conspired to produce the explosive development undergone by the Internet. The result of this interplay is a new marketplace hosting not only individual mass communication and telecommunications services but also the most diverse forms of electronic trade.

181 At this juncture it is worth pointing out that the much over-used concept of "multimedia" here only represents a breakthrough from a telecommunications perspective. From the perspective of communication science "multimedia" have existed since 1929 when Warner Bros. brought out the first talkie.

Source

Fig. 4.3: ECC 1999, based on Picot 1998a, p. 3

4.1.4 Virtual Markets as a Result of Digital Infrastructure

"Now I assign the Internet the highest level of importance."[182] This private memo written by Bill Gates in 1995 indicates the significance of the digital infrastructure being used in increasing measure as a marketplace. As in the real economy, these electronic markets,[183] often designated "virtual marketplaces" or "marketspace,"[184] are places where supply and demand meet one another. The basic difference lies in the fact that either part or all of the transaction takes place electronically with the aid of the appropriate information and communications systems.

Electronic markets before the Internet

Various forms of electronic market existed even before the Internet infrastructure came into being. In the seventies Airline companies such as American Airlines, for example, began constructing internal reservation systems, which are now also accessible to rival companies and integrated with the reservation systems of other services such as car hire firms or hotel chains.[185]

The Internet expands the frontiers of electronic markets

In general, the Internet and the constant advance of information and communications technologies permit market potential to be realised which had previously not been available or only rudimentarily so. The increasing connection of telecommunications and other networks to the Internet also means continuous growth in the electronic marketplace.

Fig. 4.4: The Internet as an electronic market

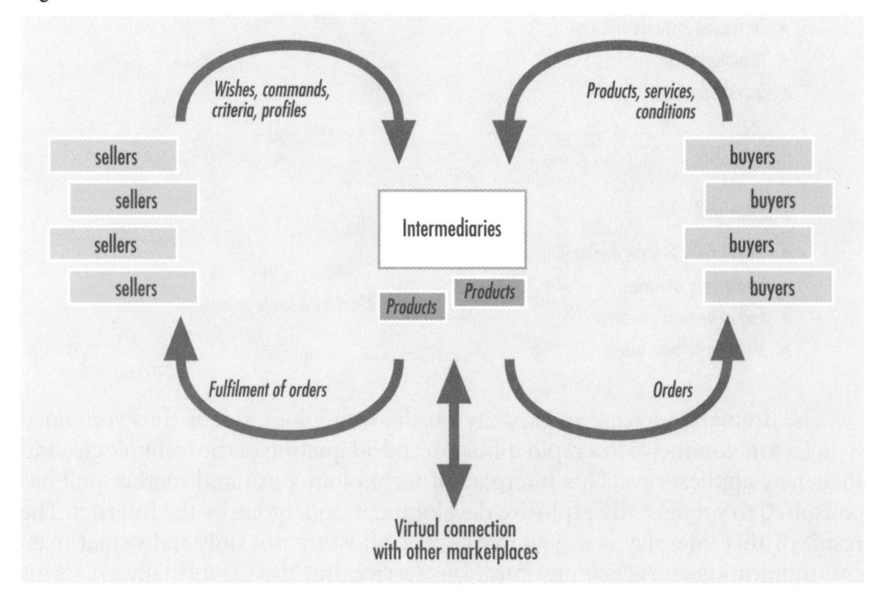

182 See Edstrom/Eller 1998, p. 200
183 See Malone/Yates/Benjamin, 1987
184 See Rayport/Sviokla 1994
185 See Picot 1998a/1998b; Wigand/Picot/Reichwald 1997, p. 259ff.

Source

Fig. 4.4: ECC 1999, based on Mougayar 1998, p. 143

"Now I assign the Internet the highest level of importance."
Bill Gates

On the whole, the electronic marketplace can host any type of economic activity (henceforth designated e-commerce).[186] Two general forms of electronic market transactions can here be differentiated: electronically supported market transactions and completely mediated market transactions.

Up until now, the prevailing form of electronic market has consisted of electronically supported market relations. This refers to a market form where either the information and decision phase or the agreement and implementation phase of transactions are electronically reproduced. An example of an electronically supported process of information acquisition and ordering might be the purchase of a computer from a direct order firm such as Dell (www.dell.com) or context related suppliers such as Zdnet (www.zdnet.com). After receiving the order, the supplier in question delivers the product to the customer in the traditional manner.

Only two types of pure electronic commerce

The complete mediation of transactions is up to now only possible with two categories of transaction. The first case consists of information goods such as media content or software products which can be distributed through the Internet on account of their digitalizability. The second case comprises transactions where all that takes place is a transfer of rights of disposal in the form of information and there is no physical exchange of material goods. This is possible especially with electronic stock exchanges, since all that is being supplied and demanded here are stocks and shares or other rights of disposal. These two categories of electronic market transaction can be illustrated by the following diagram:

Fig. 4.5: Electronic market transactions

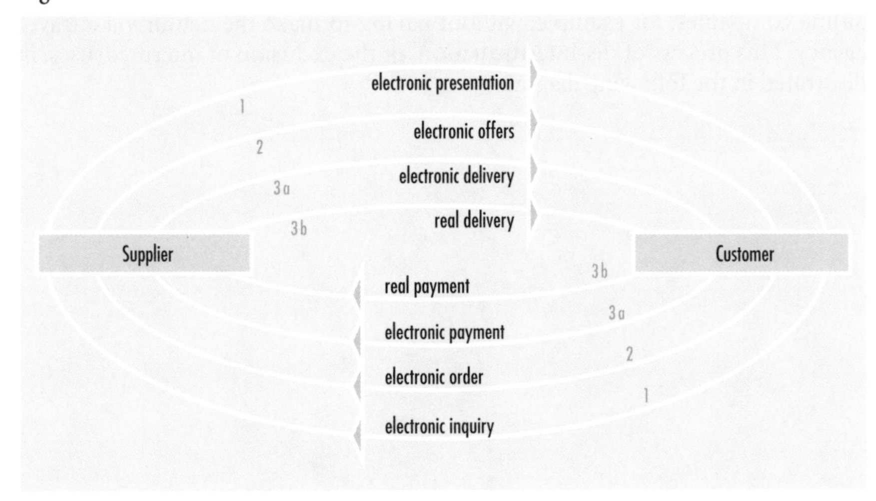

186 See Wigand/Picot/Reichwald 1997, p. 263f.

Source

Fig. 4.5: ECC 1998, in Anlehnung an Mougayar 1998, S. 143

Electronic presentation and inquiry (cycle 1) refers to the phase of information acquisition and decision making, while electronic offer and order (cycle 2) denotes the transaction's agreement phase.

These support forms of electronic market are being put into practice more and more frequently in the Internet. Noteworthy is that the most diverse mixtures of electronic transactions are here possible. One possibility, for example, is to call up information electronically on a certain product, but then make the actual purchase in the conventional manner. Another is to order a specific service electronically through e-mail, though the sales dialogue has taken place face to face.

Most types combine physical and electronic transactions

As has been seen, the complete electronic realisation of the implementation phase is only viable with information goods or with trade in rights of disposal (cycle 3a). For most goods, the physical delivery and payment are still carried out in the traditional way (cycle 3b).

Dis-intermediation

Most market transactions, however, do not take place directly between supplier and customer, but rather, as in a real economy, through intermediaries who arrange the exchange of services between the various market participants. These intermediaries can in principle take part in any phase of a market transaction. Travel agencies, for example, are intermediaries in that they match the wishes of their customers with what is being offered by airline companies. Electronic markets present the possibility of executing this intermediary function by means of electronic information and communications systems, rendering the classical suppliers superfluous. A customer can thus book a flight directly with the airline companies, for example, without having to make the detour via a travel agency. This process of dis-intermediation, or the exclusion of intermediaries, is illustrated in the following diagram:

Fig. 4.6: The dis-intermediation of intermediaries

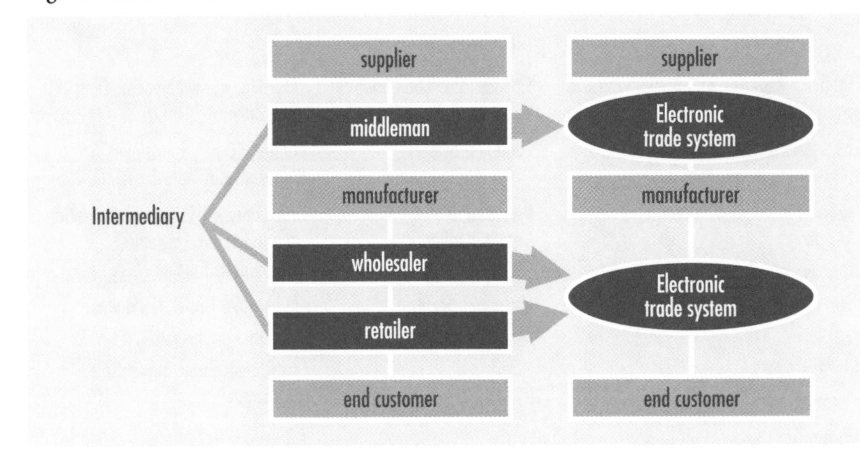

What is beyond dispute is that the new marketplace Internet will bring about a fundamental change in distribution systems. It is not clear, however, just how extensive the dis-intermediation of current intermediaries will be in this new infrastructure. The greater market transparency and the increased ease with which transactions can be carried out in electronic markets means that long-established value chain structures are tending to break up. Every intermediary is thus in principle faced with the threat of being eliminated from the trade chain.

Re-intermediation

This dis-intermediation can nonetheless be overcome if established intermediaries succeed in bringing about their "re-intermediation" in the electronic markets through the provision of high added value. This can be achieved by means of an increased personalisation of their services. Travel agencies planning a new holiday itinerary, to return to our example, could select a hotel that takes into account the specific preferences of the customers in question, as described by Nicholas Negroponte:

> *"They got to know their clients and could personalize their recommendations. 'Nicholas, since you like the Okura in Tokyo and the Peninsula in Hong Kong, you'll love Raffles in Singapore'. And I do."*[187]

Along with the undisputed risks and challenges for established intermediaries, a network economy also opens up a wealth of possibilities for new intermediaries. The dynamic growth in the connected networks and users increases the need for structuration and thus also the demand for new intermediaries. For this reason, the growth in new intermediaries could end up exceeding the dying out of the old ones.

187 Negroponte 1997, p. 208

Source

Fig. 4.6: Picot/Reichwald/Wigand 1996, p. 321

Fig. 4.7: Dis-intermediation: established and new intermediaries

Established intermediaries	New intermediaries	Examples
Financial brokers and underwriters, in-person outlets or banks	Finance transaction brokers	E*Trade, e.Schwab online, DirectIPO, DCT-Online
Specialist trade in books, computers, cars	Context providers	C/Net, ZD.net, SportslineUSA, Amazon.com, Auto-By-Tel, CDNow
Travel agencies, in-person shopping	Product and service mediators	Travellocity, TISS.com, CompareNet, edmunds.com, ConsumerEdge, CouponNet, Junglee,
Direct marketing profiles	Customers profiles	Cybergold, Firefly, NetStakes
Yellow pages, address directories	Electronic directories	Four11.com, BigYellow, switchboard.com, inquiry.com, InfoSpace

New intermediaries

The functions of these new intermediaries can take the most diverse of forms. The greater part of the offers made are as aggregators. This means that a large number of partial services are fitted together to provide the consumer with an integrated service. The aggregators' core competency consists in well-founded market knowledge – specialist knowledge of the function and quality of the products they are offering, specialist knowledge of customer preferences, and specialist knowledge of the complementary services necessary for complete customer satisfaction.[188] The partial services as a rule comprise an attractive range of content, communications and transactions possibilities grouped around a particular thematic focus. An example of the context providers described earlier is CNet.com, where the emphasis lies in computers.

"ShopBots" and "Intelligent Agents"

Along with these "full range" intermediaries, there are a number of specialists that concentrate on individual areas and are on occasion included in the "full range" offers. One example here would be "ShopBots" or shopping robots, which are software agents such as Junglee or Jango that automatically locate low-cost alternatives to particular services. The sort of service performed by these ShopBots might be to visit the Web pages of some 30 on-line CD traders, in each case enquiring into the price of a given CD. The results are gathered and listed for the customer in terms of value for money.[189]

The economic consequences of these new intermediaries are far-reaching. In the real economy sellers enjoy a head start in terms of information, based essentially on the low market transparency for buyers. With a bit of skilled price discrimination between individual target groups, suppliers can thus raise their overall producer surplus.[190]

In the network economy there is a shift in emphasis within this constellation, leading to the phenomenon of reverse markets. This means that there is a displacement of the previous information imbalance between suppliers and

188 Intershop CEO Stephan Schambach in a conversation with the ECC in April 1998.
189 See Kelly 1998, p. 121f.
190 See Hagel III/Armstrong 1997, p. 17f.

Source

Fig. 4.7: ECC 1999, based on Mougayar 1998, p. 147

customers to the clear advantage of the customers. With the aid of the above intermediaries it becomes possible for buyers in electronic markets to ascertain the lowest price for their desired product while also keeping transaction costs low. The greater market transparency thus leads to a stiffening of competition among suppliers. On the whole, buyers in electronic markets have the opportunity to make a clear gain in customer surplus, picking out the best offer on each occasion, while sellers must reckon with lower profit margins. The following diagram illustrates the shift in the relation between customer and producer surplus. In the second diagram, the slanting line in electronic markets represents the increasing realisation of lower costs, raising customer surplus.[191]

More transparency means more competition

Fig. 4.8: Increased customer surplus in electronic markets

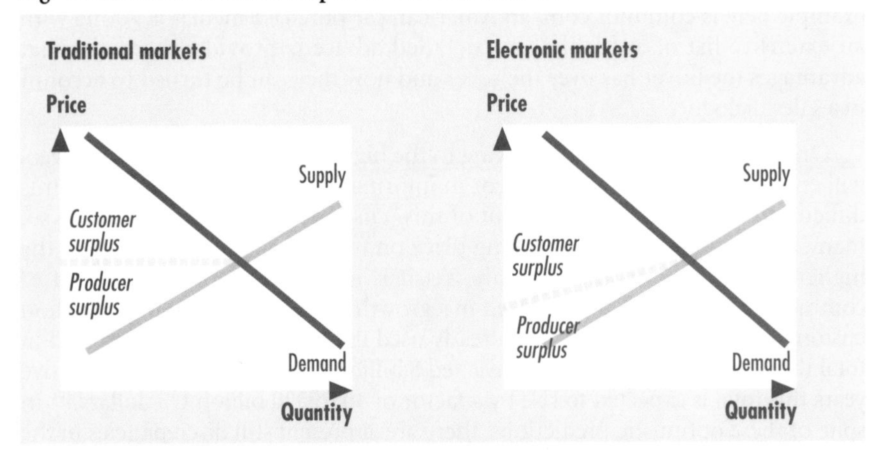

Given the continuing growth of the electronic marketplace Internet, it does not make sense for suppliers to fight shy of this new market transparency. Indeed, there are two possible strategic alternatives for achieving economic success in these new market circumstances.

Strategies for new circumstances

According to Hal Varian, economist at UC Berkeley, it is necessary for companies to go about selectively developing counter strategies: "if the frictions go away, it's in your own interest to recreate them."[192] In other words, the comparison and swapping of suppliers facilitated by high market transparency and low transaction costs can be prevented by means of appropriate strategies. An example of this are programmes that promote and reward customer loyalty. As a result of such programmes, there is not only a tighter bond between the customer and the supplier concerned, but also a degree of price-insensitivity.

These customer programmes, exemplified by frequent flyer programmes with airline companies, constitute one of the most important means of customer bonding and will come to assume outstanding significance in a network economy.[193] As they are not only technically simple but also cheaply effectuated in

191 The use of neoclassic supply and demand curves at this point serves solely to represent the split between consumer rent and producer income. An interpretation in terms of the formation of an equilibrium price is not warranted here. It is rather the case that in electronic markets there are distinct prices realised in different transactions.
192 Hal R. Varian in a conversation with the ECC, April 1998.
193 See section 5.1.4

Source

Fig. 4.8: Hagel III/Armstrong 1997, p. 25

the Internet marketplace, they will prove a ubiquitous method of differentiating between suppliers and gaining customer loyalty. One example of a successful loyalty programme is Amazon.com, who give customers a 3.5 per cent loyalty discount on each purchased book for every book recommendation submitted to Amazon.com's homepage and thus made accessible to all other Amazon.com customers. Cumulative loyalty programmes make particularly good sense as they create a degree of customer "lock-in." A customer of an online bookstore who is promised a 30 dollar discount for making 20 book recommendations, and who has made 15 recommendations already, is unlikely to switch to a rival supplier.

Supplier strategy

For intermediaries the strategy is exactly the other way round. Their concern is to give the customer as much information as possible about what is on offer. An example here is edmunds.com, an American car purchase mediator. Along with an extensive list of available cars, detailed advice is provided here on what advantages the buyer has over the seller and how they can be turned to account in a sales dialogue.

Intermediary strategy

On the whole, it is clear that owing to the higher market transparency buyers will enjoy a substantial improvement in information with respect to sellers. It is difficult to predict the precise extent of this "customer empowerment." Even so, many transactions are already taking place on reverse markets, especially in the higher price range of travel or cars. Yet it is not just these two areas but e-commerce as a whole that finds itself in a growth phase. By 1997 some 10 million customers in North America had already used the Internet for shopping, and in total the worldwide turnover had reached 8 billion US dollars. Over the next five years this total is expected to rise by a factor of 40 to 320 billion US dollars.[194] In spite of these optimistic predictions, there are at present still discrepancies in the speed of development between the end consumer and the business customer segment. Although it was private users that caused the Internet to take off, the use of the Internet for commercial transactions by private households is still at an early stage. Growth is slow but steady.

Slow development in the private sector

The business customer segment is clearly making all the running here, as the potential for lowering costs inherent in the distribution structure was recognized from the outset. CISCO for example, one of the world's leading providers of network solutions, was as early as 1997 able to secure some 33 percent of its total turnover with 10,000 registered customers and trade partners through the Internet. The business customer segment is at the beginning of a phase of massive growth. The following table gives an overview of the most frequent categories of service being traded:

Fast development in the business sector

194 See Picot 1998a, p. 12ff.

Fig. 4.9: Estimated turnovers: the continuous and unstoppable growth of e-commerce (in billion US dollars)

	1996	1997	1998	1999
Computer hardware	107	1.154	1.838	3.482
Travel	230	860	1.458	2.683
Books	34	149	287	639
Computer software	72	196	299	594
Presents/flowers	45	117	231	400
On-line brokerage	38	117	187	377
Cars	11	51	137	320
Music	27	51	125	320
Clothes	38	86	137	263
Food	34	86	112	205
Other	491	1.044	1.421	2.135
TOTAL	1.128	3.910	6.230	11.416

All in all a consideration of e-commerce or electronic markets reveals that the question is not whether but how fast the further developments will take place. The Internet thus makes possible trade and commerce on an electronic infrastructure that can be used independently of time and space.

4.2 Consequences of Technological Evolution – An Economic Revolution?

The emergence of a new infrastructure gives rise not only to changes in existent sales and trade structures, but also has consequences that are primarily economic in nature. A new market model brings a whole new set of rules with it, applicable especially to firms from the media and communications sectors. Cost structures, revenue types and value chain structures are likewise subject to a radical transformation. A knowledge of these new market conditions is absolutely indispensable for businesses hoping to devise and implement effective strategies.

Cost structures, revenue types and value chains are subject to radical transformation

Network effects in general are nothing new. The media and communications sectors for example are traditional markets that are characterised by intra-sector networking. Nonetheless, the significance of network effects and their further ramifications have increased sharply as their sphere of influence has expanded with the developments described above. There is thus now also inter-sector networking taking place, discussed above under the concept of the convergence

Source

Fig. 4.9: BancAmerica Robertson Stephens 1998, p. 1

of the media, telecommunications and IT industries. On top of this, more and more other areas of the economy are coming under the influence of networking, with the result that the new network economy is no longer applicable solely to the three media and communications sectors but is now becoming increasingly valid as a general economic theory providing fundamental insights into all areas of the economy.

Network theory becomes applicable to more and larger segments of the economy

4.2.1 New Rules of the Game: A New Economic Market Model

The new network economy denotes the functioning of economic mechanisms in markets characterised by network effects. The existence of network effects alters the market mechanisms of economic theory, making it necessary to devise a new market model. In what follows, two main categories of network effects will first be described. Building on this, it will be shown that the existence of network effects invalidates the classical market rules of traditional microeconomics. New mechanisms appear in their place, where the dominant factors are now positive feedback, the lock-in effect and the overriding importance of standards.

Negative and positive externalities

Network effects have also been designated network externalities. Externalities in general denote a situation in which one person's behaviour has an effect on the welfare of another person (or other persons),[195] and they may be either positive or negative. Environmental pollution is an example of a negative externality. In the economic sense, externalities – also known as external effects – describe "side effects on a third party from individual consumption and production activities which are neither compensated for in the market nor accumulated as costs for the individual in another manner."[196]

Network externalities denote the effects of a person's participation in the network on the other participants. In general these are among the positive externalities. A person's participation in the network can have both direct and indirect effects on the other participants, whence the corresponding distinction between direct and indirect network effects.

Rising network values with increasing numbers of participants

In the case of direct network effects, the value of a network service rises with the number of users. Metcalfe's Law describes this correlation as exponential. Communications services such as telephone, fax or e-mail provide examples of direct network effects. The more users are connected to the network in question, the more potential communication partners the network has for each user and the greater the network's value. In such markets, the original value of the service or the underlying technology is pushed into the background. A user no longer buys just the physical product but rather the access to the network that the product provides. The derivative value, i.e. the size of the network, eclipses the generic product value as the decisive purchase criterium. The product's functional properties are thus a necessary but no longer a sufficient condition for the purchase of a network good.[197]

195 See Mankiw 1997, p. 10
196 Wigand/Picot/Reichwald 1997, p. 35f.
197 This opposition between technological superiority and inferiority in customer preference lies at the heart of the frequent discussions relating to "Windows vs. Mac-OS."

Indirect network effects exist first and foremost with system products. The purchase of system products is characterised by two separate decision phases. First a decision is made regarding system architecture, for which the appropriate system components are then additionally bought as required. The choice of components is thus already limited by the choice of architecture, as the components have to be compatible with the system architecture. In choosing an operating system, for example, indirect network effects dictate that attention must be paid to the existence of the appropriate user software, as an operating system without user software is worthless.

Indirect network effects and systems products

With indirect network effects, the potential utility of the product – and thus also its value – depends on the availability of complementary services. The more people have decided on a particular system architecture and in so doing joined the "virtual" network, the greater will be the efforts made by suppliers of complementary products to develop system-specific products. The size of a virtual network thus influences the availability of complementary services. The correlation between the size of the network and its value is here only indirect.

In markets with network effects, classical economic laws are often stood on their head. According to the traditional economic perspective, the increasing availability of a good entails a fall in the individual good's value (negative feedback).[198] Unique goods thus tend to be seen as particularly valuable, while mass products lose in value. A good's value is based on its scarcity. The effect of negative feedback can be illustrated by a shift in supply and demand curves.

Negative feedback for typical products

Fig. 4.10: Negative feedback as traditionally understood

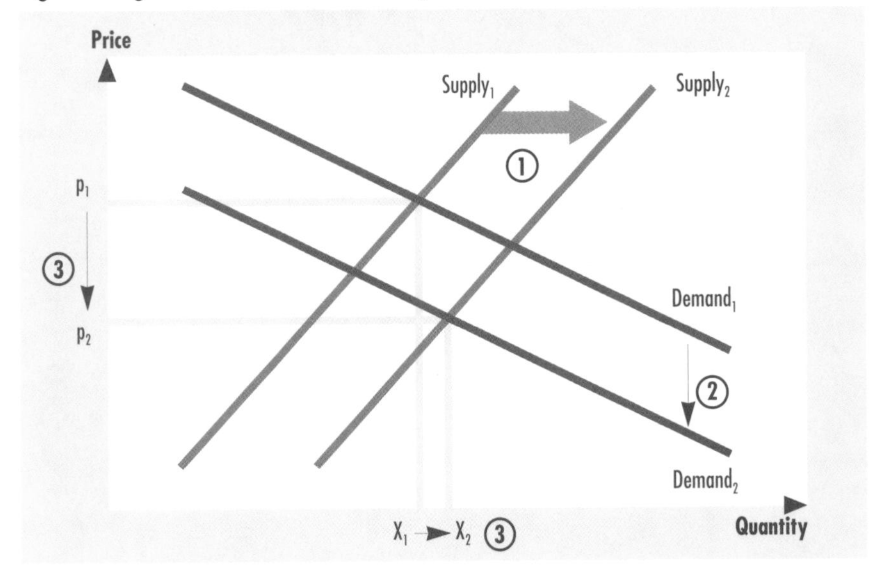

198 Negative feedback should not be confused with the law of diminishing marginal returns. The law of diminishing marginal returns prevalent in neoclassical economics refers to the reduction in a good's value for a single person as that individual's consumption increases. Negative feedback by contrast denotes what happens when a good's value sinks on account of the increasing consumption of the good by other people.

Source

Fig. 4.10: ECC 1999

"Until Bill Gates and his fellow shareholders made a killing at Microsoft, increasing returns were thought to exist only in textbooks." [199]

Here an expansion in supply is represented by a parallel displacement of the supply line (from A1 to A2). This then causes a drop in the good's value on the demand side (from N1 to N2). These two displacements produce a new equilibrium price of p2 lying beneath the former price p1 and a rise in the quantity traded from x1 to x2. The increase in supply has thus given rise to a decrease in value of the individual good in the market.

Positive feedback for network goods

In the new network economy, this correlation clearly no longer applies. Network effects dictate instead that with increasing availability the value of a particular good will go up. What was formerly negative feedback turns into positive feedback in the new network economy. Quantity supersedes scarcity as a source of value. Positive feedback, often referred to as "increasing returns" and resulting from the direct and indirect network effects described above, is a dominant feature of the new market rules.

"Until Bill Gates and his fellow shareholders made a killing at Microsoft, increasing returns were thought to exist only in textbooks."[199]

Owing to network effects, the bigger the networks, the more attractive they become. This then induces more users to get connected to the network, in turn bringing about yet more direct and indirect network effects. Growth spawns further growth. This rule lies at the core of the concept of positive feedback.

Fig. 4.11: The positive feedback cycle

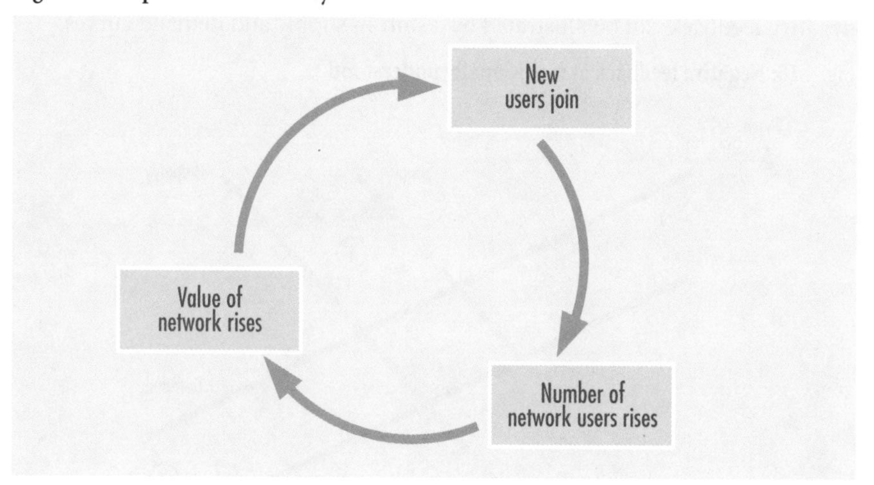

199 Hagel III/Armstrong 1997, p. 6

Source

Fig. 4.11: ECC 1999

Positive feedback and competition

In terms of the competition between technologies or companies, this means that with a growing market share the confidence of consumers in the technology or company concerned will also grow, in turn again engendering increased market shares. The reverse side of positive feedback is that these spirals can also be downward-oriented. Every inch of ground lost by a competitor saps the confidence of potential customers. Reduced attractiveness leads to steadily declining market shares. "Positive feedback makes the strong get stronger ... and the weak get weaker."[200]

Fig. 4.12: The effects of positive feedback on competition

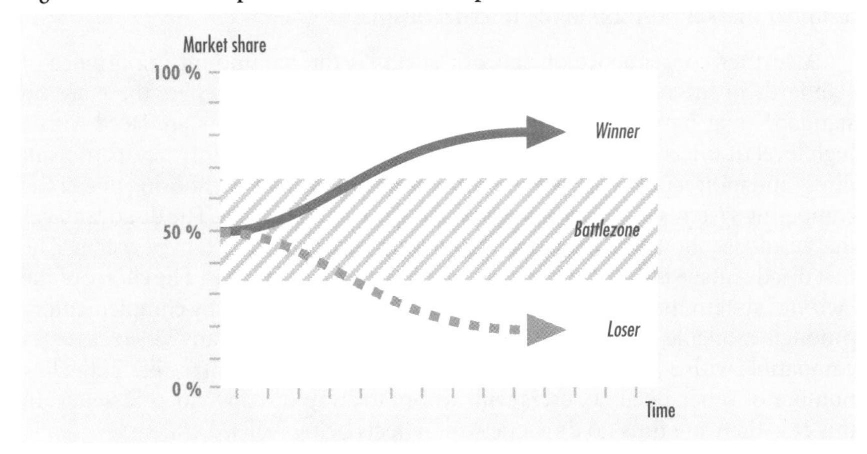

Announcements and expectations as a source of positive feedback

The value of a network product depends not only on actual developments in the number of users, however, but is also substantially influenced by expectations regarding future developments in user numbers. If expectations as regards the success of a particular network product are positive, then the offer becomes progressively more attractive, with the result that the positive expectations become self-fulfilling. This aspect of positive feedback makes expectations management a crucial factor for competitors in a network economy. Timely pre-announcements can not only speed up success in the market but even be what makes it possible in the first place. Product pre-announcements have been successfully used by Microsoft and IBM in particular, with the concept of "vapour -marketing" being coined to denote this type of marketing manoeuvre.

Positive feedback effects underlie the existence of the temporary monopolies that flourish in the markets of the network economy (for example Windows, VHS). In classical markets the significance of an individual company will only grow to the point where the company's economies of scale reach their limits: oligopolistic market structures come into being based on these supply-side economies of scale. In the network economy individual companies can come to attain virtually monopolistic market positions for limited periods of time. These temporary monopolies are founded on the advantages of increased network size

200 Shapiro/Varian 1998, p. 138

Source

Fig. 4.12: ECC 1999

for customers. Demand-side economies of scale do not get weaker once a certain point is reached, however, but continue to increase. If everyone is taking part in a network, this is all the more reason for getting connected yourself.

Positive feedback and monopolies

Even so, the monopolies in a network economy only survive until they are threatened by new technologies with high expectations for the future pinned on them, at which point the competition for market success starts all over again. This is shown by the example of Microsoft, who saw their quasi-monopolistic position in the software market under threat owing to the rapid development of the Internet. As a result, Microsoft is now doing battle with different rivals to build up a similar market position in the Internet business.[201]

The role of standards

A further consequence of network effects is the paramount importance of standards in the new network economy. In market phases where there are no standards that have yet gained general acceptance, consumers are faced with a high level of uncertainty in purchase situations. This uncertainty has its roots in direct and indirect network effects. Customers run the risk of choosing one of the competing systems that ends up not catching on in the market. The disadvantages that result are shown by the video recorder market and the Betamax system. The first disadvantage is the failure to realize indirect network effects. The choice of the "wrong" system means that in the long run there will not be any complementary products available. Nowadays, for example, there are hardly any video cassettes compatible with a Betamax video recorder. A second disadvantage lies in the low number of other Betamax users with whom to swap or copy video cassettes. In this case there are thus no direct network effects being realised either.

Potential business partners and consumers hesitate before a standard has been settled

Many consumers react to this decision making quandary by delaying their decision. The high level of uncertainty prompts them to put off their planned purchase decision until one of the systems has asserted itself as a standard, thus ensuring that they will benefit from network effects in the future. Suppliers in the pre-standard phase thus face the special challenge of building up their installed base of users as rapidly as possible so as to achieve the critical mass. From then on, the product's development gains a momentum of its own by building up confidence in its long term market success and thus becoming progressively more attractive in the eyes of customers.

Definition of "Lock-in"

Indirect network effects form the basis for so-called lock-ins. Lock-in situations occur when the costs of switching architecture are greater than the value arising through making the switch. Switching costs refer to more than just the costs of acquiring the new system. At issue above all are "sunk costs" in that former investments in the complementary products of the old system become worthless through the change of system. Take a company's decision to adopt a particular operating system. Once the company has gone for the operating system in question, it will invest in complementary software applications, staff training and perhaps also in system-specific hardware peripherals. Use of the system further entails the creation of the corresponding files. Switching one's operating system

201 See section 3.3.1

means that extensive conversion measures are required for any of these complementary products to be used again. On top of this, appropriate investments in the new system have to be undertaken in the form not just of product costs but above all expenditure on staff training and guidance. The significance of these switching costs has been described by the American economist Peter Drucker: "No new system can displace an established system unless it outperforms it by a factor of ten."

Economic consequences of successful lock-in

Lock-in situations are in principle attractive for a supplier. If suppliers succeed in "capturing" customers within their own system architecture, then they will enjoy something akin to a monopoly position in future purchase situations. Even so, this advantage is countered by the problem that long term lock-ins can produce a high level of dissatisfaction among customers, on occasion even inducing them to switch in spite of the costs. This danger is particularly great when suppliers openly abuse their monopoly position, as comes to light in poor product quality, excessively high prices and an unwillingness to innovate. Unobtrusive lock-ins are thus preferable, as they are usually more stable.

New Rules for the New Economy

by Kevin Kelly, Wired Magazine, Executive Editor

Increasing Returns – Self-reinforcing Success: Strategies

Check for externalities. The initial stages of exponential growth looks as flat as any new growth. How can you detect significance before momentum? By determining whether embryonic growth is due to network effects rather than to the firm's direct efforts. Do increasing returns, open systems, n^2 members, multiple gateways to multiple networks play a part? Products or companies or technologies that get slightly ahead – even when they are second best – by exploiting the net's effects are prime candidates for exponential growth.

Coordinate smaller webs. The fastest way to amp up the worth of your own network is to bring smaller networks together with it so they can act as one larger network and gain the total n^2 value. The Internet won this way. It was the network of networks, the stuff in between that glued highly diverse existing networks together. Can you take the auto parts supply network and coordinate it with the insurance adjusters network plus the garage repair network? Can you coordinate the intersection of hospital records with standard search engine technology? Do the networks of county property deed databases, U.S. patents, and small-town lawyers have anything useful in common? One thousand members in one network are far more powerful than one thousand members in three networks.

Create feedback loops. Networks sprout connections and connections sprout feedback loops. There are two elementary kinds of loops: Self-negating loops such as thermostats and toilet bowl valves, which create feedback loops that regulate themselves, and self-reinforcing loops, which are loops that foster runaway growth such as increasing returns and network effects. Thousands of complicated loops are possible using combinations of these two forces. When internet providers first started up, most charged users steeper fees to log on via high-speed modem; the providers feared speedier modems would mean fewer hours of billable online time. The higher fees formed a feedback loop that subsidized the provider's purchase of better modems, but discouraged users from buying them. But one provider charged less for high speed. This maverick created a loop that rewarded users to buy high-speed modems; they got more per hour and so stayed longer. Although it initially had to sink much more capital into its own modem purchases, the maverick created a huge network of high-speed freaks who not only bought their own deluxe modems but had few alternative places to go at high speed. The maverick provider prospered. As a new economy business concept, understanding feedback is as important as return-on-investment.

Protect long incubations. Because the network economy favors the nimble and quick, anything requiring patience and slowness is handicapped. Yet

many projects, companies, and technologies grow best gradually, slowly accumulating complexity and richness. During their gestation period they will not be able to compete with the early birds, and later, because of the law of increasing returns, they may find it difficult to compete as well. Latecomers have to follow Drucker's Rule – they must be ten times better than what they hope to displace. Delayed participation often makes sense when the new offering can increase the ways to participate. A late entry into the digital camera field, for instace, which offered compatibility with cable TV as well as PCs, could make the wait worthwhile.

It's a hits game for everyone. In the network economy the winner-take-all behavior of Hollywood hit movies will become the norm for most products – even bulky manufactured items. Oil wells are financed this way now; a few big gushers pay for the many dry wells. You try a whole bunch of ideas with no foreknowledge of which ones will work. Your only certainty is that each idea will either soar of flop, with little in between. A few high-scoring hits have to pay for all the many flops. This lotterylike economic model is an anathema to industrialists, but that's how network economies work. There is much to learn from long-term survivors in existing hits-oriented business (such as music and books). They know you need to keep trying lots of things and that you don't try to predict the hits, because you can't.

Two economists proved that hits – at least in show biz – were unpredictable. They plotted revenue of first-run movies between May 1985 and January 1986 and discovered that "the only reliable predictor of a film's box office was its performance the previous week. Nothing else seemed to matter – not the genre of the film, not its cast, not its budget." The higher it was last week, the more likely it will be high this week – an increasing returns loop fed by word of mouth recommendations. The economists, Art De Vany and David Walls, claim these results mirror a heavy duty physics equation known as the Bose-Einstein distribution. The fact that the only variable that influenced the result was the result from the week before, means, they say, that "the film industry is a complex adaptive system poised between order and chaos." In other words, it follows the logic of the net: increasing returns and persistent disequilibrium.

From: Kelly, Kevin: New Rules for the New Economy. 10 Radical Strategies for a Connected World. New York, N.Y.: Viking Press,1998, p. 34-36
With permission from Kevin Kelly

4.2.2 Media and Information Goods: The Near Zero Marginal Cost Issue

In recent years there have been decisive changes in the cost structure of media products, with far-reaching strategic implications for business enterprises. Generally speaking, the production of media goods, as of any other product, gives rise to both fixed and variable costs. What is notable is that media content cannot be understood as a mass product, since every time content is produced something unique is created owing to the one-off combination of the human and material resources involved.[202] As the one-off production process is independent of the number of copies sold, it gives rise to what are known as fixed costs or "first-copy costs." These are often designated "sunk costs," because they are as a rule incurred up front before production begins, and the content produced can rarely be reused in a different form if they flop with customers.

First-copy costs of media products are high

These fixed costs[203] are incurred with every form of media, whether print or electronic, and differ only in how great they are. Every author who spends a year sitting at a desk producing a piece of written work entails fixed costs for the customer, irrespective of whether it is a book or a filmscript that is being produced.

Further expenditure is necessary on marketing and promotion to attract the attention of the desired target group. As it is largely independent of the quantity of items finally sold, this too counts as fixed costs. Given the escalating information-overload faced by consumers today, attention has become a scarce commodity, provoking disproportionately high price rises in recent years.

It is only with the reproduction and distribution of the media good after its production that the unique product becomes a mass product. Here however there are considerable differences in the variable costs incurred from one media form to another. In the printing sector, these amount to substantially more than half of the total costs of a media product on account of the high printing and distribution costs.[204] In the electronic media sector by contrast, the marginal costs are very low in comparison to the fixed costs that have already been incurred. These cost structures can be illustrated by taking the German media market as a paradigm case.

Variable costs differ between media: very high in printing, very low in broadcasting

202 See Altmeppen 1996, p. 265
203 It should be pointed out here that the printing and distribution costs of newspapers and magazines do not count as fixed costs, as in theory at least the use of appropriate services makes this fixed charge seem avoidable.
204 See Sennewald 1997, p. 59

Fig. 4.13: Production, reproduction and distribution costs as a percentage of total costs of German media

	Production cost	Reproduction cost	Distribution cost	Reproduction and distribution cost in total
Newspapers	20.0	39.5	19.0	58.5
Magazines	29.5	28.1	6.6	34.7
Public service television	55.9 (without staff)	0	9.2	9.2
Commercial television	68.9 (without staff)	0	7.1	7.1

The digitisation of content has reduced the reproduction and distribution costs for media products even further. In the case of Encarta, Microsoft's digital encyclopaedia, for example, the reproduction and distribution costs of the CD-ROMs amounted to some 1.50 US dollars by comparison with 250 US dollars for the book version.[205] The reproduction and distribution of digitalised media products is especially efficient through the Internet: marginal costs here are as good as zero.[206]

Digitisation has reduced reproduction and distribution costs

This marginal cost structure applies not only to digitalised media products, however, but to all digitally reproduced and distributed services. The first-copy costs of producing the Netscape Navigator thus consisted of some 30 million US dollars in development costs, while the marginal costs for additional copies were only around one dollar each.[207] Following Shapiro/Varian, it is thus more usual to speak of information goods.[208] This refers to all information that is available in digital form, be it a stock market report, a software program, a game or a business report.

Netscape Navigator: first-copy costs 30 million US dollars, additional copies one dollar each

What emerges on the whole is that media and information goods have a high portion of fixed or first-copy costs and therefore tend to enjoy economies of scale. In other words, the average costs of production for media and information goods fall dramatically as the level of production rises. There is however considerable variation in the actual scale of degression depending on the media concerned and the marginal costs thus involved. Print products only benefit from economies of scale once a certain production level has been attained, while broadcasting and above all information goods enjoy scale effects even at low production levels.

This cost structure applies not only to media and information goods but likewise characterises various types of material goods. The construction, for example, of complex systems or infrastructures such as a stock exchange system or a telecommunications network generates high first-copy costs, while the costs for subsequent use are low.

205 See Downes/Mui 1998, p. 51
206 See Bakos/Brynjolfsson 1996
207 See Kelly 1998, p.54
208 See Shapiro/Varian 1998

Source

Fig. 4.13: Altmeppen 1996, p. 266

"Until now, the biggest lie on the Internet hasn't been about alien abductions. It's been: Don't worry, the web will make money."

It is interesting, however, that the resulting economies of scale are given an additional boost in a network economy with media and information goods. It is thus an essential prerequisite for companies operating in media and information markets to take this cost structure into account in order to push through successful strategies. This applies in particular where considerations of price strategies and product policies are at issue.[209] Until now, the biggest lie on the Internet hasn't been about alien abductions. It's been: "Don't worry, the web will make money."[210]

4.2.3 New Revenue Strategies

An understanding of the new cost structures and possible revenue types is the essential precondition for on-line media to be able to implement appropriate revenue and price strategies. The following section will first discuss revenue models for on-line media, differentiating between present models and those anticipated for the future. Secondly, it will look at a new form of price strategy for which success is only possible in a network economy: follow the free.

Consequences of the virtual marketplace for media companies

The emergence of the virtual marketplace has several consequences for the revenue strategies of media companies. Firstly, there is a higher revenue potential than in traditional markets. Geographical boundaries prevalent in the past cease to be relevant, as the marketplace extends globally to all those connected to the Internet, and the phenomenal growth of the number of people being connected (and to be connected) adds to the attraction of this factor. Secondly, with technical restrictions ceasing to apply there are no longer any limits to the deployment of the diverse revenue types. These together form a sort of "construction kit" that can be used by on-line media for the refinancing of their products. This section will first provide a brief discussion of the most important revenue forms, before introducing and analysing possible business models or in other words the weighting given to the distinct revenue types.

Advertising: attracting eyeballs is also important in the Internet

The most important source of revenue has been and will continue to be financing through advertising. The generation of reach is thus one of the most crucial goals for on-line suppliers in their attempts to attain the highest possible turnover in the complementary market for advertising. The absolute volume of expenditure available for Internet advertising is as yet relatively low. Although expenditure on the part of advertisers is rising continuously, it is still only a very low portion of total advertising expenditure. The table below illustrates the development taking place in the United States market, as the Internet's diffusion as a form of mass media is more advanced there than in other countries.

209 See sections 5.1.2 and 5.1.3
210 Advertising text for Pandesic, Wired 2/98,p. 3

Fig. 4.14: Advertising turnovers by media in the United States (in billion US dollars)

	1996	2000*	2002*
Newspapers	38,2	47,4	50,1
Magazines	9,0	10,9	11,9
Online	0,3	4,4	7,7
TV	36,0	42,9	46,9
Cable TV	6,4	9,9	12,2
Radio	12,3	15,9	17,8

The table shows clearly that Internet advertising is growing rapidly. Even so, as a percentage of the total its absolute volume is still in single figures and as such too low to serve as a revenue source for many suppliers. This applies especially since subscription fees are hardly accepted and most on-line publishers thus have to rely almost entirely on advertising as a source of revenue. Up to now, the big players in the advertising industry – manufacturers of consumer goods such as Procter & Gamble – have proved restrained in their on-line advertising expenditure. The biggest advertisers are still IT companies from the hardware and software sectors, media companies and financial service providers.

Internet advertising dominated by technology, media and finance

For this reason new (old)[211] forms of advertising have been created. "Sponsorship" in particular is a popular model currently being used by many on-line media. Exclusive sponsorships are here arranged for individual theme sections or chat-rooms for a certain period of time: Forbes Digital Tool (www.forbes.com) for example has thus sold the advertising space for its finance section exclusively to the American financial service provider Fidelity.

Sponsoring

On the whole a market structure is emerging – as in the classical media industry – in which most of the attention and therefore also most of the advertising income is grabbed by just a few content providers. In 1997, for example, ten Web publishers – portals and on-line services such as Yahoo and AOL – obtained some 60 per cent of global Internet advertising revenue.[212] News and information services in 1997 managed to capture some seven per cent of total turnover, while entertainment-oriented services ended up with just three per cent of advertising expenditure.

The second revenue form available to on-line media is the transaction. A Web publisher such as SportsLine USA, for example, thus sells sports articles that have been bought on its own account. This revenue source is included under e-commerce and will slowly but surely catch on as an accepted part of consumer behaviour. For on-line publishers such as Disney (www.disney.com), for example, transactions turnover through the sale of merchandising articles already constitutes roughly 50 per cent of total turnover.[213]

211 These forms of advertising have long been common in the printing sector: split-advertising, promotions etc.
212 See Jupiter Communications 1998, p. 15
213 See Geirland 1998, p. 155

Source

Fig. 4.14: Jupiter Communiactions 1998, p. 10
* estimated

Commisssion for e-commerce ...

The third revenue form is the commission. It is used above all in the case of so-called partner programmes, which refer to the attempts of e-commerce suppliers, up until now predominantly from the book or music sector, to find "affiliates" or partners. These are on-line providers sought out either for their strong reach or their affinity with the supplier's target group. The procedure is as follows. An on-line medium such as USA Today (www.usatoday.com) puts on its Web page a banner ad from the book supplier Barnes & Noble (www. barnesandnoble.com). By clicking into this banner ad it is possible to order a book directly from Barnes & Noble. For every successful sale Barnes & Noble then pays approximately seven per cent of the sales revenue back to USA Today. Further examples are given in the table below:

Fig. 4.15: Examples of partnership programmes

	Commission for partners	Number of partners	Examples	Proportion of revenue through partners
Amazon.com	15%	8,000	Atlantic Monthly, Village Voice, Star Chefs, Puppy Net	n.a.
Barnes & Noble	5–7%	40+	CNN, ESPN, SportsZone, USA Today, Wired	n.a.
CDNow	12%	2,000	Pathfinder, JazzOnline	10%
Virtual Vineyards	7%	n.a.	American Express, Excite, Match.com	10–15%

... and for special-interest media

Especially in view of the low advertising revenues, commissions are a viable revenue source for on-line media from the special interest sector. Their low reach and the resulting low advertising income can be compensated at least partially by catering to a specific target group and in this way gaining commissions. For suppliers of the most diverse of goods, the key thing here is above all the context related content of the on-line publisher. The Web page of an on-line publisher specialising in computer equipment such as ZDnet.com thus provides a much greater stimulus to buy a computer whether on impulse or otherwise – than a banner ad on the Web page of a search engine such as Yahoo. Commissions are also one of the most important sources of revenue for the on-line service AOL, which demands not only an "upfront payment" but also a single-figure percentage of the relevant sales revenue from suppliers within its service.

Hi ho, hi ho – down the Datamine we go ...

The fourth revenue type for on-line publishers is datamining.[214] This refers in general to the processing into knowledge of data that are obtained through user-transactions of whatever kind (telephone conversations, questionnaires, competitions).[215] The refinancing of these suppliers (address listings, direct marketers) is achieved through the provision or the resale of the data to third

214 The term "data warehousing" is often used instead of "datamining", but the two concepts have different meanings. "Data warehousing" refers to the mere storage of data, "datamining" to the evaluation and processing of the data stored.

215 See Laube 1998, p. 80ff.

Source

Fig. 4.15: Jupiter Communications 1997, p. 8

parties wishing to use the information to address new customers or optimize the range of services they provide. In addition to the simple provision of data, direct marketing firms frequently also offer a number of consulting services relating to the conception of the campaign, for instance.

Datamining

Newspapers and magazines, predominantly in the United States so far, are likewise already using data concerning their readers as an indirect source of revenue. The American magazine publisher Meredith for example, whose publications include Homes & Garden and Golf for Women, use their database as a strategic weapon to aid the introduction of new titles: "Our database covers about 60 per cent of households in America. Along with Reader's Digest we have one of the two really mass market publishing databases that exists. ... We use it extensively on our business ... one most vividly is to help us reduce the cost and risk of introducing new magazines."[216]

The "one-to-one" future, in other words the strategically targeted attention given to customers by suppliers, is one of the most talked about expectations for the future of the network economy. The advantage for users is said to consist in not receiving any unwanted communications such as junk mail but only offers for services that are actually needed or of direct interest. For suppliers the advantage will be the greater economic efficiency in group targeting, with a minimalization of losses through scattering effects. Forms of on-line datamining realized so far, however, hardly extend beyond direct marketing activities already available in the material world and in some cases are still lagging somewhat behind: one-to-one is as yet still more of a promise than a reality in the internet economy.[217]

Attractions of one-to-one marketing

One reason for this is the inadequate analysis of the data. There are indeed media-specific advantages to be gained through the direct technical registration of user activities. The results of this "tracking" are stored by the Web pages concerned in so-called log files, or digital logbooks. These data include primarily:

Specific advantages in tracking Internet usage

* use made of the main page and individual pages (visits, pageviews) per unit of time;
* user's domain origin;
* operating system (Macintosh/Windows etc.);
* browser (Netscape/Explorer etc.);
* data relating to means of advertising (click-through etc.);
* links generating traffic for the Web page in question.[218]

Yet up until now the extreme market dynamism and intense competition have meant that many of the on-line publishers have been excessively preoccupied with the management of day-to-day business. As a result the data have just been stored but not integrated or processed. The problems have been outlined by Craig Donato, marketing manager of the portal Excite:

216 PaineWebber 1997, p. 294f.
217 See section 5.1.4
218 See Bayers 1998, p. 132

"The question is, how do we turn 12 million registrations into revenues and earnings?"

"Excite has had enough trouble just keeping up with its phenomenal growth: In February [1998] the company served 28 million pageviews each day, up from 4 million per day a year earlier. At this pace, Excite collects 40 gigabyte of data in its log files every day. If we tried to look at things at that level, we'd go insane – we'd drown in information."[219]

Even so, knowledge of service-users can be one of the greatest assets for the generation of further revenue sources such as advertising. AOL for example has thus been able to ascertain that from 1996 to 1997 the average visit time rose from 21 to 41 minutes, that 80 per cent of users surf in the AOL service and not the Internet, and that AOL now enjoys a greater prime-time reach than MTV.[220]

Combining Internet usage data with personal data from other sources

Along with these usage-related data, on-line suppliers are interested above all in their users' personal data. In the United States in particular, the generation of what are known as ZAG-data (i.e. data relating to zip code, age and gender) are of great significance. The reason for this is a market research procedure popular in the United States that permits further statements pertaining to salary bracket, consumer habits etc. on the basis of these data. The table below shows the main forms of information generation in use up to now:

Fig. 4.16: Forms of information generation

Scope of information	High			Low
Type of information gathering	Billing	Registration as precondition for use	Registration necessary for receiving value-added content newsletters, e-mail, etc.	No registration
Examples	Wall Street Journal, AOL	PointCast, Juno, New York Times, Talk City	CNET, ESPN, SportsZone, ZDNet, Yahoo, Netscape	All others

Datamining strategies

After the information has been generated, the big question is how it can be used to make money. Or as the Yahoo founder Jerry Yang puts it: "The question is, how do we turn 12 million registrations into revenues and earnings?"[221] There are several revenue strategies available in this context. The following table shows areas in which datamining can be used to promote effective and efficient group targeting with the generation of additional revenue:

219 Bayers 1998, p. 132
220 See Leo 1998, p. 61
221 Alden 1998, p. 2

Source

Fig. 4.16: Jupiter Communications 1998, p. 38

Fig. 4.17: Possible areas of datamining application for on-line suppliers

Area of application	Examples	Objectives	Type of desired revenue increase
Content	Mass Customisation (my.yahoo.com) Daily E-Mail Newsletter (CNet.com)	Intensifying relationships to website users Increasing reach	Advertising
Advertising	Banner targeting by OS, browser, domain	More effective targeting for advertisers	Advertising
Product supply	Website personalisation, eMail notification	Increasing buy rate	Transactions
Transmission of stored data	Sale of data to mailing services	More revenue	Datamining

As the above list makes clear, datamining only generates direct revenue for on-line suppliers in the case of the sale of the stored data to commercial suppliers such as address listings. Apart from this, datamining is above all a way of improving content or product supply and as such it can lead to increased advertising or transaction earnings.

Subscription fees

The fifth form of revenue consists of subscription fees. This revenue form is one of the most important sources of earnings for media in off-line markets. Owing largely to the history of the Internet, the new network economy has so far been dominated by the widespread idea among users that content should be accessible for free. There have nonetheless been a number of attempts to introduce subscription fees as a form of revenue. Yet most of the on-line suppliers who have tried demanding subscription fees have experienced a drastic decline in users. A well-known example of this happening was at the expense of the American financial daily The Wall Street Journal (WSJ). Since April 1996 there has been an electronic version of the WSJ with a readership that reached up to 650,000. From autumn 1996 the WSJ charged yearly subscription fees of 49 US dollars, with a special rate of 29 US dollars for those also subscribed to the print edition. The consequence for the WSJ was a substantial drop in on-line readership to just 35,000, though towards the end of 1998 this was climbing back up to 150,000 paying subscribers again.[222]

What emerges on the whole is that as a revenue form subscription fees are best suited for two target groups. Firstly they can be used by suppliers of specialist information, whose users stem primarily from the business sector. These include database suppliers such as Lexis-Nexis (www.lexisnexis.com) or the services

222 See Schuler 1996, p. VII; Loizos 1998, p. 36

Source

Fig. 4.17: ECC 1999; the list makes no claims to completeness, but attempts to give diverse examples of possible areas of applications for datamining.

Difference between special-interest and general interest content

provided by the German Weka-Gruppe (www.wekanet.de). Secondly, on-line publishers can also charge fees for the sort of special interest content that generates high personal interest levels among users. These include above all sex/erotica and sport.[223] General interest suppliers by contrast find it difficult imposing this form of revenue. Lack of success had in 1998 already led two dailies, "The New York Times" and "San Jose Mercury News," to discontinue their subscription fees.[224]

Business models for on-line media

Combined revenue strategies

Following on from this enumeration and characterization of the individual revenue types, this section will provide a summary of the most common business models for on-line media. By the end of September 1998 there was still only one supplier already making a profit in the Internet: ZDnet. For the rest, the yearly overheads – amounting to some 3 million US dollars for the bigger on-line media suppliers – were greater than the revenue brought in.[225]

The following table gives an overview of the current business models:

Fig. 4.18: Business models of on-line publishers

	Content	Business model	Users	Method of financing	Number of staff
AOL aol.com	Content Communication Marketplaces	Advertising Commission Subscription	12,5 million	Stock market (NASDAQ)	n.a.
CNET cnet.com	Computer	Sponsorship/Advertising Transaction Commission	4 million	Stock market (NASDAQ)	586
FEED feedmag.com	Culture & Lifestyle (New York)	Advertising	50,000	Venture capital	6
FORBES Digital Tool forbes.com	Network Economy	Sponsorship/Advertising	40,000	Forbes New Media	23
SALON salonmag.com	Culture & Lifestyle (San Francisco)	Sponsorship/Advertising Syndication Subscription	140,000	Investors	38
SLATE slate.com	Politics	Sponsorship/Advertising Subscription	140,000	Microsoft	30
WALL STREET JOURNAL INTERACTIVE wsj.com	Finance	Advertising Subscription	150,000	Wall Street Journal	120
ZDNet zdnet.com	Computer	Sponsorship/Advertising Transaction Commission	5 million	Ziff Davis	268

In summary, it looks as though advertising is going to continue to play a dominant role in the weighting of revenue types,[226] even though the absolute volume of advertising available is still too low. For the foreseeable future it will only be possible to impose subscription fees for specialist services. Commissions

223 See Karepin 1998, p. 64
224 See Barlas 1998, p. 2. The New York Times only charged subscription fees for users outside the United States. Its domestic offer was accessible for free from the outset.
225 With its on-line offer "Pathfinder," for example, Time Warner was making losses of some 5 to 10 million US dollars per year. The investment made by Knight-Ridder amounted to approximately 20 million dollars a year. See Sennewald 1998, p. 104
226 Further revenue types for on-line publishers that we have not explicitly enumerated include syndication and archive access. Syndication means the resale of editorial material to other editors. Archive access refers to the sale of individual articles from the archives. Business Week (www.businessweek.com) for example offered five articles for two US dollars, ten for five dollars and 100 for 50 dollars.

Source

Fig. 4.18: ECC, based on Loizes 1998, p. 38; as of April 1998

and transactions are possible revenue sources for theme-based specialist suppliers. Datamining has so far only rarely been used as a means of generating income but it does offer on-line publishers creative opportunities for bringing in additional earnings.

On balance, the future of on-line publishing seems promising. The basic problem is that it will probably not yield sufficient revenue for the next three to five years. For many on-line publishers this will be too long.

4.2.4 The Erosion of Traditional Value Chains by Multimedia Value Networks

The changes brought about by the network economy relate not only to new economic laws and companies' cost and revenue structures, but equally to the value chain structures underlying the markets. As was described in section 3.4, the convergence in the media and communications sectors has led to substantial interpenetration among the media, telecommunications and information technology markets. This convergence has given rise to a new joint market, the composition of which is marked to varying degrees by the three media and communications sectors.

Convergence creates a new joint market

The basis for these developments is provided by innovations in the domain of information and communications technology. As the marketplace, the digital infrastructure that has evolved as a result is the precondition for the fundamental changes in the provision of information, entertainment and communications services.

- The various services provided in the individual media and communications sectors have been qualitatively improved through steady technological advances and the constant application of new technologies (such as digital television, ISDN telephony).
- Connections between infrastructures have made possible the integration of previously distinct services, which can thus be provided in parallel by means of a shared infrastructure (for example TV viewing and information research without changing medium).
- The connection of infrastructures and the connection of services provided have given scope for the development of innovative services that go beyond the integration itself (such as Internet telephony).

The connection of the media and communications sectors has thus created a new market offering the most diverse of opportunities for the unilateral or reciprocal exchange of information. This multimedia market provides the whole range of information, entertainment and communications services. The distinct services constitute a new multimedia value chain that contains the following value chain stages.

Distinct services constitute a new multimedia value chain

Six stages in the multimedia value chain

* The first value chain stage provides content of all types. These include the classical media content such as films, programmes, books, reports and articles. They further cover computer games, databases and the Web pages of businesses, private individuals and various organisations.
* The second stage (packaging) incorporates services relating to the preparation and arrangement of content. These embrace the functions provided by broadcasters, publishers, information services (such as weather and sport), news agencies, as well as on-line Service Providers and Internet portals.
* The third stage refers to the transmission of the produced and aggregated content to the receiver, as well as the transmission of communication content between communication partners. This stage contains all network operators and transmission providers using one or several of the following networks: Internet, telephone, cable, satellite and terrestrial. Transmission through networks linked via the Internet is here at the fore.
* The fourth stage provides the customer with the navigation service. As a value chain stage, navigation denotes a manipulation of the infrastructure, in particular the use of hardware and software components to facilitate and improve orientation in and control of the physical infrastructure. These components include operating systems, browsers and intelligent agents.
* The fifth stage contains value added services (VAS) such as billing, installation and training as well as multimedia consultancy.
* The sixth stage in the multimedia value chain is the final one and thus the one closest to the customer. It includes all devices that are necessary for the reception of content that has been produced, aggregated and transmitted, and thus necessary for use of the information, entertainment and communications services. These reception devices cover all multimedia receivers such as MM-televisions, screenphones, mobile telephones, MM-PCs and personal digital assistants (PDA), as well as others.

The six stages of the multimedia value chain are summarized in the table below:

Fig. 4.19: The value chain for the multimedia market

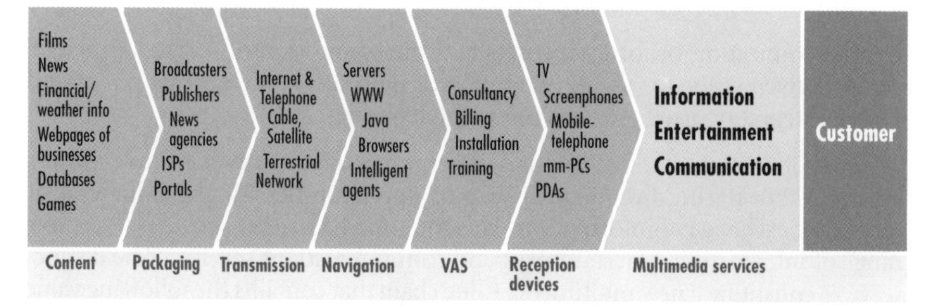

Source

Fig. 4.19: ECC 1999

This linear representation of the value added system of multimedia services as a chain rarely corresponds to what actually takes place in the real world, since the provision of multimedia services seldom follows the linear order of the value chain activities. In order to give an adequate picture of the flexible links between the individual segments of the various value chains from the media and communications sectors, a circular or even a network-shaped representation is more appropriate than a linear chain. The customer here moves from the end of the value chain to the centre of the value network, surrounded by the whole range of value added functions. The result is the following multimedia value network, which does justice to the great flexibility in the provision of multimedia services:

Linear representation in a value "chain" insufficient to describe more complex realities – a new type of a value "network" emerges

Fig. 4.20: Multimedia value network

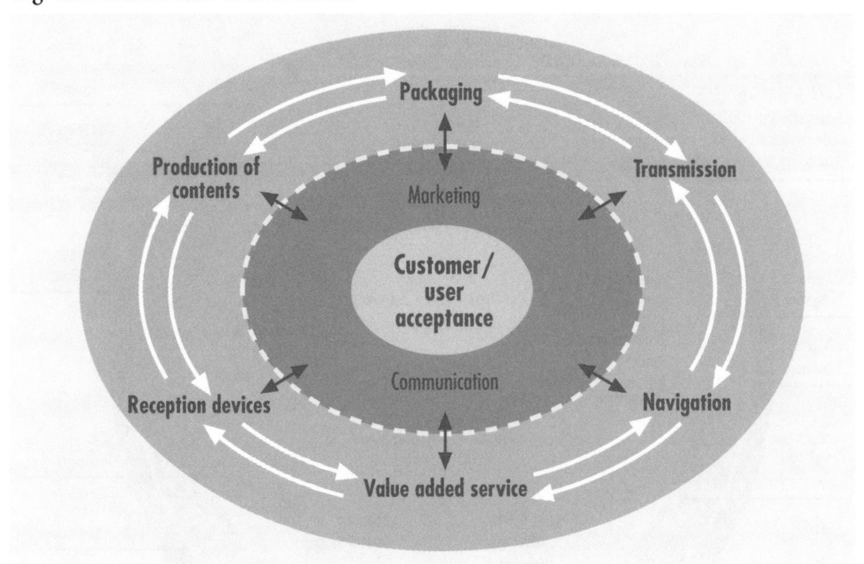

The value chain system for the multimedia market is characterized by the absence of the classical limitations that existed for earlier value added processes. This produces on the one hand a greater variety in the services provided, but on the other hand also a more diverse supply of associated services ("media richness"). For supplier and customer alike, the great attractiveness of this multimedia market lies in the enormous flexibility it offers in the provision of information, entertainment and communications services.

Enormous flexibility of the new multimedia markets

An even more elaborate and interesting concept has been developed within the EU CONDRINET project.[227] It centers around the final customer, and considers four main groups of end users: residential, business, government and education. The core industry activities are defined as (and grouped into) creating, aggregating, hosting, connecting and interfacing. Around this core a number of "enablers" (infrastructural preconditions for the value added process) are forming

227 See www2.echo.lu/condrinet/

Source
Fig. 4.20: ECC 1999

Integrating infrastructural preconditions as "environmental enablers"

concentric rings. "Enabling technologies and services" include device and user interface technologies, networking and hosting technologies as well as professional services, the business and transaction systems and authoring and content management technologies. The outer ring describes those factors which influence all of the different elements within the circle as "environmental enablers": the legal and regulatory framework, the logistics system, standard organisations and financial capital. This concept allows the inclusion of all the various and different stages and elements of the "value chains" in an integrated model. It also visualises the fact that there are no fixed boundaries between these stages, and rather describes a dynamic system undergoing permanent change.

An interesting model combining a high degree of sophistication with mirroring the flexibility of the multimedia markets

Fig. 4.21: CONDRINET model of the value system

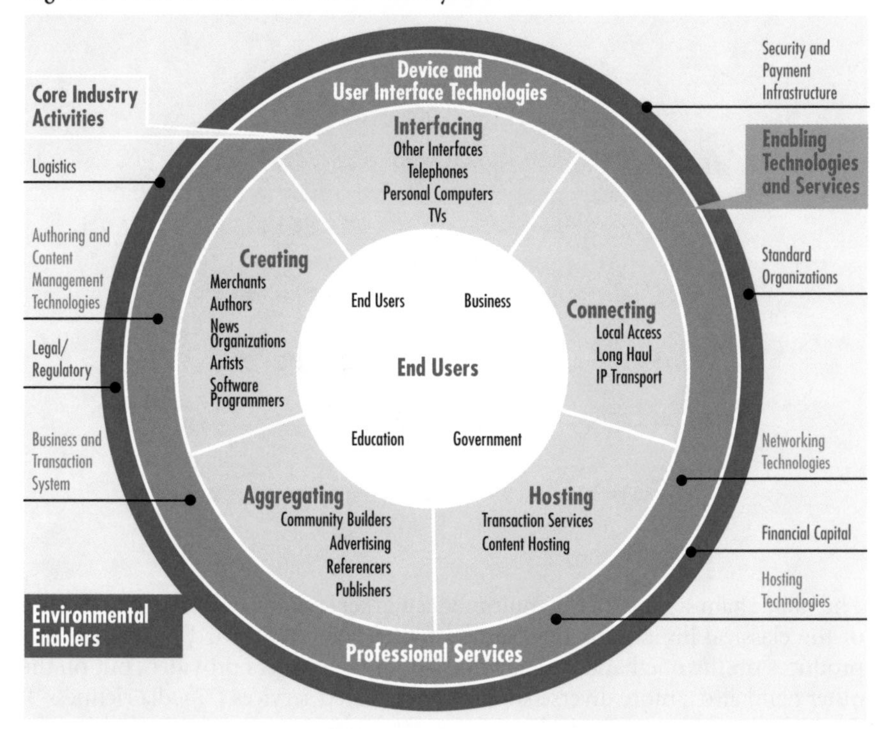

An example of the flexibility that exists in multimedia value chain processes is the project TeleChoice run jointly by the American telephone company US West Communications together with NextLevel Communications.[228] Those customers of US West taking part in the project have been provided with a VDSL modem from NextLevel, and this has been used to connect the television to the US West telephone network. Via television and modem customers received 120 digital television channels, 50 music stations in CD quality, Internet-access with a transmission speed almost 1,000 times greater than with ISDN, as well as a high-quality

228 Hohensee 1998, p. 148

Source

Fig. 4.21: Europäische Kommission 1998b

telephone service. By the end of 1998 there were expected to be more than 400,000 customers participating in the project. This combined use of telephone network (telecommunications), data transmission modem (information technology) and television (consumer electronics/media) would not be possible in a non-integrated media and communications framework. This example shows the potential inherent in the multimedia value network.

New types of competition

The great variety of possible combinations for linking the individual stages from the distinct value chains leads to a whole new set of conditions of competition for the players in the multimedia market. Film producers such as Warner Bros. find themselves in direct competition with Web pages of business enterprises and manufacturers of print content for users' scarce resources of time, money, acceptance, competence and attention. Internet portals such as Yahoo mean new competition for television stations such as NBC. Cable network operators such as TCI have to contend with telephone companies such as Bell Atlantic and suppliers of Internet telephony such as VocalTel for telephone services. Microsoft's dominance in the software market is under threat from Netscape. PC-manufacturers such as Compaq have to face up to mobile telephone suppliers such as Nokia or screenphone manufacturers such as Alcatel. And this proliferation of competitive relations is but the beginning of the emerging network economy.

Given these market conditions, many companies are trying to expand the range of services and the corresponding core competencies they have been offering up to now. Since buying into AOL in 1995, Bertelsmann in particular has been converting itself into an integrated multimedia concern. Along with the purchase of the Berlin firm Pixelpark, a number of multimedia projects such as "Avanti" have been initiated. Their "mediaways" project – a joint venture between Bertelsmann and debis – is seeking to secure entrance into the market for Internet telephony.

New multimedia value network emerges in addition to traditional media value chains, and does not replace them

It is worth emphasising that the convergence of the separate markets to form a single multimedia market does not mean that the independent media and communications sectors will be completely lost in the new market. The multimedia value chain does not instantaneously replace the earlier media, IT and telecommunications value chains. It is rather that a new multimedia value chain emerges in addition to the others, bringing new services with it. The resulting increase in the range of possible applications means added value for the customer. The result is a shift in demand such that the traditional value chains are "rewarded" with a smaller flow of payments than the multimedia value chain. A steady process of transformation sets in, leading to a continuous erosion of the traditional media and communications value chains.

The speed of this process of erosion and the success enjoyed by the new value chain structures depend first and foremost on the added value brought by the new possibilities. Attractiveness of content, quality of performance and reception, as

well as cost and user-friendliness are all factors that must show a clear customer advantage.

Crowding out of traditional media by Internet media

The shift in demand towards the multimedia value chain has been in evidence since the early nineties and is speeding up steadily with the expansion of the digital infrastructure and the corresponding development in the range of services provided. A classical example of the decline undergone by the traditional value chain is provided by the Encyclopaedia Britannica. The renowned reference work in 32 volumes has been marginalised within a matter of years by CD-ROMs containing encyclopaedias of inferior quality and lesser scope. Another example of the changes taking place is the constant increase in the use of electronic mail or e-mail, which will slowly but surely reduce the turnover of the classical postal delivery service. A further topical example is the relation between television and Internet use, with consumers tending increasingly to devote their scarce resources of time to the Internet.

Electronic books

Similar developments can be anticipated for the future. The traditional value chain for the book market is thus having to confront competition through the development of innovative appliances like the "RocketBook." This portable computer is equipped with a special display for reading digital book content and has a memory capacity sufficient for the content of ten books. New value chain structures are thus made possible for the distribution of book content. This entails not so much the disappearance of the traditional value chain for books but rather a slow process of erosion through redistribution and overlapping.

Examples for self-cannibalising of incumbent companies

As one of the biggest publishing companies, Bertelsmann has reacted by extending its interests to the manufacturer of the RocketBook, the Californian enterprise NuovoMedia. In this way Bertelsmann is seeking to reduce the risk stemming from the incipient decline in the classical book trade. Similar considerations have been behind the recent activities of Deutsche Telekom. Their participation in the Israeli company VocalTel is a sign of support for Internet telephony even though this poses a threat to their own classical telephone business. The ulterior motive underlying these steps is that if the erosion of your own core competency can no longer be prevented, then the best thing to do is grab a share of the multimedia market as quickly as possible. For companies in the media and communications sector, the strategic conclusion to be drawn is thus:

"Cannibalise yourself, before somebody else will do it."

Chapter 5

Navigation Aids for the Internet Economy

5.1 Strategic Consequences for Media and Communications Companies

5.1.1 New Competition Strategies – Creating Business Webs

An Internet economy is characterised by the availability of an increasing number of goods and services that are best understood as system products. This means that when customers judge the product value it is not the individual part that counts but the product as a whole (in other words the system product). The individual subsystems thus stand in a relation of complementarity to one another.

System products rather than individual elements are reference points for customer evaluation

Up to now this phenomenon has appeared above all in the information technology industry. Although the hardware, the operating system, the user software and peripherals such as printers, modems and CD-ROM drives can be seen as distinct component parts, from the consumer's perspective they all exist within a system context. A similar constellation can also be observed in the Internet economy. Subsystems such as Internet access, e-mail, news and information services, as well as shopping services all combine to provide the customer with a valuable integrated system.

A result of the increased presence of system products in an Internet economy is a broadening of existing perspectives as regards competition strategies. Instead of independent individual products, the focus is much more on the production of complementary system goods and services. An understanding of network effects and positive feedback thus yields a decisive competitive advantage for system providers.

Focus on complementary system of goods and services, positive feedback and network effects central for new strategies

In this sort of market context, companies have the choice of either producing the complete system themselves or concentrating on just one element within the system. Given the higher efficiency that comes from the division of labour, the best solution in many cases is to spread the production of system products among a number of firms. Companies that choose to produce the whole range of system products themselves are said to be vertically integrated. IBM is an example of a heavily vertically intregrated concern.[229] Such companies are contrasted with horizontally specialised concerns such as Microsoft, Compaq or Intel.

229 See section 3.3

Fig. 5.1: The market structure for vertical integration and horizontal specialization

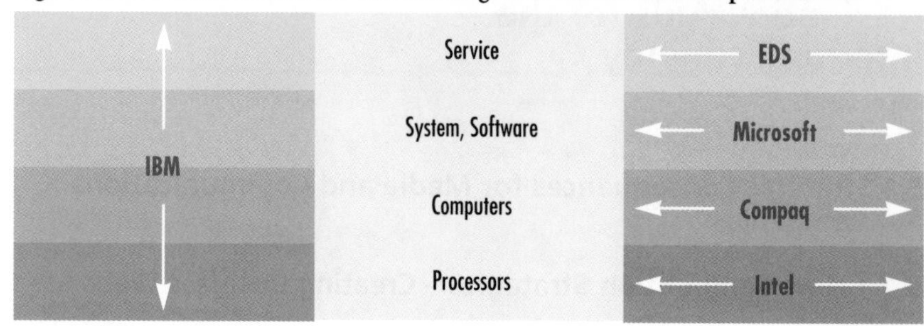

Fragmented value chains typical for Internet economy

The Internet economy is typified by a fragmented value chain structure and an increasingly dynamic market with a rapidly rising rate of technology and product innovation. In such circumstances it becomes very difficult for any single company to produce the integrated range of complex system products. Flexible responses are only possible when companies concentrate on just one or a few elements from the value chain system concerned.

This explains, for example, the splitting up of the telecommunications concern AT&T into three smaller companies in 1995. By the same token, the poor 1997/8 figures for the chip division of Siemens and the plans for splitting it off as a separate listed company as a result of intense competition by horizontally specialized American and Japanese companies can be regarded not just as a sensible measure but as a more or less inevitable response to the circumstances.

Traditional strategic triangle

The new market conditions entailed by an Internet economy are of considerable importance in devising competitive strategies. Formerly, competitive strategies were geared towards the "strategic triangle," where the company's own performance was compared with the competitor's as seen from the perspective of the customer.

Fig. 5.2: The strategic triangle

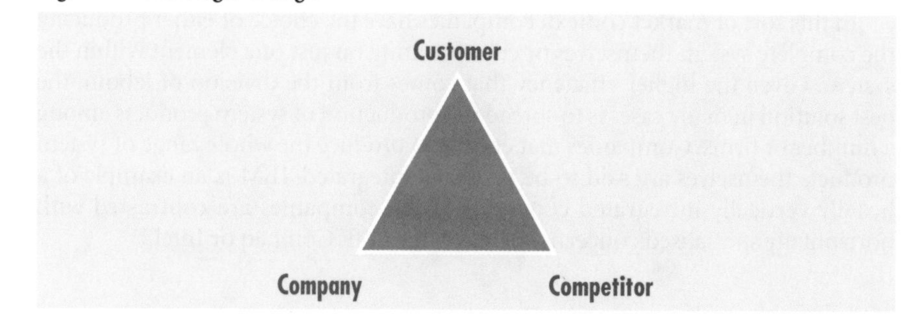

Source

Fig. 5.1: ECC 1999

Fig. 5.2: Ohmae 1986, p. 71

"Focus not just on your competitors, but also on your collaborators and complementors." [230]

Within this strategic triangle (according to Porter) a number of generic strategic paradigms are open to the business concern. One possibility is to go for competitive advantages by lowering prices while maintaining the same quality level, i.e. through cost leadership. A second possibility is to go for so-called differentiation, i.e. quality leadership at a relatively constant price level. These two strategic paradigms can be directed both at the market as a whole and at individual market segments (i.e. focusing on key areas). However, although they continue to be of central importance for companies in the Internet economy, they are no longer sufficient for market success.

Business webs – a strategic answer to a changing systemic environment

This is because for specialized companies the production of individual system components leads not just to greater adaptability to a complex environment but above all to participation in a value network. This results in a change of strategic perspective, as not only competitive cost-effectiveness but also cooperation with other value chain partners comes to assume fundamental significance.

One strategic approach to this systemic environment can be found in the concept of "business webs" elucidated by Hagel III.[231] Business webs are groups of companies that participate in the same value chain system independently of one another and thus exist in a relationship of mutual complementarity. The market success enjoyed by these companies is coupled with one another, since the customer only receives integral results from the system product created by the value network in its entirety, and it is this system product that must win through against rival products.

Shapers and adapters

The element linking the system architecture of a particular business web is in most cases based on a specific technology, but can also be constituted by a customer segment. The first category is referred to as a technology web and the second as a customer web. An essential feature of all business webs is the hierarchical relation between "shapers" and "adapters." A shaper controls the central element of the business web, which is accepted by the adapters who align their complementary products with it.

Technology web

Technology webs are as a rule based on a technological de facto standard. A certain technology here forms the fundamental focal point of the system product in question, and the rest of the components have to be geared to this. What is particularly important here is the interface compatibility, which is essential to the trouble free coordination of the various complementary products with one another and with the core technology.

The best known example of a technology web is the "Wintel" partnership in the PC industry.[232] Here, it is the microprocessor from Intel and the operating system from Microsoft which are the key technologies that have to be coordinated with one another for the purposes of optimising performance. The system was given an open structure, so interface specifications for complementary products were freely accessible. The result was that a range of adapters grouped around the

230 Shapiro/Varian 1998, p. 7
231 See Hagel III 1996
232 See section 3.3

central processor/operating system architecture from the two shapers Microsoft and Intel, aligning their hardware products and software applications with the central "Wintel" architecture. As the business web became more and more successful positive feedback set in. The business web benefited from an increasing number of business web partners, providing the customers with indirect network effects.

In general, business webs based on technologies are more stable than customer webs. When the core of the technology web is a proprietary technology in the hands of a single concern, a change proves expensive for customers and technology web partners alike. Lock-in effects are produced as a result, making long-term success highly probable for the technology web.

Customer web

With customer webs it is a particular customer segment that is the focus of attention. The business web then produces a set of services within a system product for this customer group, focusing for example on the satisfaction of needs linked to a specific theme. This may consist, for example, in arranging and offering information, entertainment and communication services or shopping facilities relating to a particular theme. A number of companies are currently trying to create this sort of customer web by building up virtual communities.

An example of a customer web is the sports service available on the WWW from the American television channel CBS. The services offered by Sportsline USA (www.cbs.SportsLine.com) range from news, databases, match reports, address directories, video sequences, interviews, advice, chatrooms, ticket sales, lotteries for trips to sporting events, to the sale of fan accessories and other sports articles. The companies offering the services thus focus on the sports fan as a customer group. Sportsline USA is the shaper, therefore, and its thematic specialisation gives it access to a target group and control of customer information that is highly valuable to other enterprises. Through its control of customer access, the shaper in customer webs has considerable influence on the business activities of the business web as a whole.

Loose business web ties – high degree of coordination

Regardless of whether the business web takes the form of a technology web or a customer web, this "loose" combine requires a high degree of coordination, at first sight apparently giving vertically integrated companies the upper hand. The coordination of interfaces between the individual system components is easier to achieve within a single company than within value networks. Often – though not always – integrated system products from a single firm such as Apple do better than products stemming from coordination between a number of concerns.

In dynamic industries, forming business webs is not just an attractive strategic alternative but virtually a must if a competitive edge is to be achieved. The following reasons help explain this:

* It is by concentrating on core competencies that firms are able to strengthen their specialisation. The combination of these core competencies in a business web leads in the long run to increased quality in the system product.

Five advantages of business webs

* By sharing the risk involved, uncertainty is considerably lessened for companies in these complex and highly dynamic markets.
* The sharing out of value added activities means that network effects can be exploited to generate a high level of flexibility. In addition to this, the modularisation of the system product brings a great potential for innovation and allows for rapid market penetration.
* The combination of the business web partners provides system products with access to more extensive resources.
* Within the business web there is room for a fruitful interplay of competition and partnership ("coopetition").[233]

Cornerstone of business web strategies: "win-win" situation

Yet these advantages are offset by certain risks. In general, business webs offer little in the way of security to the companies concerned, since the value chain partners for the most part lack any full contractual arrangement concerning the involvement and cooperation of the various companies. Instead, the cornerstone of the business web is the possible "win-win" situation: the individual companies are only successful if the value network as a whole is successful. Sales of a company's own products are directly dependent upon the success of the system product in its entirety. Given the interdependencies within the business web, it is crucial for participants to support the whole value network. As the aim common to all the companies concerned is to maintain and stabilise the business web, group interests move to the fore.

> *"In the network economy a firm's primary focus shifts from maximizing the firm's value to maximizing the network's value."*[234]

Competition within business webs

This alignment of interests is in most cases counterbalanced by competition within the business web. The value added created by a business web is distributed among the individual companies according to the position they occupy as shapers or adapters. In the "Wintel" system, for example, it is Microsoft and Intel that get proportionately the highest revenue. This is supported by the other companies as long as they too enjoy a sufficient slice of the cake. Given that their core competencies are generally interchangeable, the adapters lack any explicit competitive advantage of their own and thus find themselves in an intense struggle for the distribution of the value added.

A further problem can arise if the position of the shaper is not made clear from the outset. In such cases there will be a lack of investment in the construction and expansion of the network infrastructure as the individual firms fail to take responsibility for it. An especial danger is the problem of free-riding, where third parties profit from a business web without themselves shouldering their fair share

233 See Browning/Reiss 1998, p. 112
234 Kelly 1998, p.64

of development costs. In extreme cases, problems of this order can pose a serious threat to the emergence or existence of a business web. If a shaper leaves the value network because other companies fail to provide sufficient support, then the chances are that the business web will break up.

Expansion of business webs by takeovers and investments

A number of strategies are open to business web shapers in the attempt to increase the value of their network. Both Microsoft and Intel, for example, invest in promising young companies as a way of securing their own position in the business web for the future.[235] These takeovers and investments are further seen as a means of ensuring that the existing business web also expands into new market segments.

Enlargement of market segments

It is for this reason that Microsoft is trying to establish a "Windows Light" version – Windows CE – as the dominant operating system in markets outside the PC realm. Microsoft is looking to establish Windows CE not only in handhelds but also in decoders used for the reception of digitally transmitted TV programmes and Internet services. To this end it has signed contracts with B.T. and with TCI, the biggest American cable network operator (recently taken over by AT&T):

> *"The rumor around the cable industry says that Microsoft is paying TCI $100 per copy to put five million copies of Windows CE in its digital cable boxes. That's $500 million, folks, which just about equals TCI's cost for going digital in the first place. Funny, isn't it, how those numbers work out? And remember that TCI also has a second OS in the box -- a version of Sun's JavaOS, apparently doing the real work while Windows CE pays the way."*[236]

These comments from the author and journalist Robert X. Cringely also testify to TCI's unwillingness as a gatekeeper to put its control over the customer segment at risk by opting for just one operating system. As the shaper of a customer web (comprising the cable customer segment), TCI finds itself in a typically ambivalent situation. On the one hand the aim is to integrate as many adapters as possible into the business web in accordance with their core competencies. Given that TCI, as a cable network operator, has no core competence in the spheres of decoder production or the manufacturing of operating systems, the logical thing to do was to leave these to third parties. Access to the business web thus had to be as open as possible. The greater the number of different subsystems that are produced for any particular business web, the more attractive the system product as a whole becomes for the customer. On the other hand, a shaper must try to secure and defend its central position through appropriate strategies.

System products thus mean that competition in an Internet economy is marked by an extremely high level of complexity. It is possible, for example, for a particular company to enter into partnership with different business webs that may be in competition with one another. Microsoft is a shaper in the Wintel web,

235 See Rich 1998, p.1
236 Cringely 1998: "I Cringely"; Vol. 1.6.1, www.pbs.org/cringely/archive/may2898_text

for example, yet at the same time provides the rival Apple Macintosh system with applications software.[237] All in all, the result is a fruitful interplay of competition and cooperation. This "coopetition" promotes the dynamic development of the markets in question.[238]

Portals – fighting for control of the entrance door

These dynamics come to light especially in emergent markets where an as yet vague value chain structure means that clearcut business webs have yet to come into existence. 1998 saw the start of a bitter struggle for the market segment of so-called "portals" – i.e. the starting pages for Internet usage. A number of companies are striving to impose their own business web. In this case it is customer webs that are being created, since a well-known brand name means control of a large customer segment highly attractive to other enterprises as well.

An example of this is Yahoo (www.yahoo.com), formerly a pure search engine, which since mid-1997 has step by step been trying to build up a complete business web. Firstly, a number of other specialists were integrated into the web through cooperations: from content providers to telecommunications enterprises providing Internet access through to service providers of all kinds. The result is that Yahoo now offers a diversity of functions to rival the on-line service AOL (www.aol.com). Since early 1998 Yahoo has thus taken up a position not just as a content aggregator but also as a portal.

AOL has made a swift response. At the beginning of July 1998 it announced a broadening of its Internet services, underscored by a relaunch of its homepage. In the field of content there was the introduction of the personalised newsletter "MyAOL", corresponding to already existent services such as "MyYahoo." In the field of context AOL acquired the software ICQ ("I seek you" – www.icq.com). The function of the "Buddy List" up to now accessible only to AOL members has thus become available to users of the AOL homepage as well, who can ascertain straightaway which of their friends are on-line and directly communicate with one another.

Similar activities directed towards the creation of a portal can be observed in the form of Netcenter (www.home.netscape.com) from Netscape and Start (www.start.com) from Microsoft. The upshot is a multitude of portals competing with one another for the consumer's attention. The eventual success or failure of these business webs will here depend not just on the proprietary name, but above all on the quantity and quality of the cooperation partners involved.

What emerges on the whole is that competition is no longer something that exists between individual companies but rather between business webs of the most diverse types. The key to competition is to win the customer over with attractive integrated products, entailing a shift to a higher level of aggregation. A business web will be successful and competitive if its system product is superior to what is offered by rivals.

237 Admittedly there was also a political (or rather a regulatory) dimension involved in this cooperation between Microsoft and Apple. It is suspected that Microsoft is currently supporting its rival in order to avoid the consequences of anti-trust action when it comes to Apple's eventual demise.

238 See Browning/Reiss 1998, p. 112

Fig. 5.3: Competition between business webs

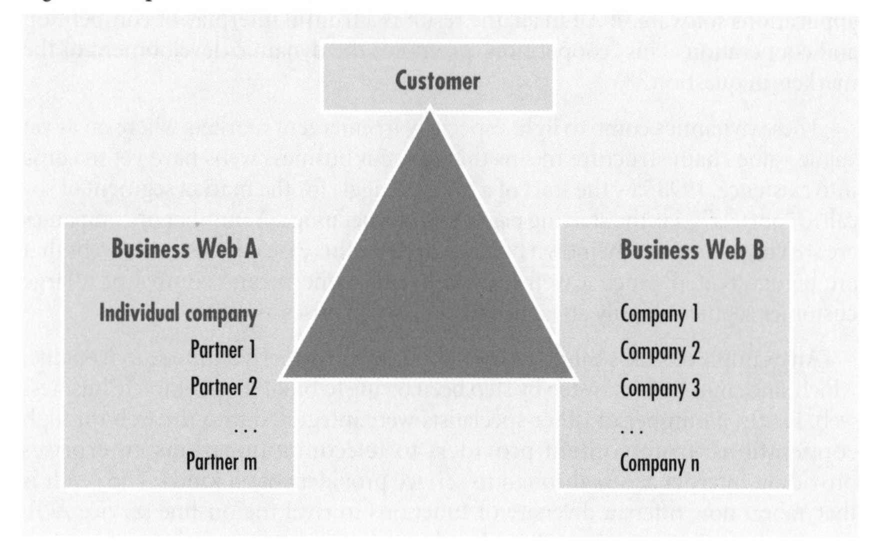

Competition shifts from individual companies to business webs

To summarise, a business enterprise's strategic focus will be both narrower and broader than has previously been the case: narrower, since firms will be limiting themselves to their core competencies, and broader, since the creation of partnerships will be viewed as a key strategic element.

5.1.2 New Product Strategies – From Windowing to Versioning

Windowing in the movie industry

Feature films such as "Godzilla" are transmitted not only in the cinema, but also through video, pay-TV and television. This use of four distinct channels of distribution – known as windowing – is possible because consumers have different preferences as regards how up-to-the-minute the film they watch should be.[239] While one part of the target group feels a pronounced urge to see the latest Roland Emmerich film as soon as it comes out and is willing to spend $ 8 at the box office to do so, another part waits until it can be picked up for $ 3 from the video store.

As this example shows, consumers regard how up-to-the-minute a film is as an essential service criterion. It is thus possible for suppliers to use an appropriate grading of consumers' willingness to pay according to whether distribution is by cinema, video, pay-TV or free-TV in such a way as to maximise revenue. If, for example, a film were only available as a video purchase, this would exclude two customer groups: on the one hand those who would have been prepared to pay a higher price to see the film in the cinema, and on the other those hoping to watch it as part of what is offered on either pay-TV or free-TV. The outcome would be substantially lower revenue.

239 See Owen/Wildmann 1992, p. 26, as well as section 3.1

Source

Fig. 5.3: ECC 1999

"If you add a fancy new feature to your software or information product, make sure there is some way to turn it off."

Behind windowing there lies a more general principle of product design known as versioning.[240] This refers to a media or information product being offered in a number of different versions. Versioning is a form of product differentiation, a common strategic approach for gaining a competitive edge in the consumer and capital goods industries.[241] For media and information goods up to now, product differentiation has only been possible in the form of windowing, as apart from staggered distribution through successive channels (cinema, video, pay-TV and free-TV) uniqueness of content did not allow for differentiation. Now that media and information products are available in digital form, however, product differentiation can be deployed on a large scale, as it becomes easier to differentiate the dimensions relevant to how the product is valued by different customers.

Versioning – the more general strategy for information products

The basic thought behind this strategy is to design a product line from which the consumer can automatically select the product with the greatest value for his own needs. To do this, the supplier must first identify product characteristics that are valued differently by different consumers in terms of the appeal they hold. Possible product characteristics upon which versioning can be based may include such dimensions as delay, comprehensiveness, speed of operation or added value. By choosing various gradations of the product characteristics in question, the supplier can thus design a product line. Shapiro and Varian describe how best to design an information product:

Product and service characteristics

> *"If you add a fancy new feature to your software or information product, make sure there is some way to turn it off! Once you've got your high-value, professional product, you often want to eliminate features to create a lower value, mass market product"*[242]

It is essential for the supplier to make sure that the various gradations are perceptibly different from one another. Otherwise the danger exists that the premium product within a particular product line will not be recognised as such and the willingness to pay the higher price will be insufficient. The delivery period offered by Amazon.com, for example, is split into three clearly distinguishable categories: 1-3 days, 7-21 days and 4-6 weeks. Amazon.com is keen to make sure that a customer from the middle category does not receive the books after four days, for otherwise the willingness to pay for the first category will be low since it makes so little difference. This constellation serves to ensure that consumers really do select the version with the highest subjective value for themselves. All in all, such "self-selection" means that consumers' willingness to pay can be skimmed as a way of maximising revenue.

Clear perception of the difference between versions essential for customer's willingness to pay higher price for better versions

The principle underlying how versioning functions can be illustrated by looking at "financial information" as a paradigm product. The characteristic taken by consumers to differentiate the product value is clearly how up-to-date the information is. Professional users require real-time information, while private

240 See also Shapiro/Varian 1998, p. 39ff.
241 See Kotler/Bliemel 1992, p. 461
242 Shapiro/Varian 1998, p. 48

investors are as a rule able to cope with a slight delay in managing their portfolio. It is for this reason that the American financial information provider "PAWWS Financial Network" can use a graded time delay to version its product line. For professional users it charges a monthly price of roughly $50 for real-time portfolio information, while the same information becomes available to private users twenty minutes later at a cost of just $8.95. If PAWWS were only to offer one of these two versions, this would mean that the customers' willingness-to-pay was not being optimally skimmed.

Example for time-delay type versioning: stock market information

A second example is the distribution of information products. The two possibilities available here are in general off-line delivery and on-line delivery. Because the readability of information products such as reference books is a product characteristic that is evaluated very differently according to particular requirements, the product line offered can be versioned on the value of this dimension. For example, the reference book about the digital economy – "Unleashing the Killer App"[243] – is available in its complete form on-line for free download on Web page www.killer-apps.com. The layout is reader-unfriendly, as each page has to be called up individually. If readers really are seriously interested in it, they will give up trying to read it on-line after twenty pages at the most and buy the book as hard copy for $24.95. The free availability of the book in its on-line version is thus more likely to boost sales of the hard copy than to cannibalise them. Indeed, both the "National Academy of Sciences Press" and the "MIT Press" have found that having their books available on-line has approximately doubled the hard copy sales.[244]

Example for convenience type versioning: online book and printed book

Suppliers of media and information products are faced with a whole range of possibilities for versioning. Three central dimensions – time, quantity and quality – can here be distinguished, as shown by the following table:

243 See Downes/Mui 1998
244 See Shapiro/Varian 1998, p. 66

Fig. 5.4: Possible dimensions for versioning

Dimension	Characteristic	Potential variation	Example
Time	Actuality	Immediate or later access	PAWWS Financial Network (real-time stock market information versus 20-minute delayed information)
	Duration of infomation availability	Long-term and short-term use	Lexis/Nexis databases (on-screen and/or as download)
Quantity	Comprehensiveness	Professional and private use	Dialogweb and Datastar (degrees of database comprehensiveness)
Quality	Image resolution	High or lower resolution images	10MB for gloss images 600K for matt images
	Legibility	Monitor or hard copy	"Unleashing the Killer App" (online version free and book for $24.95)

It should be emphasized that versioning is highly dependent upon the information product concerned, in other words it is very product-specific. Even so, versioning can also be used to design product lines over a range of diverse product categories, as shown by the example of the database ERNIE from the management consultancy Ernest & Young.

Consulting and databank versioning

As a general rule, management consultancy demands considerable effort providing reports and documentation, in theory making it easy to mediate these services. This is not in practice completely feasible, however, as most companies realise they have a problem without understanding precisely what it is, and a problem analysis thus has to be conducted before the problem itself can be tackled.

As a result of this, differences arise in the assessment of consultancy services. For some firms, consultancy only makes sense if it includes a prior problem analysis, an individual solution and/or an ensuing implementation. These firms will usually fall back on the classical services offered by management consultancies.

For firms whose need for consultancy is smaller because they already know what the problem is and/or they have lower financial resources Ernest & Young offer ERNIE. This is a database available through the WWW that for a yearly price of 6,000 dollars can be used by companies to provide them with an approach to their specific problems. Its content includes modularised answers and paradigm cases. The varying willingness-to-pay of companies in relation to their distinct

Source

Fig. 5.4: ECC 1999, based on Shapiro/Varian 1998, p. 45

consultancy needs has thus been broadened beyond the conventional differentiations in consultancy services.

When versioning, special attention also needs to be paid to the number of versions offered. Often, having three classes will be found to yield maximum revenue potential, for when there are just two options available the problem is that undecided customers in most cases tend to go for the cheaper version. For this reason a third, clearly distinguishable "gold" or "maxi" or "premium" version should be created, so that when in doubt customers automatically select the middle – formerly premium – version. This explains, for example, why McDonalds offers three distinct sizes of soft drink.

Three versions seem to be the optimum number of choices

In general, the advantage of versioning for media and communications enterprises lies not only in the possible designing of product lines but also in bringing this about extremely cheaply. Digitisation means that not just reproduction but versioning itself can be achieved at a very low cost.[245] The comprehensiveness offered by a database can thus be restricted or expanded at will by just a few programming commands. Given these numerous advantages, it thus seems likely that versioning will establish itself as an integral strategic approach for media and information products in an Internet economy.

Versioning of digitised information is extremely cheap to organise

5.1.3 New Price Strategies – Follow the Free

The most astonishing and most talked about phenomenon in the Internet economy is taking place in the realm of price strategies. A new software company decides to give away its only product for free. The consequences are enormous. Within less than six months the browser Netscape Navigator reached a worldwide market share of some 80 percent. This figure is in itself not particularly astounding, given that beforehand no real browser market existed. Even so, the potential inherent in the venture succeeded in winning over investors, with the result that when Netscape was floated on 8 August 1995 the ensuing explosion in its market value wrote economic history. An enterprise that was more or less giving its only product away saw the price of its shares rise in a matter of weeks from a starting value of 28 US dollars to over 170 dollars: "The stock market created its own version of artificial reality."[246]

Netscape writes economic history

This distribution model was so successful that shortly afterwards Microsoft too went over to giving away its Internet Explorer for free. This price strategy – designated "follow the free" – combines the special cost structure of digitalised information products with the phenomenon of network effects. It demonstrates the rules of the new Internet economy and their strategic consequences in their purest form.

It is worth stressing that "follow the free" is not completely new as a price strategy but is built upon existing strategic approaches. Along with price

245 See section 4.2.2
246 See Lewis 1997, p. 1

differentiation,[247] marketing science in general draws a distinction between penetration pricing and skimming strategy. Penetration pricing refers to a supplier entering the market with particularly low prices. The goal of this strategy is to achieve a high level of market penetration and thus lower costs as quickly as possible through economies of scale and experience curve effects.

Penetration pricing

The use of penetration pricing is most effective when the product in question has no particular USP ("unique selling proposition") and is exposed to intense competition from the outset. The ensuing cost-lowering effects here produce a price advantage that is difficult to compete with. Japanese consumer electronics manufacturers have thus managed to conquer the European and American markets with low prices which in turn have resulted in a further expansion in production and thus facilitated more price cuts.

Skimming strategy

By contrast with penetration pricing, a skimming strategy is a way of entering the market with high prices. High starting prices can only be imposed provided the product has a clearcut USP in the eyes of consumers. The broad aim is to skim or cream off the early users' willingness to pay. Afterwards prices are steadily lowered, thus to a large extent opening up the market. A classical example of skimming strategy is book publication. Initially a hardbound version comes out at a relatively high price, but after a certain time has elapsed a cheaper paperback version is published.

Penetration pricing and network effects

In devising price strategies in an Internet economy, attention should be paid not only to these classical approaches but especially to the significance of network effects and their ramifications. When a company such as Microsoft gives a product away for free, it is not without a reason. This reason can be summarized in the formula "mind share leads to market share": giving the product away is intended as a way of bringing about rapid market penetration, which in turn generates additional attractiveness and – through positive feedback – a further acceleration of the process of market penetration. "Follow the free" is thus an extreme form of penetration pricing, possible particularly on account of the very low variable costs in the development and distribution of information products.

The central question for firms in a network economy is how to refinance the frequently very high "first-copy" costs of the products distributed for free. The Netscape Navigator, for example, cost some 30 million US dollars to develop and design.

"Can you imagine a young executive in the 1940s telling the board that his latest idea is to give away the first 40 million copies of his only product? (Fifty years later that's what Netscape did.) He would not have lasted a New York minute."[248]

Refinancing the prior investments in a "follow the free" strategy typically occurs in two stages, separate in time and with distinct objectives. In the first step, the free distribution of the product and the ensuing network effects build up an

247 See section 2.2
248 Kelly 1998, p. 56

installed base of users. The aim here is to create lock-in effects that will bind customers long-term to the firm's products.

Two phases of "follow the free" strategy

It is not until the second step that the investment in the installed base bears any financial fruit. In the Internet economy the rule is that "locked-in customers are valuable assets."[249] Possible starting points for the generation of revenue flows are the sale of complementary products or product versions that are newer (upgrades) or more powerful (premium). This is illustrated by the following examples:

Examples

* Network Associates (formerly McAfee) began by giving away its antivirus software free of charge. In this way it was able to win a third of the market for virus protection software, and from this strong market position substantial revenue was earned through the sale of upgrades. A further positive side effect of giving products away is the free participation of users in product development. Through the large installed base and the strong customer links a number of viruses were detected which could then be eliminated in upgrades. These indirect network effects improve product performance and so stabilise the company's market position.
* In the field of mobile telephony, it is through the subsidised distribution of mobile phones when a contract is signed that companies invest in building up an installed base of users. In this case revenue is generated through the sale of the complementary "telephone service". The investment is thus financed by call charges.
* In the course of rapid market penetration the creation of an installed base can yield numerous possibilities for realising network effects. The increasing attractiveness of the offer then justifies the imposition of a normal price for a product that was initially free (for example, digital audio broadcasting).
* A high market share and a correspondingly high public profile can lead to network effects in the sale of other products from the same company. Netscape for example sells business customers powerful server software for Intranet solutions. Moreover, the fact that Netscape's home page is as a rule accessed automatically as a "portal"[250] and that this generates considerable attention among surfers is exploited to generate income from advertising and commissions.

A further possibility for using the installed base is the sale of access to users, exemplified by the long-established use of advertising to finance media products as a common variant on "follow the free." Broadcasting financed by advertising, for example, uses indirect network effects that exist between recipients and advertisers.[251] TV stations financed purely by commercials provide their programmes for free, but it is only through a broad circulation that the offer becomes attractive to the advertising industry, allowing the market to cover the costs for programme production and actually make a profit.

249 Shapiro/Varian 1998, p. 113
250 Mobile telephone suppliers had the advantage of being able to access an already existing basis – the telephone network in existence – with their product. Even so, participation in a given single mobile network usually has considerable price advantages for calls within this network; this makes it appropriate to speak of network effects.
251 See section 3.1

Along with the targets of rapid market penetration and the ensuing refinancing that this facilitates, the Internet economy also holds other, non-financial incentives for distributing a product for free. One of the best-known examples of such a free product is Apache. The server software Apache dominates the Internet server market with a worldwide share of some 50 percent.[252] The peculiar thing about this is that Apache is not a commercial product and receives no support through marketing. As it was developed – like the programming language Linux – by a virtual network of volunteer programmers, there are neither directly classifiable first-copy costs nor reproduction and distribution costs. Its developers see Apache as an investment for the future. As their names are given in the home page, a reputation is built up which may result in job offers or professional support for the developers. This particular objective is nonetheless something of an exception in "follow the free" terms. In general, it is the success of a company that is at issue.

Non-commercial incentives for free distribution of a product

On the whole, the initial sacrifice of income means that "follow the free" constitutes a considerable investment in the future, albeit with high entrepreneurial risks inevitably involved. Even so, as a price strategy it is an effective means of asserting oneself in the high-speed competition for positive feedback, lasting competitive advantages and so for revenue.

Penetration pricing a high risk strategy

5.1.4 New Communication Strategies – One-to-One Marketing

Networks such as telephone or fax facilitate the exchange of information, serving – whether in human or purely economic terms – to improve the quality and quantity of relations. But for this the networks always require real human beings. The central determinants in an Internet economy will thus always be interpersonal relations.

Interpersonal relationships significant for market success

The great significance of stable (customer) relations has been recognized for years. With the aid of network technologies, however, these relations can now be raised to new heights. Networks such as the Internet are more than just a means of interpersonal communication; they make the unrestricted exchange of information more rapid, more intensive and more international. They are what make many relations possible in the first place.

Networks thus stimulate the relationships between suppliers and consumers. They are particularly crucial because it is not only companies that invest as suppliers in their relations with customers, but customers likewise invest in their relations with companies, conveying them their preferences, opinions, personal data and interests. Once established, such relations between supplier and consumer ensure increasing and above all stable revenue, as a change would entail high, often excessively high, transactional costs:

252 See Mieszkowski 1998, p. 38. A further reason for Apache's success is the widespread attitude among the Internet community that activities without a commercial background should be supported.

"Relationships between people are a central category in the Internet economy."

"When the florist sends a note reminding you of your mother's birthday, and offers to deliver flowers again this year to the same address and charged against the same credit card you used with the florist last year, what are the chances that you will pick up the phone and try to find a cheaper florist?"[253]

Internet allows new quality of relationships

Airlines and the retail trade likewise cultivate customer relations of this kind with their frequent flier or frequent buyer programmes. The aim is to create long-term customer links, preparing the way for a genuine one-to-one relationship between company and customer. Such relations are a fundamental characteristic of an Internet economy, since customers can for the first time be genuinely integrated into the company network by means of a relation-optimising feedback loop. Three instruments help contribute to this:

Three main instruments

1. Cheap, powerful databases, which allow companies to recognise customers on differentiated, personalised terms and treat them accordingly.
2. Interactivity, which permits customers to communicate directly with the company and receive the corresponding response without delay.
3. Mass customisation, now that technology makes it possible for companies to offer personalised products, i.e. different customers receiving different products.

Integrating these three factors into the production process helps suppliers optimize output and at the same time further increase customer loyalty. One-to-one communication is nothing new in itself. Direct marketers have been using it for ages. Yet the Internet lends it a new quality, allowing dialogic communication to be achieved much more efficiently now that companies can give personalised, real-time attention to each customer's individual needs.

Collaborative Filters and Feedback Loops

One starting point for an intensified relation between customer and supplier is the extrapolation of customer preferences from interactions in the past. A more far-reaching instrument, however, is provided by the concept of collaborative filters. Such filter software creates the infrastructure for users to relate their preferences to the preferences of other users. One of the first such web-based recommendation systems was developed by the "Firefly" company,[254] bought up in early 1998 by Microsoft.

Feedback loops optimise inter-personal relationships

Firefly functions as follows. A user accesses the music site MyLaunch (www.mylaunch.com), for example, and names at least ten of his favourite CDs. To give a broader definition of his preferences, he can add as many more as he fancies. These data are then taken by the Firefly software and compared with those of the roughly half million other MyLaunch members who have already named their preferences. The responses given by those users whose preferences coincide most closely with the newcomer's are sought out, and the user is then notified of CDs that he did not name himself but were positively evaluated by other users

253 Peppers/Rogers 1997, p. 180
254 www.firefly.net/

with a similar taste. In return the others are notified of CDs named by the new user but not mentioned by themselves.[255] There is reason to suppose that users are more likely to like the CDs chosen by other users with a similar taste than those chosen for discussion in a magazine for example.

Fig. 5.5: Collaborative filter

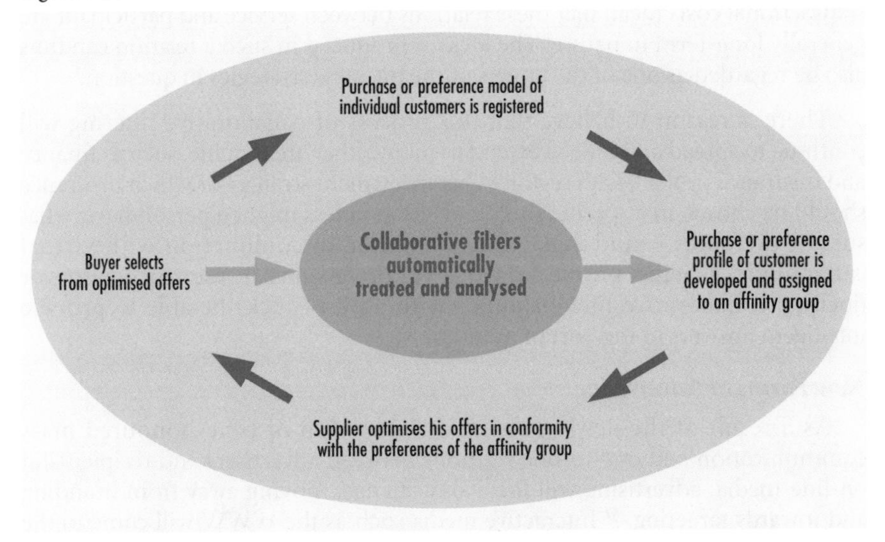

The appeal of systems such as Firefly lies not only in the recommendations it is able to convey but in its formation of preference or affinity groups. "Communities of taste" come into being, something that previously would have simply been impossible. Firefly provides similar services in books, films and Web pages to a total of three million users.

Communities of taste(s)

Booksellers such as Amazon.com or Barnes & Noble / Bertelsmann likewise use collaborative filters in order to optimise what they offer their customers: instead of preferences it is here spending patterns that are analysed. On the basis of orders made by the customer in the past as well as by other customers with similar spending patterns, the customer is sent recommendations by e-mail. These offers are regarded as Amazon's most potent marketing tool. Customers are made optimised offers, develop ties with the company and thus ultimately help increase Amazon's turnover.[256]

Amazon can thus be viewed not only as an on-line bookseller, but also as a company that sells relations. Everyone who logs in to Amazon receives access to an affinity generator that gets to know the customer better and better as time passes – rather like a real shop assistant in a real shop. On the basis of collaborative filters of this sort, feedback loops develop which turn the enterprise into a

255 Without referring the individual user to the other actual persons themselves. See Kelly 1998, pp. 119-21
256 See Kelly 1998, p. 120

Source

Fig. 5.5: ECC 1999, based on Lewis 1997, p. 170

learning enterprise able to direct its marketing more and more precisely at the individual needs of its customers.

Long-term lock-in

Yet one thing should not be forgotten here: all participants who have once submitted their preferences to such a system have to go to considerable lengths if they should then wish to join any other system of collaborative filters. The high transactional costs mean that these relations between service and participant are generally long-term in nature. The lock-in produced in such a relation can thus also be regarded as one of the targets of the business strategies in question.

Collaborative filtering for communities in health, finance and insurance

There is reason to believe that this process of collaborative filtering will continue to spread and indeed extend to many other areas of life, such as finance and insurance or the health sector. What investment strategy or which insurance should be chosen in a specific situation? What illness might a person have, what sort of therapies would make sense for him? In conjunction with virtual communities, whether private individuals or professional investment advisors or doctors, collaborative filters should (with a bit of luck) be able to provide intelligent answers to this sort of question.

New Forms of Advertising

From push to pull – a radical change in advertising strategies

As a result of the newly possible combination of time-honoured mass communication and one-to-one relations between advertisers and recipients in on-line media, advertising will inevitably change, moving away from branding and towards targeting.[257] Interactive media such as the WWW will come to the fore, spurring on the change from push to pull marketing. Potential recipients will enjoy a new sovereignty, allowing them to select all the content and thus also the advertising they see. Instead of inundating recipients with unasked for advertising material, interactive media will have to place greater emphasis on offering viewers the opportunity to use their advertising content. The demise of classical (TV) advertising will nonetheless not come about quite as quickly as some prophets think; the habits of TV viewers are not going to change overnight. Yet the new potential offered by the WWW and digital television do allow a whole range of new models for advertising to be discerned.[258]

257 The problem here is that at present neither genuine branding is possible on TV nor genuine targeting through the Internet.

258 In addition to this, a number of demands can also be made on the classical field for improving its advertising. In a few words:
- Advertising enlargement - "So big, you can't miss it."
- Advertising enrichment - "So good/funny, you got to see it."
- Advertising enlightenment - "So smart (or useful), you don't mind."

New Models of Advertising: Four Challenges to Advertising Customers and Agencies

Which models for advertising make sense in a net-based (interactive) media landscape? In what follows, four challenges are presented for advertising models that go beyond what is in use today.[259]

Consumer empowering – ***"Smart users know where to go"***

The Internet media environment shows a clear difference from the reception situation as it has been up until now: the user can select advertising as and when he likes, instead of having to zap helplessly from channel to channel as with television. Flickering banners are admittedly a temptation to click through, but here it is the consumer who decides independently. Yet if this is the case, then advertisers should respond appropriately. Customers should be given full freedom of choice as to when they receive which advertisements, and when not! The "pull" element in Internet shows just how important consumer power is. Accordingly, users should be made the offers that are best and most useful to them. Added value is here the order of the day. The "push" concepts that were much discussed in 1997 – transferring the idea from broadcasting to the Internet – seem by contrast to have become obsolete again apart from a few newsletters that can be subscribed to for free.[260]

Let customers decide

Financial Benefits – ***"Pay for Eyeballs"***

The above notion is also the base for the idea of paying users. Instead of paying media for transmitting advertisements no one watches, companies would be better off paying consumers directly for actually looking at their advertising. This already occurs and functions in the Internet. Under certain circumstances it can even work out better value than classical advertising. In the long run of course, it would be important to develop a scale of prices for attention that varies with socio-economic background.[261] As well as this, there is the problem of what the consumer's intrinsic motivation is for receiving advertisements in this way. Purely financial interests would possibly lead to a "devaluation" of advertising. Yet the basic idea remains: pay customers for receiving advertising and not the media for merely transmitting it!

Pay them for their attention

Mass Customisation/Personalisation – ***"So personal, I want to see it"***

A further proposal, stemming from the concept of advertising enrichment, is the personalisation of advertising. If databases with the aid of collaborative filters make it possible to build up psychological profiles of the individual recipients and to match these up with specially suited advertising, then every viewer will optimally receive their own adverts or personal offers emphasising precisely those aspects that are likely to appeal to them. If the industry[262] actually succeeds in pulling this off and producing advertising

259 These advertising models are at present probably better suited to advice-intensive than image-oriented products. Also: all these new advertising forms are dependent upon the advertising targets and require both expenditure and planning if they are to be successful.
260 The channel bar of Microsoft's Internet Explorer is thus no longer automatically loaded in versions 5.0 or later.
261 See Noam in ECC 1997, p. 43f.
262 Privacy aspects also play a role here.

Target individuals, not groups

that is a mass product yet tailor-made for the individual customer, then not only can it gain a great deal more attention, but much more than this – it can win over valuable customers who stick with their brand, make or firm because this firm knows them. For this reason, personalise your advertising!

Branded Content – "No way out"

As with product placements, there is more and more content (above all in the Internet) that is no longer produced by professional media (as content providers) which then distribute and refinance it through advertising, but is instead directly produced and financed by advertisers. Instead of spending large sums of money on developing and placing advertisements, therefore, it is becoming more and more attractive for advertisers themselves to produce interesting content that goes beyond advertising. This content needs not even be directly connected with the company. The spectrum of offers can range from a Tamagotchi in the shape of a purple cow developed by the chocolate brand Milka, to the 24-hour motor race at LeMans being transmitted on the Internet by BMW (focusing exclusively of course on the BMW pit), or the ATP tour broadcasting its own tennis matches live on the Internet. The motto for this new form of advertising thus runs: disseminate content not adverts!

Content attracts attention – not advertising

5.1.5 Strategic Development of Content in the Internet Economy

Three trends in content development: Meso-media, regionalisation, games

Three trends can be pinpointed in the development of new content. Firstly, there is the triumph of meso-media, i.e. products located somewhere between the mass media and one-to-one communication. Secondly, the video and computer games industry deserves closer examination as a possible trendsetter, since it is here that the development of (interactive) content is played out in advance. Thirdly, and in conformity with the first point, the regionalisation of media content is also assuming more and more significance for the question of new content.

The Triumph of the Middle Level

There are mass media such as television, radio and print that reach millions of people at once with a single message. And there are micro-media such as the telephone, letters, fax and e-mail, all of which facilitate one-to-one communication between individual people. The middle level – comprising smaller groups or circles with fewer than ten thousand people – manifested itself relatively rarely in the world of classical electronic media, as technically it tended to be a loss-maker. Up to recently there was practically no technology outside the print media that made it possible to provide media content in an economical way to smaller groups of around five thousand people. Electronic media failed to furnish the necessary economic basis.[263]

263 See Kelly 1998, p. 98. Some cable channels such as CSpan in the United States or n-tv in Germany nonetheless at times register less than 100,000 viewers per day.

In the Internet economy, media-suppliers for smaller and medium-sized groups are highly valuable. They are crystallisation points within networks, offering perspective, access, orientation and support in a network world where it is easy to lose one's bearings. They represent a level that has so far been missing in our world of electronic communication. The development of a segment of "Very Special Interest" programmes is something that has been made economically viable by the Internet.

Internet makes "meso-media" economically viable

Alongside the competition for existing advertising funds, there is reason to believe that on-line broadcasting, or "webcasting," will actually open up new revenue potentials, making wholly new offers possible for producers. One promising new idea here is a form of differentiated on-line subscription, where a content provider organises a small group of subscribers willing to spend, for example, $ 25 a year for specific content or offers. The supplier only needs to be able to get 10,000 subscribers worldwide in order to generate a yearly turnover of $ 250,000 as a producer – enough to produce a weekly programme that can be cheaply distributed through the Internet and/or called up on demand by subscribers. Special interest channels will thus probably spring up in great number and undreamt-of diversity in the Internet, as the Web's infrastructure not only offers an economical platform for production and distribution, but in the long run also provides an effective basis for the hitherto expensive business of billing such closed user groups.

Webcast subscription for "very special interest" programming

New Perspectives for Content in the Markets for Computer and Video Games

Attempts to develop interactive television have stemmed predominantly from the TV concerns. Many ventures, such as Time Warner's Orlando project, were not able to meet the high demands, even though – or perhaps because – some of the investments involved were so substantial. Interactive television as the concerns had envisioned it proved a failure. Instead, 1993 saw the more anarchic Internet following in the footsteps left by the interactive dreams.

For the last five years at least, developments in the market for computer and video games have been largely absent from public awareness. Yet this is a media world that exemplifies just what form interactivity and entertainment can actually take.

Video games show the future of interactivity in entertainment

As early as 1996, roughly a quarter of all European households already had video game consoles. Although the absolute figures are sinking slightly, household penetration is on the rise. In addition to this, almost every PC equipped with a CD-ROM drive can today be used as a games computer at little or no additional expense. It comes as no surprise, therefore, that in 1996 the total expenditure on interactive entertainment software for games consoles and PC games in Europe was 3.44 billion US dollars, higher than the turnover from the distribution of cinema films.[264]

264 European Audiovisual Observatory 1998, p. 130. In the 1999 Report, the European Audiovisual Observatory changed the basis for video game consoles: only advanced consoles (like Sony Playstation) and software sales for these as well as for PC leisure are taken into account.

Fig. 5.6: Video game consoles in Europe

	Consoles in thousands			Penetration in %		
	1994	1995	1996	1994	1995	1996
Germany	11,781	11,645	11,446	21	22	24
UK	12,977	12,391	11,518	34	36	37
France	7,321	7,515	7,583	21	24	26
Italy	2,838	3,083	3,274	10	12	13
Spain	3,251	3,281	3,312	22	24	26
EU5	38,168	37,915	37,133	n.a.	n.a.	n.a.
EU 15	44,443	44,445	43,676	21	22	24

Computer games such as the PC game Blade-runner based on the film classic or the virtual archaeologist Lara Croft have not only attained cult status with their fans but are also paving the way for a new form of entertainment, combining playful, interactive and receptive situations in a computer-run "world" with a steadily increasing realism that is now approaching feature film quality. From computer games of this sort scenarios for new forms of media entertainment can be extrapolated into the future.

Video games and digital tv

The possibilities already in existence for entering the Internet through a low-cost games console, as well as the introduction of the more powerful DVD (digital versatile disc), will further enhance both the level of realism and the degree of networking among players. The huge number of young people today growing up with such games will in future be thoroughly at home with this kind of product, via digital TV as well. A further key factor is that console manufacturers such as Sega, Nintendo or Sony are well aware of how to handle both the demands and the willingness-to-pay of users. There are also indications of a further convergence between games consoles, PCs and digital television receivers that may well help video games to their definitive breakthrough within the next generation.

Source

Fig. 5.6: European Audiovisual Observatory 1998, p. 127

Sega, the once and future king of video games, has a plan to dominate personal computing, too, and it just might work

by Robert X. Cringely

There was a joke – more of a riddle, really – that PC industry people used to tell about the largest software company in Redmond, Washington. Everyone thinks of Microsoft, of course, but that was the joke: Until a couple years ago, the largest software company in Redmond, Washington, was Nintendo of America, with about $6 billion in sales. Nintendo's U.S. headquarters, just across from Microsoft, looked back then like it had a license to print money.

The video game business is so much like the PC business and yet so different. Both industries have hardware and software. Both are built around the microprocessor and a video display. Both have been embraced by millions of American families, and that's where the similarity ends. We're personal computer people, why should we care about this video game stuff? Because it is about to force a restructuring of the PC business. (...)

One thing that makes the video game different than the PC business is its domination by the Japanese. Let the Department of Justice try to dictate to a bunch of guys in Japan, hah! Even more important, though, is the fierce competition in the game market and the regular change of leadership. Nintendo dominated the market for 8-bit games. Sega dominated 16-bit games. And Sony has mastered the 32-bit game business. In each case the brand seemed supreme only to be overturned completely with the next change of hardware platform.

What's after 32-bit games ought to be 64-bits, but for the moment only Nintendo is operating in that space and its Nintendo 64 machine hasn't made the big inroads the company expected against the Sony PlayStation, king of 32-bit gaming. And now Sega plans to leapfrog 64 bits entirely, jumping to what the company claims will be a 128-bit game machine. (...) Sony is the big gun right now and the reason is games. Sony's games are the best and there are more of them. Sheer size accounts for some of this, but it is also important that Sony has more types of intellectual property to leverage than the other companies. With records, movies and TV shows to cross license, Sony is a powerhouse of game generation. (...)

Windows CE, which not only Sega has endorsed for the Dreamcast, but Sony has said will be inside its new game system that will debut in 1999. Needing to be at the heart of all that is digital, Microsoft has bought its way into the game systems, just as it has the digital cable TV boxes. (...) And now Windows CE is going to be on two of the three major game platforms, too. But this is less interesting than the platforms themselves. (...) We've got a very

fast rendering engine, a modem and a CD-ROM drive all running on an OS that already has its own Java virtual machine. This sounds like a network computer to me, but this network computer will come into the family at the bidding of a 12-year-old with discretionary income, which is much more of a sure thing than can be expected of Oracle's NCs or even Microsoft's own WebTV.

Play all this against the economics of video game hardware, which is completely different than the economics of PCs. Video games are literally a razor and razorblade business. At best, the game maker breaks even on the hardware and makes all the profit on the software. Introductory pricing is always aimed at breaking even by the middle of the product cycle, so Sega will deliberately lose money on the Dreamcast systems for at least a year. There isn't a PC vendor in the world that deliberately aims to lose money on a system. (...)

What Sega is getting revved-up to do (with Sony right behind it) is force that digital convergence we've long talked about. A cheap box with a fast processor and a CD-ROM drive will enable some serious Web surfing through a portal that will inevitably be co-branded by Sega and Microsoft or Sony and Microsoft. We'll still be buying our stuff through Internet Explorer, just on a video game. And with the IE graphics cached on the CD and pumped out through a rendering engine that's easily twice as fast as any Pentium yet built, the Dreamcast will make the best of that little 33.6 modem. Even with such a thin straw to sip through – for now – the Dreamcast will be a very fast Web surfing machine.

And because Microsoft owns a part of it, Gates and Company won't need to make an effort to protect their PC franchise. In the video game business, they'll just be another software developer and the Feds will pay no attention. Two years and 20 million units from now, that could all be different, of course. At least that's the plan.

From: "I Cringely"; Vol. 1.6.1, www.pbs.org/cringely/archive/may2898_textl
With permission from Robert X. Cringely

Regionalisation and Localisation

Up until now, regional interests and media needs have been looked after above all by classical media such as local newspapers and freesheets. As a countertendency to the wave of globalisation that is at present holding sway in the fields of information and entertainment, users will ask more and more for regional and local information and products. As such information is easily accessible in the real world, it is quite understandable that users will demand to retrieve more of this type of local information over a large-scale data network as well.

That part of the Internet going by the name WWW thus faces the prospect of a growing regionalisation of its content. This tendency can be clearly illustrated by a four-field matrix relating to the focusing of media content.

Fig. 5.7: Focusing of media offers in the Internet

		Regional Focusing	
		Narrow	Broad
Content Focusing	Narrow	On-line entertainment and events guide (www.tip-online.de) *Freesheets*	On-line sports service (www.sport1.de) *Special interest channels*
	Broad	Digital city *Local newspaper* *Local TV*	Search engines/Portals/Communities (www.netcenter.com) *National TV channels* *Satellite channels*

Two types of increased focusing in the Internet

In the Internet too, therefore, more and more regional and local "centres" will come into being, corresponding either to the existing topographical realities or to entirely new interest groupings. The world as it is – divided up and fragmented in a multitude of ways – is in this sense reproduced in the Internet, since from a linguistic or cultural point of view any excessive homogenisation or one-sidedness cannot remain stable in the long run. This means not only greater regional content (on a linguistic level), but more information stemming from a particular city, a particular district, a particular street. In the process there will be both new groups forming and old structures finding themselves reproduced on-line. Information about the regional weather is no less important here than the question (for example) of how much a pork cutlet costs at the two butchers down the road.

The real world will be mirrored in the future of the Internet

In accordance with this trend, the on-line service America Online (AOL) in July 1998 announced its intention to step up the element of regionalisation in its service (which had already been working with regional offers in the USA since 1995) in an attempt to increase its number of subscribers and advertising customers. From the program version 4.0 of AOL access software onwards, which appeared in late 1998, customers have been greeted with a starting page particular to their own region. AOL is in this way not only attacking the advertising market of the local newspapers, but above all moving more into line with the needs of its users.[265]

265 In Germany alone there were some 60 volunteers, so-called AOL scouts, producing regionalised offers for 57 German cities. As payment they receive free AOL access.

Source

Fig. 5.7: ECC 1999

Summarising Conclusion

There are three lines of development that can be discerned as regards new content in the Internet economy:

Three lines of development

1. The Internet opens up opportunities for forming new interest groups that can be provided with media content on an economically attractive basis. This content may be highly diverse or even marginal, but the point is that it taps into potential that has hitherto been left unexploited, namely medium-sized interest groups of fewer than ten thousand people. Such groups have not up to now been provided for separately by electronic media.
2. It is the games industry – whether on-line in the Internet, through games consoles or the home computer – that is currently developing the most successful concepts for designing interactive entertainment. The branch can thus be considered a trend-setter in the development of new content and formats in a networked media world.
3. The regionalisation and localisation of content is a further trend essential to the network economy. Instead of ongoing internationalisation, content and offers relating to one's immediate vicinity are becoming more and more relevant.

> *"The principles of the net, such as increasing returns, were seen as special cases, anomalies within the larger, "real" economy of steel, oil, automobiles, and farms. What did such weirdness have to do with, say, making cars, or selling lettuce? At first, nothing. But by now every industry (shoe retail, glass manufacturing, hamburgers) has an information component, and that component is increasing. There is not a single company of consequence that does not use computers and communication technology."*
>
> Kevin Kelly 1998, p. 70-71

5.1.6 Perspectives for the Internet Economy

Rules of Internet economy will spread across all industries

The development of the information and communication-based Internet economy is already pushing ahead at full tilt. Just as industrialisation reduced the extent of agricultural work and manual labour, so the industrial jobs of today will be transformed by digitisation and networking. At the end of the nineties, only 18 percent of all those in employment in the USA are still working in the industrial sector. And of this 18 percent three quarters are in turn involved in information-based activities such as administration and research, design and sales, marketing and law, where it is bits being dealt with, not atoms. Only a very small portion of today's working population is still actually active in industrial production, while the vast remainder of society is being slowly but surely transformed into a communication and knowledge society, which – with the aid of networks – will further develop into an Internet economy.

As nearly all companies react to the developments in information technology and integrate them into their business processes, the rules of the Internet economy are of direct relevance for most companies. The whole economy will thus evolve more and more into a networked economy. Even today cars are designed, developed and manufactured in a networked production process based on network technologies. Some car manufacturers are already reducing development times by passing data for a new car round the world "parallel to the course of the sun" – from one development office in the USA after eight hours working time to the next office in Asia and then on to the next in Europe. More and more jobs are being taken over and transformed by network technologies, until the whole economy ends up networked.

International, immaterial and interconnected communication is the mainspring of this development

Communication between and among participants is the mainspring of this development. Networking leads to the growth of a new dynamics of communication. Digitisation and networking not only improve and accelerate communication possibilities and the targeted exchange of information; they become the very determinant of the economy itself, centred as this is on communication or the exchange of information. There are three fundamental conditions essential to the functioning of a networked economy:

* it is international;
* it is based on immaterial ideas, information and relations;
* it leads to a comprehensive interconnectedness between all those involved.

The underlying idea that the atomistic worldview of the twentieth century is being replaced with the triumph of networks dealing above all in bits and not atoms leads to the simple recognition that the Internet (or the networks) is becoming the basis of our economy.

An analysis of the processes and requirements characteristic of such networks thus throws light on the rules determining how an Internet economy functions. Kevin Kelly, Executive Editor of the Californian magazine "Wired," has drawn up ten "rules of thumb" for a new network economy. Although these "rules" make no claim to be absolute or exclusive, they certainly present interesting perspectives on the sort of behaviour that will prove economically successful in an Internet economy.

New Rules for the New Economy

by Kevin Kelly, Wired Magazine, Executive Editor

1. Embrace the Swarm. As power flows away from the center, the competitive advantage belongs to those who learn how to embrace decentralized points of control.
2. Increasing Returns. As the number of connections between people and things add up, the consequences of those connections multiply out even faster, so that initial success aren't self-limiting, but self-feeding.
3. Plentitude, Not Scarcity. As manufacturing techniques perfect the art of making copies plentiful, value is carried by abundance, rather than scarcity, inverting traditional business propositions.
4. Follow the Free. As resource scarcity gives way to abundance, generosity begets wealth. Following the free rehearses the inevitable fall of prices, and takes advantage of the only true scarcity: human attention.
5. Feed the Web First. As networks entangle all commerce, a firm's primary focus shifts from maximizing the firm's value to maximizing the network's value. Unless the net survives, the firm perishes.
6. Let Go at the Top. As innovation accelerates, abandoning the highly successful in order to escape from its eventual obsolescence becomes the most difficult and yet most essential task.
7. From Places to Spaces. As physical proximity (place) is replaced by multiple interactions with anything, anytime, anywhere (space), the opportunities for intermediaries, middlemen and mid-size niches expand greatly.
8. No Harmony, All Flux. As turbulence and instability become the norm in business, the most effective survival stance is a constant but highly selective disruption that we call innovation.
9. Relationship Tech. As the soft trumps the hard, the most powerful technologies are those that enhance, amplify, extend, augment, distill, recall, expand, and develop soft relationships of all types.
10. Opportunities Before Efficiencies. As fortunes are made by training machines to be ever more efficient, there is yet far greater wealth to be had by unleashing the inefficient discovery and creation of new opportunities.

"New Rules for the New Economy" published by Viking/Penguin 1998
With permission from Kevin Kelly.

In what follows, various aspects of the new network economy will be discussed in the light of the theses put forward by Kelly.

Basic Premise: the Internet Economy is Decentralised

Today there are already some six billion chips at work in a whole range of objects from toasters to air-conditioning. This process will probably continue until virtually every object fashioned by the hand of man is equipped with a simple, disposable – since cheap – chip that performs easy tasks or sends out short messages. Networks are thus already coming into being in the most diverse of areas: articles of clothing are provided with chips so that their size and composition can be read off automatically. Cows have chips implanted under their skin allowing them to be allocated the optimal amount of food according to their condition. An average car already has more chips and a higher processing power than an average PC.[266] Work is already being done on how to incorporate browsers and further interconnections into the next generation of cars in a way that makes sense both for customers and the car industry.[267] Just as aeroplanes can be regarded as chips with wings, cars are likewise turning into chips on wheels.[268]

More computing power in a car than in a PC

Developments of this order are impelled a stage further when these chips are then networked up with one another. For just by exchanging a few pieces of information with one another inanimate objects are suddenly transformed into animated network nodes, capable of intelligent – in other words self-optimising – solutions. This is the motor of the Internet economy: a swarm of single, simple chips in communication with one another through a network produces intelligent, more economical solutions.

Networking distributed intelligence

The computer manufacturer "Silicon Graphics" effectively dissolved its entire administration programme by providing every single one of its workers with access to its Intranet, called "Silicon Junction." Here all the company's administrative processes, forms and internal procedures were kept in largely automatised form. In this way it was possible for day-to-day activities within the firm to be noticeably speeded up.

Research into artificial intelligence (AI) provides an example of the power inherent in the decentralised swarm. An AI experiment was designed where the aim was to use several small robots to gather a number of balls distributed evenly over a square surface and push them together into a single heap. The optimal solution proved not to be a complex centralised control system, but a simple programme where each individual robot moved around apparently "aimlessly," stopping and changing direction whenever three balls had accumulated in its shovel. The balls quickly gathered in a heap in the middle of the test surface.

The power of the (decentralised) swarm

Another example: in Berlin a number of glass-recycling containers have been fitted with a sensor chip and a transmitter that signals to the head office whenever

266 Kelly 1998, p. 72
267 For example a check on the most important engine data without the car having to go into the garage.
268 See Kelly 1997, pp. 140-197

a container is full. From the data received, the office then every day works out the best route to be taken by the lorry that empties the containers. Beforehand, recycling firms were perpetually struggling with the problem that containers were either too full, making emptying them an especially awkward business, or they were much too empty for the journey there to be worthwhile. The system is so successful that the experiment is now being extended to the city as a whole.

The future of computer technology thus lies in its disappearance. Through its invisible and ubiquitous presence and the exchange of information it facilitates, intelligent solutions can be developed. To this end, as many objects as possible need to be animated and connected in order to produce efficient results.

The issue of standards: Lock-in Effects and their Consequences

Standards make or brake the Internet economy

In the Internet economy standards are in general becoming more and more significant. "Interfection" is the term for the process of using existing standards from one business web, applying them to another network and then revolutionising this other network. Just as TCP/IP – the standard in data communication – is currently revolutionising voice transmission, so business webs can successfully deploy in completely new fields the standards they have developed.

Open systems enlarge potential

The temptation here is for companies to offer closed systems, intended to help impose the desired standard by securing the supplier the exclusive right of disposal for the system in question. Yet such closed systems are often less competitive than open systems represented by several suppliers, above all in their flexibility in response to the demands of the market. A lot of firms here fail to recognise that a small piece of a very big pie is usually more profitable than the greater part of a small pie. As demonstrated by the operating system Linux, the Web server software Apache or Netscape releasing its source code (on www.mozilla.org) for Navigator, system openness is thus one of the most important preconditions for growth in an Internet economy, in the end helping to exploit greater potentials than would have been possible given an insistence on closed systems.

The greater the number of networks with which a particular product or service can be connected up, the more powerful and profitable the product becomes. Owing to its great flexibility (for both data and voice transmission, as well as connection to the Internet), the telephone network is an essential service with a correspondingly high optional value. A situation here arises, however, that makes it increasingly difficult to alter standards once they have been established – a so-called lock-in.

Lock-In

by Carl Shapiro and Hal Varian

Visionaries tell us that the Internet will soon deliver us into that most glorious form of capitalism, the "friction-free" economy. How ironic, then, is the event that will usher in the next millennium: the dreaded Year 2000 Problem, a testament to the enormous rigidities that plague the information economy. Far from allowing for frictionless choice, information technologies are notoriously subject to switching costs and lock-in: once you have chosen a technology, or a format for keeping information, switching can be very expensive. In extreme cases, these switching costs can be so prohibitive that the user is virtually locked in to his current technology.

Lock-in arises whenever users invest in multiple complementary and durable assets specific to a particular information technology system. You purchased a library of LPs as well as a turntable. So long as these assets were valuable, i.e., the albums were not too scratched and the turntable still worked, you had less reason to buy a CD player and start buying expensive CDs. More generally, in replacing an old system with a new, incompatible one, you may find it necessary to swap out or duplicate all the components of your system. These components typically include a range of assets: data files (LP records, COBOL programs, word processing documents, etc.); various pieces of durable hardware; and training, human capital. Switching from Apple to Intel equipment involves not only new hardware; the software has to be updated as well. And not only that, the wetware – the knowledge that you and your employees have built up that enable you to use your hardware and software – has to be updated. The switching costs for changing computer systems can be astronomical. Today's state-of-the-art choice is tomorrow's legacy system.

Lock-in can occur on an individual level, a company level, or even a societal level. Consumers were locked-in to LP libraries, at least in the sense that they were less inclined to purchase CD players because they could not play LPs. Many companies were locked-in to using Lotus 1-2-3 spreadsheets, because their employees were highly trained in using the Lotus command structure. Today, at a societal level, most of us are locked-in to Microsoft's Windows desktop operating environment.

The great fortunes of the information age lie in the hands of companies that have successfully established proprietary architectures which are used by a large installed base of locked-in customers. And many of the great headaches of the information age are visited upon companies who are locked in to information systems that are inferior, orphaned, or monopolistically supplied.

"Significance precedes momentum."

Here are some critical points to keep in mind when thinking about lockin:

* Customer lock-in is the norm in the information economy, because information is stored, manipulated, and communicated using a system consisting of multiple pieces of hardware and software, and because of the importance of specialized training to use specific systems.
* Switching cost must be evaluated relative to revenues on a per-customer basis. Even small switching costs can be critical in mass markets such as the telephone industry or consumer electronics.
* Total switching costs include those borne by the consumer to switch suppliers and the costs borne by the new supplier to serve the new customer.
* As a rule of thumb, the present discounted value to a supplier of a locked-in customer is equal to that customer's switching costs, plus the value of all other advantages enjoyed by the incumbent supplier based on lower costs or superior product quality.

"Information Rules: A Strategic Guide to the Network Economy", Harvard Business School Press, 1998, by Carl Shapiro and Hal R. Varian. Web site: www.inforules.com.

With permission from Carl Shapiro and Hal Varian

Significance Precedes Momentum

It is difficult to ascertain clearly the decisive factors that determine the success of a network (or its standard). While the first supplier of a network service as a rule enjoys competitive advantages ("It's better to be first than it is to be better."),[269] the further course of events is always "path-dependent" – which is why in spite of possible lock-in effects individual networks can quickly also become obsolete. Who now knows the operating system CP/M?

Exponential growth of standards' diffusion

A characteristic feature, as in biological systems, is exponential growth. Having been scarcely perceptible for a number of years, a network can then suddenly "explode" in its number of participants and its (economic) significance, in the process becoming externally visible. The best example here is the development of the Internet itself. For decades it was a small, obscure network for a handful of academics. Following certain improvements (HTML, browsers), the number of participants exploded. Like the Internet, companies dependent on network effects such as Microsoft or technologies such as the fax may well simply "make up the numbers" for a long time before their market suddenly takes off with a bang. These are examples of "biological growth" in technological systems. The significance of the development of such network markets always precedes its

269 See Ries/Trout 1993

discernible momentum. It is in general difficult to predict in advance and as a rule only lends itself to analysis in retrospect.

Let go at the Top – after Success Comes Devolution

Companies are highly inventive when it comes to defending their technologies. Whole varieties have been on the point of perfection when, like dinosaurs, they have become extinct, superseded by entirely new technologies. Electronic valves, for example, had been developed to the level of an elaborate high-performance technology before being replaced at one fell swoop by the much more powerful and durable transistors. Computer chips today have a similar product life-cycle, described by Moore's Law as an innovation-cycle lasting from 18 to 24 months at the most. After this time the chip will have become largely obsolescent and can only be sold on the cheap. Up until recently, Intel was always able to defend its high profit margins by having new generations of chips brought onto the market before competitors could copy them. Rivals now appear to have caught up with them.

Fast innovation cycles mean giving up acquired potentials

The dynamics of innovation also holds considerable problems for the companies concerned. Successful enterprises thus tend to stop paying sufficient attention to potential rival products from other fields. In his book "Only the paranoid survive," the Intel CEO of many years' standing Andy Grove thus emphasizes that it takes extreme attentiveness and quick reactions to secure a company's long-term survival. Especially when it comes to moments of decisive strategic importance, such as the invention of the transistor for the valve industry or the establishment of the Internet for the software industry, it is essential for market leaders to make the necessary changes to their products and strategies in order to be able to survive in the market.[270] Otherwise today's winners will very rapidly become tomorrow's losers.

Today's winners can be tomorrow's losers

Along with the skilful construction of a business enterprise, Kelly accordingly claims that there is a need for a form of creative destruction, even going as far as the appointment of a "Chief Destruction Officer."[271] However, this destruction undertaken for the sake of creating something new only rarely has any success in old structures, especially since most industries are organized with the specific aim of optimally defending their territory. For this reason it is often the case that established firms entering a new market spark off more innovation than the old market participants. Frequently indeed the decisive push for innovation even comes from complete outsiders. It was not IBM who invented the PC but (amongst others) two students who founded Apple Computers. It was not the American cable concerns with their plans for an information superhighway who invented the WWW, but a series of researchers in the European nuclear research centre CERN in Geneva. It was not the car manufacturers who first stood up for the construction of a small car, but the watch-and-clock concern SMH.

Creative destruction

270 See section 3.3
271 Kelly 1998, p. 82

"Intranets dismantle hierarchies."

In the Internet economy with its rapid innovation cycles and product life cycles, the decisive competitors will thus come from other branches and fields. Stability, it seems, no longer has a place in such a high-speed economy. The only thing stable is the continuous change. And it is precisely by fostering the constant flow, the imbalances and disparities, that a contribution can be made to stable development. Between the poles of chaos and order, firms will thus in future find themselves operating more and more in the realm of flux.[272]

Operating in the realm of flux

New Work Organisation enabled by Information Networks

Relations between several people can be organized in various ways, for example hierarchically or as a network. Up to now hierarchies have frequently served as the form of organisation in business enterprises since they successfully even out differences in information. In a networked world, however, hierarchies seem to make less and less sense. As the real-time dissemination of information through internal and external networks becomes progressively easier, groups and teams increasingly become the dominant form of organisation within a networked working world, with information now available to all members simultaneously. Intranets thus help bring about the dismantling of hierarchies.[273]

Changes in work environments and methods

In addition to this, an Internet economy can actually promote new forms of work that operate beyond the classical industrial models. Many of the knowledge-intensive branches, especially in the fields of information, communication and entertainment, will in the long run presumably tend to adopt project-oriented methods of production, for speed and flexibility of output are higher when organised in this way than in large-scale industrial structures. [274]

5.1.7 Hang on a minute... – Critical Comments

The Problem of Critical Mass

Critical mass refers to a condition of imbalance where the number of network participants drops back down if there are not enough of them to turn the spread of a network into a self-generating process.[275] Up to now discussions about how to overcome such critical mass problems have focused mainly on network externalities.

Four starting problems of network goods

Yet the much-discussed teething troubles of network goods involve considerable coordination difficulties that even the marketing activities of experts and suppliers cannot completely dispel:

1. There is no incentive for a consumer to buy a network good if he cannot be sure whether a sufficient number of further customers will materialise.
2. The teething troubles of network goods are directly connected to customers' willingness to pay. For this reason it is essential that the initial price for the network good should be geared to the number of willing-to-pay users that

272 Kelly 1998, pp. 104-109
273 See Kelly 1998, pp. 112-114
274 See Kelly 1998, p. 105f.
275 Rupp 1996, pp. 13-15

permits the critical mass phenomenon to be overcome. Subsidies can here prove helpful, allowing an attractive price to be charged.

3. Constantly falling prices lead to "leapfrogging," which impedes the initial spread of the good. Potential buyers wait for a long time before making a purchase in order to skip whole cycles of technology.
4. Market penetration problems for network goods can also be solved by arrangements formed between participants or by international committees.[276] Yet new problems can come about in the process if there is a clash of interests between the participants or if the transactional costs that arise are (too) high.

Limits to Growth for Network Goods

Under certain circumstances the network as a whole may experience a decreasing marginal utility once (either temporarily or permanently) a certain size or spread has been reached. This limit may curb, if not completely counteract, the attractiveness of the network and thus also the incentive for participating in it or further extending it.

Optional rather than actual value

Up to now in the Internet economy the argument has been that network externalities mean direct network effects for each additional participant. In fact it is usually just the optional value of the network not the actual value that increases. In the end – and here above all regional or national as well as linguistic factors come into play – network externalities too reach a limit, defined as when each additional participant no longer produces a real increase in utility for the other participants. The increase is at this point only marginal, and later only optional or even negative if additional users no longer have any effect on the utility enjoyed by the rest of the participants or if they intensify a technological lock-in.

Smaller increases of network value with larger number of participants

According to network economics theory, the value of a network such as the telephone network rises with its number of participants. The smaller the likelihood of a telephone user actually making contact with the additional new users, however, the smaller the increase in value for the user becomes. An example here might be the (as yet theoretical) idea of fixing the whole of China up with telephone connections. For the average Central European, these connections are of relatively little importance, as there is hardly anyone with a sufficient reason for ringing up an average Chinese person. Only if there were to be an increase in individual tourism or a marked expansion in trade with China might there be a certain optional increase in utility – provided both parties spoke a common language. The marginal utility of developing the telephone network in China thus tends towards zero for Europeans. To the extent that it provides the pure potential for getting directly in contact with any Chinese person at all should the need arise for whatever reason, such a network nonetheless does have optional value.

276 For example MPEG, ADSL Working Group, W3B-Consortium etc.

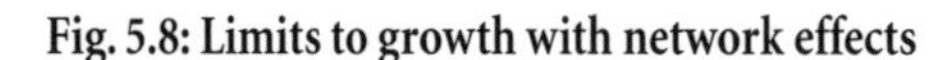
Fig. 5.8: Limits to growth with network effects

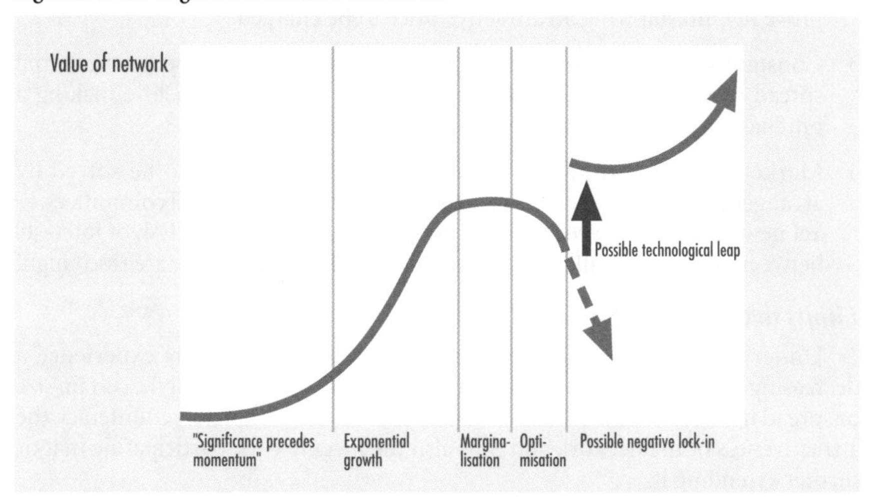

Much greater problems, however, are posed by the lock-in situation, which can even end up reducing the value of a network good. This would be the case, for example, if a more powerful telephone technology is to be introduced worldwide as the new standard but this is prevented in the afore-mentioned instance of China because the country is unwilling to meet the costs of conversion. At this point the network externalities are being cancelled out and reversed by lock-in effects. In such situations, however, it often happens that a technological leap occurs (for example from analogue to digital telephones), allowing a new network to skip the limitations in place beforehand. In the case of the telephone network, an example would be the development of IP-based telephony after ISDN.[277]

Reversal of network effects

The above example makes it clear that a derivative of the product-life-cycle familiar from traditional economics can also be described for network goods in the Internet economy. The difference is that growth curves for network goods take an exponential course, as do possible decline curves. The limiting elements stem – as seen above – mainly from regional, national or linguistic factors that hinder or prevent the further spread of the network good and thus put an outer limit on its potential for growth.

Limits to growth also true for networks

A Problem with Network Externalities – Natural Monopolies

Network externalities often lead to natural monopolies. The market for operating systems is today dominated by Microsoft, the market for network routers by Cisco and the market for databases by Oracle. A functioning Internet economy means at best that the winner takes most, at worst that the winner takes all, as in the case of Microsoft.[278]

277 See section 3.2
278 See Kelly 1998

Source

Fig. 5.8: ECC 1999

"A lot of dumb chips networked up do not necessarily result in intelligent behaviour."

What is paradoxical about these near-monopolies is that they are often considered desirable by the users themselves: from the customer's perspective inter-operability and compatibility are valuable qualities in a networked world. Standards are honoured because they raise the network's value. The greatest problem with such network externalities, however, is above all the possible check they may put on innovation and the concomitant suppression of optimal solutions. Along with network-winners there are thus almost always network-losers too. These are the companies that may have even brought better products onto the market but that have proved unable to assert themselves against the network standards.

A paradox of the Internet economy: Network effects and "natural" monopolies

Limits to the Power of the Leaderless Swarm

A lot of dumb chips networked up do not necessarily result in intelligent behaviour. The more participants there are in a network, the more important it is for there to be guidance along with the self-organisation. Especially when there is a large number of options, things can easily come to a standstill if there is no system of preselection. The Internet protocol TCP/IP needs databases to preselect the best routes for reaching individual addresses.

A need for guidance

A further problem is that the advantages of the swarm only set in when there is comprehensive networking of all participants. Networking only half the participants fails to produce the desired network advantages, since the old infrastructure models still need to be maintained. Half-heartedness thus prevents network advantages from ever coming to fruition, instead creating additional costs and increasing friction.

Networks must be comprehensive

"The 'communication revolution' has not reached the household yet."

5.2 An Integrated View of User Behaviour

5.2.1 Basic Ideas and Fundamental Principles

User perspective: clear distinction of different media

Ask the individual citizen which of the media they use for what purpose, and even in the age of the "communication revolution" you will get a relatively straightforward reply. Television, radio, newspapers, magazines, as well as computers and telephones are from the viewpoint of their users all clearly differentiated media or instruments for dealing with distinct day-to-day situations. There is very little confusion or overlap, at least in the minds of the people who use them. Interestingly enough, the analysable convergence of different economic players, the technical integration of diverse electronic systems on the basis of digital technology, and the continuous intertwining of value chain processes have found hardly any echo in the daily consciousness of consumers.

Individual consumers, in households and families, are unlikely to be concerned with more than ease of use and cost, when it comes to the use of new media and the services they increasingly offer. Convergence in the household will be a matter of an unselfconscious capacity to move between interfaces according to perceived need and desire. Consumers need neither to know about or understand the technology. Their rationality will be defined and constrained by the culture that sustains them in their everyday lives. That is enough, but it is a huge challenge for providers of such services who will value speed and simplicity of access, system responsiveness and an unfailing level of service as preconditions for full commitment to the electronic age.

Rhetoric of convergence differs from consumer experience

Experts were already predicting the integration and convergence of technical and economic systems decades ago, in the process coining such lexical monstrosities as "telematic integration." They put the increasing dovetailing of systems down to the underlying "digitisation" of data, content or services. Yet the most important player in these markets – the private customer – has continued to view the rhetoric of convergence with a mixture of bewilderment and equanimity. On the way into the new epoch, the private customer has simply been confronted with new pieces of equipment or service offers, incorporated into day-to-day dealings (if at all) only as something complementary to traditional media uses.

Comprehensive communication management – too much to ask for

The fact that television can now reach the individual household by cable as well as satellite forces the private user at most to do a few financial sums regarding the purchase of new appliances or the monthly costs entailed by the necessary connections or services. And while the modern telecommunications customer is bound to be concerned that the systems should be comparatively easy to use, this does not mean he or she will have an overview of developments as a whole. Integrated, comprehensive communications management is as a rule asking too much of the individual household. In the old days Deutsche Bundespost and later on its successor organisation Deutsche Telekom relieved customers of decisions

relating to infrastructure. Even if new network and service providers or even house owners or property management are now coming to assume responsibility for the provision of technology and channels, the matter of getting connected to the ISDN, for example, is only rarely associated by individual citizens with the question of their future media provision.

As yet, the communication revolution has only reached the household in rudimentary form. Private customers are sticking to "business as usual" in managing their finances, even though the potential for making use of new media and technology offers is staring them in the face and new value chains are entering the individual household more and more directly. Contrary to common belief, however, the household itself is not characterised by transparency, nor do its members always have a clear overview of the possibilities on offer. Instead, the rationality of household activity and the complexity of household functions tend to go begging in the bustle of random day-to-day emotional dealings: "Here I am a human being, here I can be myself." In this context there is little room for making clever strategic calculations or rationally planning one's technological media and communications relations with the outside world.

The dilemma posed by the functional diversity of multimedia applications for the private household and the myth of problem-free convergence

The development of the media tends towards ever greater diversity in the appliances and services on offer. The choice is growing by the day and markets are changing and expanding. Target groups are shrinking, making it more difficult to bundle together large markets. This affects the economic and political capacity for building up and financing extensive new infrastructures, even though these have now become technically feasible. Essential to any explanation of the development of future markets through the changing behaviour of household members is a suitable understanding of new trends in the evolution of modern systems of communications. The following model of "dedicated communication" attempts just this.

Four basic functional areas of the private household

The household can in principle be divided into four functional areas:

* organisation of daily life
* social networks
* entertainment and relaxation
* news and commentaries

Four clusters of media at the interfaces

Within these four areas, the interface between any two sectors has always seen specialised media take on a specific function. Four clusters of media thus form, numbered from 1 to 4 in the diagram below:

Fig. 5.9:Fig. 5.9: Model of dedicated communication

Social engagement – World consciousness – Quality
Dedicated communication
Personal and intensive
Organisation of daily life
① Newspaper
Teletext
WWW
WebTV
On-line services
③ Telephone
Letter
E-mail
Fax
Answering machine
Mobile telephone
News and commentaries
Individual
Media in
private household
Immediate and/or direct
The realm of individual choice
Social network
The realm of social control
② Radio/TV
Book/Magazine
Video
Record/CD
CD-ROM
④ Video-/Computer games
Chat-groups
Multi user dungeons
Averaged mass communication
Entertainment and relaxation

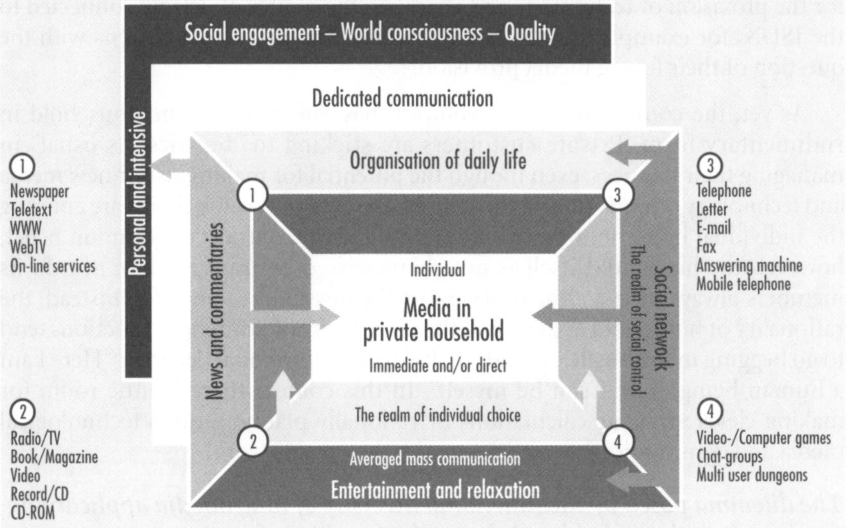

The model of dedicated communication attempts to do justice to the diverse and at times seemingly contradictory trends of individualisation of selection decisions on the one hand and the re-communalisation of lifestyles induced by communications technology on the other.

Dedicated communication

The model describes two fundamental axes of movement. One of them refers to the "implosion" from the abstract and general to the individual and personal. This can be considered as a movement inwards on the first coloured rectangle. The arrows describe the orientation in relation to the centre of the diagram. This central axis leading inwards thematizes the greater selection possibilities in the realm of individual choice for multimedia electronic products. The other axis, which runs from the foreground to the background, takes up new themes such as the growing demands concerning the quality of communications products and the desire for further increases in the level of personal experience. The second axis is produced when the diagram is conceived in three dimensions as three rectangles superimposed upon one another, with "dedicated communication" (white level) mediating between the first (coloured) and the third (black) levels as a catalyst for both development axes. The afore-mentioned developments take place simultaneously within the same space/time continuum.

Individualised selection does not rule out collective experience, and a desired trivialisation or even primitivisation of programmes on the one hand may well be associated with an equally intensely requested increase in level on the other.

Source

Fig. 5.9: Lange 1998

Dedicated communication – i.e. forms of communication not only offering greater selection but at the same time also more passionate and intensive – will in the long run come to supplement or replace traditional forms of mass communication, which in spite of their richness of information and their entertainment value have always lacked a certain personal touch through catering to the mean. The model of "dedicated communication" makes it clear on the one hand that communication geared to the masses has more or less had its day and that the differentiation of electronic media products opens up more individual selection possibilities and thus also a new realm of more personal choice. At the same time, however, there is a rapidly growing desire for more intense personal experience in new, individually chosen or individually created communities that call for wholly different, more flexible contexts of media-usage.

Communication geared to the masses more or less had its day

Continually being bombarded with messages, we no longer want to be immediately indulged or disturbed in every situation. Increased flexibility is offered by a technologically determined but individually chosen personal communication scenario. Technology should thus present us with effective buffers and possibilities for at times uncoupling from the system while also broadening our spatio-temporal connections with our social surroundings, so that – without being tied to any single medium or technology – we are nonetheless in a position to carry on relatively variable contact and communications management. This is where the genuinely new markets are to be found.

Genuinely new markets in variable contacts and communications management

The integration or convergence people have been looking for in recent years is to be understood less as a technological phenomenon than as a social construct, therefore, although the economic and technical advantages of synergy effects are certainly also viewed with approval.

The will to the Other: personality as a key to social, economic and technological convergence

The increased use of personalised media will be the social, economic and technological trend of the coming decade. The success enjoyed by the mobile telephone, for example, does not lie exclusively in the ubiquitous use-possibilities its technology makes possible, but also in the fact that it presents each individual with a world of personalised communication that unlike the letter is direct and immediate. Before the personalised communication system comes the personal communication appliance. The development towards personal and then also personalised appliances is unmistakable. The use of personal computers had already been preceded by personal television sets – the second TV set in the household. And in upper class families it was common early on for individual family members to have a telephone of their own in their personal room. While the personal appliance is characterised simply by a personal right of ownership or disposal, the personalised system also features a certain degree of knowledge about its user. It is adapted or geared to him like a good friend. Through the numbers it has in its memory, the mobile telephone enjoys a "knowledge" of its

The difference between personal and personalised appliances and systems

user's personal circle of friends, and through the type of memory in operation or the use of the quick-dial number memory it may even impart some idea of the user's communication priorities and personal communication preferences.

"Learning" appliances

The last-dialled number provides a record of the user's personal behaviour, and the mobile telephone's number is not a household address but rather a possibility for direct personal contact. The mobile travels with its user, and its "intelligence" is growing from one (technical) generation to the next, turning it into a contact- or info-man that handles our personal information and communication. Our cars, fridges, cookers and even our bicycles too – according to the wishes of some of the researchers at the MIT Media Lab in Cambridge, Mass. – will have to learn more about us in order to be able to serve us better.

New types of demand for personalised services

Media suppliers have up to now made use of market and audience research in the attempt to get as faithful a picture as possible of our immediate needs and interests. With the new intelligent and personalisable systems it will no longer be absolutely necessary to take this circuitous route. In future the cynical user will instead perhaps be able to say "My computer knows more about my behaviour than my own wife does." New markets can be opened up in this field, and the new demand for personalised services will lead to new cash-flows for the competitive suppliers of such products. It is thus a mistake to ask how to go about actively attaining more entertainment or information in the context of new communication systems.

A new trend towards rational passivity

The interactive user of the future is often misinterpreted as a universally hyperactive and highly selective person. The somewhat more comfortable criterion of choice for the calmer consumer could just as easily run: what contribution does a system make to stimulating or encouraging third parties to take me and my needs and wishes seriously and thus do something about satisfying them. New communication technologies are thus only in part instruments for active intervention in knowledge and entertainment culture. They also help third parties orientate themselves in order to be able to attend to the user's personality. This is the preparation for a new "knowing" movement of rational passivity, which lets fate simply run its course in the interests of well-understood consumers, rather like the angler who with a good bait knows how to captivate his victim using the minimum of energy.

The more hedonistic principle of passively waiting for the desired message is something that has been one-sidedly excluded from the development of new media offers owing to a misunderstood work ethic and work culture, although it is precisely here that the whole secret of success lies for experienced suppliers of traditional mass communication products. The ranks of those actively seeking information – formed largely by the "educated classes" – thus tend to underestimate the role played, for example, by news and chat groups in opening up new communications markets and it is only rarely that these are incorporated into the Internet presentations of firms or institutions with any strategic

astuteness. These new forms of communication thus include not only actively sending messages but also the special happiness of receiving messages that have been hoped for (as well as the frustration or despair stimulated by receiving messages that had not been hoped for, or were entirely superfluous, as well as the failure of the one expected and important message to arrive). Interactive systems should therefore not automatically be connected with the vision of a more active and engaged consumer. The market opportunity presented by having greater knowledge of users also lies in passively relieving them of responsibility, something so frequently sought after and enjoyed by consumers in daily life.

Economic relevance of news-groups and chatrooms

The concept of personality is also linked to visions and illusions of power to which we are all subject. Power and the certainty of its exercise become marketable aspects of media and communications technology, insofar as through money one has data, information, knowledge and in the end also people at one's disposal. Services based on communications technology create access, and in particular a certain voluntary or involuntary access to people being sought after or identified.

Personality and power

Media communication is not used merely as a means of seeking real or virtual access to acquaintances, for it lures us likewise with the adventure of the unknown and the unexplored. Stories, letters, newspapers and books have long since been making foreign continents accessible in our minds. The delays in time they inevitably entailed were in turn overcome with the supplementary presence of telephone, radio and television. The vision of telecommunications is the overcoming of distance; its realisation the creation of a new proximity. But its impetus is also what is unknown, what is distant and thus also what is interesting – whatever drags us out of the normality of daily life.

Old and new windows on the world

Through our personalised devices we "reproduce" ourselves and so broaden our contingent sphere of availability and reception. The welfare and development of our personalised appliances thus become a major impetus to our own behaviour. "Let your fingers do the walking" was a succinct slogan for telephone-based telecommunications. Now there are personal agents in the Internet who play the role of looking after our personal interests. Whereas before the invention and application of agents in the Net it was not yet possible to transfer the user's own virtual personality into the telecommunications network as a sort of learning traveller, this has now become one of the essential cornerstones of the communication revolution. Technology provides us with new artificial antennae for feeling our way into the world independently of our actual location, with less effort and fewer risks. The user just wants to be there for the harvest – in order to pick the fruits of his search in person.

Software agents looking after personal interests

The avatars appearing in the Net as electronic butlers, slaves, riders, messengers, warriors and ambassadors, for example, turn every personality into a mini-state that develops in accordance with its own rules and laws. We invest the technologies we use and the appliances we operate with our own personality in order to bring them even closer to our actual needs. Yet the better and more

Playing with personalities: Atavars

"personally" we are able to satisfy our own needs, the greater will be the boredom when we look at our mirror reflection. It is for this reason that we need the Other – other persons and other things.

Communication and de-communication

Once a personalised connection has been created, however, it is difficult to circumvent it. The closer a technical appliance comes to a person, the greater the risk of this person feeling pressurised or harassed. Connected with this trend towards personalisation in new media, it is therefore essential for there to be a compensatory element of de-communication and the possibility of periodically withdrawing from the communication system. Personal immediacy is optionally replaced again and again by spatio-temporal directness: we remain within reach – by answering-machine, e-mail or fax – and yet are not always so, and thus are less compulsively so.

Towards a new equilibrium of communications

Time and again, new social limits are placed by users' personalities on the technological principle of convergence dictating that everything seems to be inevitably flowing together or is being artificially brought together. The greater the extent to which people consciously isolate themselves, the greater the need for neighbours. The more that people form bonds together, the more urgent becomes the option to separate again. This is the communicative equilibrium that will come to play a decisive role in the description of new media and markets. People require breaks, interruptions and boundaries to become conscious of their own roles and interests. The active media-user of the future, creating his own connections, will be just such an active personality, in turn provoking new separations and new severances. This applies to the professional as much as to the private field. Active and passive use of media, of course, is not mainly an issue of different personalities, but rather one of different preferences under varying moods and circumstances: the engaged user constructs a media centred culture which has moments of activity and moments of passivity, times for involvement and times for withdrawal within it.

Professional uses continue to be the decisive socialisation field for private multimedia applications

Low level of optimising the household work-flow

Up until now the "multimedia" value chain – insofar as it has concerned transaction-based services and the marketing of products in interactive electronic distribution and mediation systems – has been located almost exclusively in the business sector and in trade between firms. While the business communications system has been founded on the continuous and often profound revolutionisation of all work and administration relations, the "rationalisation" of the household through innovations has been plodding along on a rather lower level. Organising one's time and optimising one's work-flow appear to be far less important in the private household than varying and intensifying one's personal experience.

To this extent individual citizens and consumers are perfectly innovative in daily private life and in special situations they recurrently prove to be creative as well. Yet the economic colonialisation of the private sphere has meant that the domestic services formerly provided for free by relatives, neighbours and friends (services such as care, attention, conversation and love) are now extensively marketed in return for money, and in view of this the household has shown itself notoriously reserved and sceptical towards private acceptance of the all-embracing technical uncertainties faced in daily professional life. The absence of economics and the refusal to accept the primacy of time betray a collective wish for a special quality of life in individual everyday dealings.

Values of professional life are different from those of personal life

In this context, the economic process is characterised by other spheres of perception, thought, experience and action. The economic sector must therefore, if it hopes to continue to be successful in entering this field, deal more intensively than hitherto with people's personal strategies of "collective" behaviour and their dissociation from "rational" and "functional" choices. The apparently (or seemingly) irrational behaviour in everyday live does have a rationality of its own. The failure of data and communications technologies to spread further in private life seems to have a lot to do with the fact that instruments such as the computer arrived on the scene in everyday professional life as "efficiency boosters." It is interesting that this evaluation of technology has itself been contradicted by the gap between wish and reality. Computers – those most effective of machines – are themselves often used highly ineffectively.

Efficiency not the prime objective

This lesson learnt in professional dealings could hardly escape the attention of the individual citizen in his capacity as a private person. Even when he actually does want to "rationalise" in his personal surroundings, the critical consumer will not always – and possibly in future less and less frequently – resort immediately to a solution only available in the short term. This is something learned from the excessively rapid succession of generations and the infantility of a number of "final" technological solutions. The convolutions resulting from the many provisional solutions produced by the "informatisation of the professional lifeworld" are a crucial experience that will influence the integration of professionally deployed technologies into the private sphere.

The business sector stands under the primacy of objective time structures, while the private sector follows the dramatics of subjective time. From the viewpoint of company logistics, wasting time on "useless" transactions in the private household is virtually incomprehensible. The deployment of transaction-based electronic systems thus becomes particularly important in those households that have come under a certain external pressure in the private "use" of time and with problems in the organisation of their personal agenda. This is most visible in those households which have a teleworker in their midst, for that the whole household has to become a more "rational" place.

New products in individual and personal time structures

Is there a consistent multimedia portfolio?

There is no such thing as a "consistent" multimedia portfolio

One of the central questions, which if answered could make it possible to assess the volume of future markets, relates to how much a month the individual consumer or household is prepared to spend on multimedia services. Now it has already been argued here that there is no classical multimedia service and that as an abstract construct the private multimedia budget is something artificially composed by simply adding up various flows of payment. In the minds of the consumers there is no self-consistent multimedia portfolio.

In calculating a differentiated national account, however, statisticians are confronted with the task of consistently and clearly defining and classifying the categories. From this perspective the most advisable course of action may well be to attempt to work with the model of a socially average basket of multimedia goods. From the perspective of the individual household, on the other hand, the criteria according to which some expenses should be included in this budget and others excluded are variable.

Multimedia budgets are statistical artefacts

Since the normal citizen lacks a multimedia overview, he or she is also without a coherent awareness of multimedia costs. Here at the latest it becomes apparent how impractical the concept of "multimedia" is for planning the individual household budget. A statistically "objectivised" budget generally includes expenditure on television, video, radio, stereo system, computer and telephone. Others go as far as to count what is spent on newspapers, magazines, sound-carriers and even cinema-going in such a "multimedia budget." Yet although a standardised basket of goods along these lines can provide a good idea of slight fluctuations over a longer period of time, it is the regroupings and restructurings within the budget that in the end prove to be the better indicators of the growth and decline of specialised and segmented markets.

Long-term and short-term relationships between uses of different media

The use of each particular medium thus tends to follow a logic that describes the utility value of the medium in the everyday life of the individual in question. In order to undertake a sensible allocation of means to their appropriate ends, attention should thus be paid to a general and abstract need for entertainment or information rather than to a generalised yardstick such as money. When a newspaper becomes slightly more expensive, this only rarely has any effect on television use, and the same applies vice versa. Short-term fluctuations in the price of a particular form of media seem to have little influence on actual selection decisions. If, however, there is a long-term increase in interest in a specific media form noticeable on an individual level, and clear fluctuations then arise in this field, there will indeed be reason to expect a degree of selection and de-selection given a limited budget of media and leisure time.

Even the interrelation in the choice between leisure time and media usage is not without problems of its own. A visit of several hours to an open air swimming pool can indeed be compared in financial terms with the reception of a sports programme lasting several hours or even with a long telephone call between

friends. What cannot be known, however, is what actually lies behind such a selection decision as the standard of comparison within the individually contingent range of possibility. As a general trend the signs are that the budgeting of time is continuing to gain in importance relative to the budgeting of money. This can be explained by the long phases of prosperity and the growth of private funds that have had a determining influence on the reconstruction of almost all European countries since the second world war, yet surprisingly enough it continues to apply even in the context of relatively stagnant household income. The decision to buy or the individual willingness to use a service is not always founded on the existence of a clearly identifiable "finished" product. Frequently it is anticipated "options" that are measured rather than actual "use." The list of chemists in one's local district together with the timetables for their after-hours service, the directory of all telephone subscribers, or the timetable for local and long-distance transport facilities may never be used by many of those who take on interactive use-options. Yet the point is to have these options in store and kept as up-to-date and accessible as possible in case a situation should arise in which they are actually required. The challenge in terms of a return on investment is particularly great here, since the fixed costs of providing the service hardly vary according to whether it is used or not. In order to be able to provide a clear budget for products and services for private households, an unambiguous classification of their possible uses and budgetary value is absolutely essential. The more accurate the designations chosen for the services and appliances offered, the easier it will be to identify them in a modular "bouquet" or "bundle" for the potential buyer or user as a value added component in their own right.

Options – a challenge in the Internet economy

Forms and classes of private media use – elements of an integrated offer

It makes sense to define classes of data provision and communications, entertainment and information offers that can at least be differentiated along the axes of "regularity," "frequency," or "duration" of usage, as well as the "intensity" of the attention that is paid them. The diagram below attempts an "ideal" classification based relatively arbitrarily on average assumptions, even though such evaluations can differ greatly from individual to individual and from medium to medium. The arrows in the diagram indicate trends of future development.

Time and media use

Fig. 5.10: Model of media use from a temporal perspective

		regular		irregular	
		more frequent	less often	more frequent	less often
longer duration	intensive	TV (general)	Magazine (general)	Telephone (relational) Soundcarrier (CD/MC)	Events (cinema, concert, theatre, sport) Video/book Computer user Surfing, chatting
	Superficial	Radio (general)		Sports in stadium Music in pubs	
shorter duration	intensive	TV news Radio news		Telephone (functional) TV and radio advertising Teletext	Mobile telephone CD-ROM Internet use
	Superficial	Traffic news		Magazine (adverts) Poster/Placards	Internet advertising

It remains completely open whether specialised appliances such as telephones, radios, televisions, computers or stereo systems will survive in a multimedia environment on the basis of already acquired habits of use or whether the future will see communications technology appliances that combine various functions. Up to now what has become apparent is that function-specific solutions with a clear use-profile are able to impose themselves more easily and more lastingly. In the case of certain technologies and appliances such as the mobile telephone, however, there is already a pronounced trend towards an integration of functions: the organiser, the mobile fax, and e-mail access seem to "grow into" the device as if through a natural process.

Will specialised appliances survive?

Even if there is at present still no acceptable interface for the reception of dedicated news in the mobile telephone and the costs of using mass services continue to be somewhat too high, it is in the long run perfectly feasible that a mobile phone will in future receive the data in its display not only through an expensive specialised network but also in parallel – as a cheaper option – from other sources of mass communication transmitted via satellite or terrestrially. Another particularly interesting question is whether such an integrated mobile might be able to play a special role in financial and other transactions, for example recharging an electronic purse or wallet.

Source

Fig. 5.10: ECC 1999 based on Lange 1998

For most users, the PC is still not a multimedia machine. It is something that has developed from the world of work, and this is characterised, in Europe at least, by clearly demarcated functions. Shaking off this aura of strict functionality – with its main use plainly as an instrument for processing and storing texts or alphanumeric data – will be hard work indeed for the PC in the next few years.

The "desktop potato"

The passive couch potato is often contrasted with the active computer user. Yet at least since the discovery that there is no inner order or guiding structure of meaning when surfing the Net there has been a new awareness of the Internet suction that can turn people just as easily into desktop potatos. If the computer gains new entertainment elements through games or chats, then it makes little sense to rule out TV use on the computer. Just as the TV set can gain interactive functions in the context of Internet TV solutions, so the computer too will doubtless be increasingly used for entertainment purposes.

The bumpy road to more choice: the "selectivity" dilemma, or the responsible user between acceptance and habituation

Like the selection decisions for or against particular technologies in private daily life, media use is only rarely chosen rationally. The field of traditional media use is often characterized by paradox. This is brought home by the empirical results from a comparative European study of five countries carried out on behalf of Telecom Italia.

A parodox situation between perceived programme quality and frequency of use

In Germany for example, it emerges that older people claim to be watching more and more TV, although at the same time they assert that programmes have become worse and worse in recent years. And while younger people in the same survey more frequently claimed to be watching less and less TV, they nonetheless think that if anything programmes have improved. One of the reasons for this is certainly the extent to which the opening up of new markets in the last few years has been patently tied up with a fetishisation of young people. Elderly people have been ignored as a consumer group and more or less forgotten. The balance has recently begun to be redressed in this area. This new trend towards a "silver market" for older people is connected with a further development model on the part of technology.

Technology, like the media, can be regarded as assimilated and integrated into daily life when its use or application no longer necessarily involves a conscious awareness of it on the part of the user. It must be possible for the user to grow so "used" to it that – as with TV – it is hardly even necessary any more for him to make a direct and immediate selection decision in any particular usage-situation. In the future there are thus sure to be products on offer that are simpler and easier to understand, "plug & play" solutions that even without prior knowledge can be connected directly to old and new networks alike and are correspondingly easy to operate.

Source

Fig. 5.10: Lange 1998

Two divergent audiences

Younger people have fewer difficulties with new technologies and are more curious to try out new media content and forms. They are also – within the context of their own basic media orientation – more discerning and less habitual media users. Two basically divergent audiences and consumer groups have thus come into being, each of which will have to be won over with different products or services:

Fig. 5.11: The "selectivity" dilemma

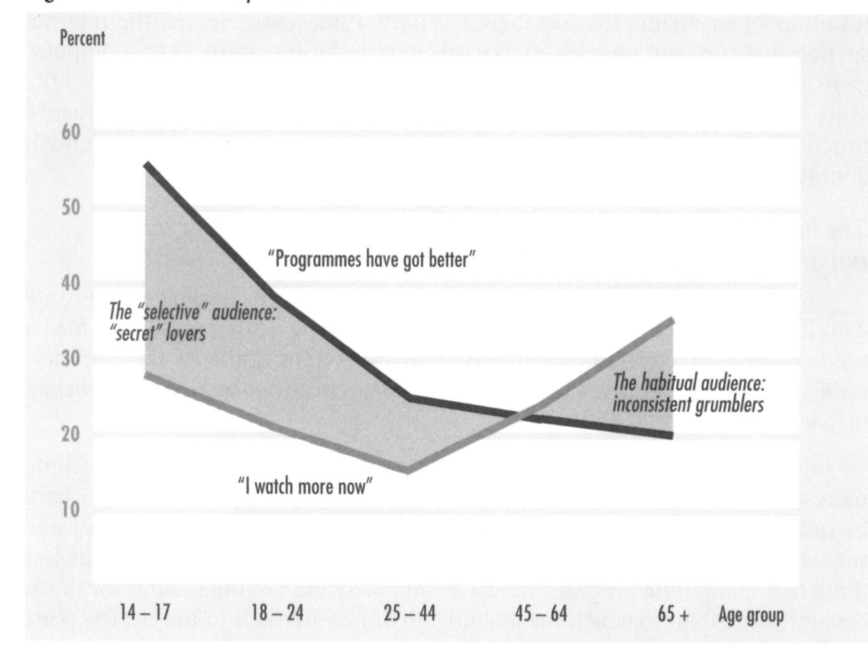

Problems for the advertising industry

The younger public tends not only to be more satisfied with the new media products for the private household offered "preferentially" with it in mind. Interestingly enough, on account of its tendency to spend more time outside, it actually uses them less often than other, older consumer groups. However, if the positively evaluated products are not necessarily used more frequently or intensively, at least in terms of time spent viewing or tuned in (for television), then this spells trouble for the advertising industry in particular, which depends on contact and attention in order to keep the financial merry-go-round going round.

As "secret" lovers, the members of this younger audience group are more than satisfied, yet there is little scope here for any increase in use associated with their passion for the product offered. This also helps explain the slow growth of the digital TV offer with the majority of consumers. It is directed at a basically younger persons' market rather than at the older people who watch more, even

Source

Fig. 5.11: Lange 1998

though the younger target group has in reality long since been perfectly satisfied with the diversity already on offer.

Fig. 5.12: The "acceptance and habituation" syndrome

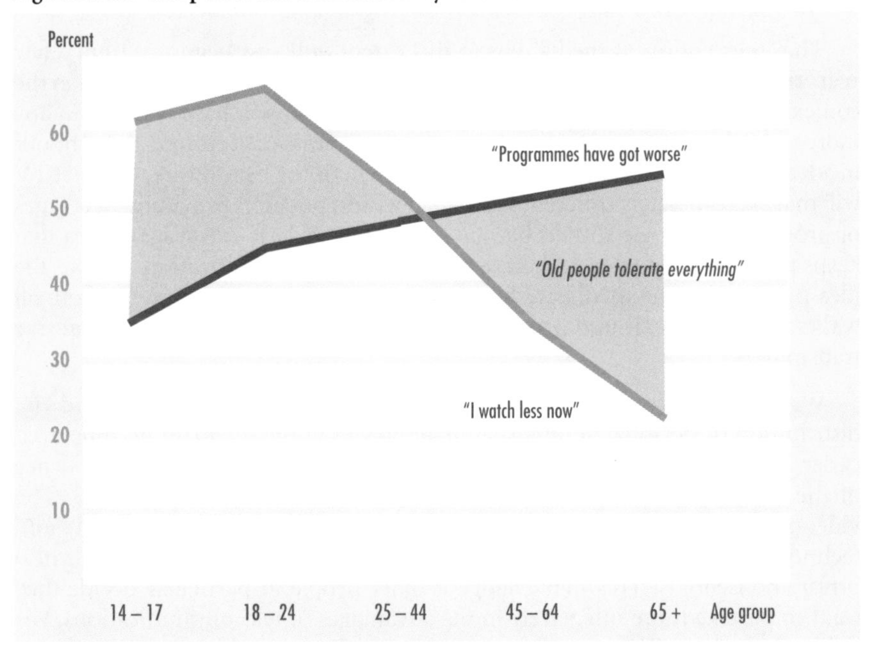

The older public, who use the old media forms intensively, have been indirectly forced to "put up with" what has been subjectively perceived as a deterioration in programme quality. For this reason, an interactive offer developed not exclusively with young techno-fetishists in mind but older "normal" people instead could well meet with great resonance and contribute considerably to shifts on the media market.

Re-integrating the society

The great wave of success for interactive services with older, lower-income and less educated people is still to come. Such a trend would be a double bonus, both contributing to the reintegration of some of the people excluded from today's two-thirds society and opening up new options for qualitatively more demanding media consumption. In view of the fossilisation and ossification of public life that has taken place where there were once distinctive neighbourhood and company cultures contributing to the community's spiritual balance, the boost that this would give to further education and social participation would certainly make it something worth encouraging.

The growing loneliness that springs from social inequality and discrimination in the education system is a further challenge to be met by the new "entertainment" products for the private household in their capacity as

Source

Fig. 5.12: Lange 1998

"technologies of communality." It is thus quite conceivable that easy access to new computer networks could prompt older people too to take part in chat or news groups, in turn impelling these groups beyond the exchange of trivialities and perhaps even giving them a new, more vital and dynamic structure.

Re-inventing media

The "reinvention of media" has to this extent only just begun. Within a few years time the computer will have virtually ceased to be perceptible as such in the context of network applications, and the television set will have taken on many more tasks than that of simply passing on programme conceptions thought out in advance by the TV stations. As a media system of the near future, Internet TV will make a completely different kind of television possible, providing new types of programme. Yet we should beware of one particularly extravagant idea that keeps rearing its head where there is fundamental media innovation at stake, the idea namely that the media are becoming so simple that all recipients from all walks of life and all age groups will end up becoming their own active transmitters.

... and what about the videophone?

We can be sure that the private production of home videos and the distribution of personal audiovisual material will in the long run become even easier. Through networking and new picture-insertion techniques, the influence of the videophone on private television will be every bit as profound. The videophone is becoming more and more acceptable both economically and technologically, even though it has yet to achieve the big breakthrough into private households. This may change if older people in particular decide they want to partake more intensively in the advantages of telecommunications. For unlike the users of PCs or on-line services in the Internet, older people are rarely willing to switch to text-based modes of communication. It would thus be an unfortunate side effect if the classical telephone were to lose value in the process. Especially for conversations involving advice or instructions, the videophone seems far more simple to operate than its rivals, above all in comparison with text-based dialogue modes. The videophone could also become an add-on for television use, just as it has already made a limited entrance into the computer world.

Alternating between teletext, Internet and normal television may also come to be carried out more smoothly in future. Yet the specific features of content-provision necessary for each individual form of media mean that it will remain a task for specialised professionals and continue to require large-scale commercial broadcasters, network operators and service providers. The consumption of finished products is convenient and useful to boot, since the division of labour it permits gives the users in question the necessary scope to concentrate on what is agreeable or necessary to them.

The trend towards individual shifts in use between the integrated media

It is at present still scarcely possible to pinpoint the actual shifts in use between the individual types and forms of media. It has been said, for example, that many Internet users in the USA today are already spending more time in their own virtual reality on the computer than classical couch potatoes spend in front of the television set. Complementary media uses exist alongside the substitution effects of new media products. Parallel uses can lead to a doubling of reach and are not necessarily the sign of a society that is segmented, segregated and lacking in integration. There is a development coming into play here that can best be described as a trend away from the "media-based public" of full-service broadcasting and towards the "event-based public" of target-group media forms.

A shift from media audiences to event audiences

The more the media lose cohesion and bonding in their interaction with the public, the more essential it is for events that have an inherent and intersubjectively shared "significance" to impose themselves in the public perception. Just as a segmented programme offer seems to neutralise or even negate the significance of the individual popular figure as a "persona" in the public perception by dissecting the potential audience group,[279] target-group media reconstruct "celebrities" through personality shows specific to the particular kind of media in question. These will have to make their mark in the hearts and minds of users in order to build up marketable media profiles and the accompanying channel loyalty.

5.2.2 Towards Media Integration in the Private Household

An Internet that is easily accessible to all users is certainly foreseeable within the next few years. Whether all consumers or at least the vast majority actually make use of this depends on factors that will be discussed below.

Internet gaining importance in private households

The Internet is expected in the future to produce closer electronic links between more and more people, enterprises and institutions. New products and services are expected to become possible in this way and new markets opened up. The marketing of products and services through the Internet and the financing of new information and entertainment offers demands a steady increase in user figures. Yet the use of the Internet by more and more people depends on the equipment being easy to operate and affordable even for consumers with low incomes. Having an electronic department store in one's own home seems an attractive proposition for saving time on travelling and getting a better market overview of the products on offer. If consumers without experience in computers are to be persuaded to make electronic bookings or other transactions from home, however, technological solutions will have to be devised involving simpler appliances with improved interfaces or application possibilities.

In terms of ease of operation, television and the telephone have up to now always been exemplary technologies. Now they are also expected to facilitate

279 This is in accordance with the impertinent chat-show motto: "Tout le monde est important."

straightforward access to electronic networks such as the Internet and the offers connected with them. New TV sets and new telephones, as well as new set-top boxes and modems, will provide "normal" citizens – even those without any prior technical knowledge – with easier access to the offers in the Internet.

Private telephone lines have now developed into an entrance for private Internet use and along with conventional telephone communication are now used by well over 20 percent of German citizens for sending faxes or e-mails as well as accessing data on the Internet. In Germany the boom in Internet use has only just begun, but the first signs of weakness in the classical telephone network are already making themselves visible. In the long run it is too slow and too expensive to get one's information and entertainment through the double-core telephone cable. Even low-data transactions, such as running an account from home or booking a flight or train journey through the Internet, demand great patience from on-line-service users waiting for the flow of data. The enthusiasm with which people are prepared to give themselves over to the vagaries of classical computer communications still has its limits.

Internet use through the PC – the starter drug for experts

PCs form too high a threshold for Internet use in mainstream households

For the time being the PC remains the most successful and most powerful tool for gaining private access to the international data services, since – as will be explained more fully below – the set-top boxes and WebTV-boxes designed for the television set are technically more simple in their conception than PCs. PCs are more expensive and more complicated, however, and thus form an artificially high threshold to entry for Internet use and on-line traffic. Even a higher level of computer ownership will not protect the PC branch from competition from simpler "on-line machines." America is far ahead of Germany in terms of private computer use, yet even there new box systems are being successfully marketed. According to a survey made by Dataquest, 43 percent of all Americans in 1997 already had a PC, while some 10.2 percent were planning to buy one in 1998.[280] This also means, however, that for there to be "full" provision with on-line services there are still more than 50 percent of the total population waiting for more acceptable solutions than the PC.

Does Germany have a deficiency in Internet use? Europe is of course behind the USA in its level of usage. Yet the deficiency lies not with the users, but with the technology providers. As long as it is so slow, expensive and complicated, why indeed should all Europeans find themselves condemned to Internet use? Is Europe perhaps once again missing the opportunity to turn this technology into something genuinely useful and convenient? We are on the threshold to an age in which interactive communication can for the first time really become an enrichment (which it hardly is at present for the normal citizen). It is thus imperative to find intelligent ways for everyone to get in on it.

280 See also www.desi.excite.com/News/971211/15.TECH-PC

One of the most exciting questions concerns the personal willingness of consumers to use the Internet and just how accessible it is. Whatever we may think about private Internet use, what is certain is that in spite of Germany's still fairly low general level of usage the Internet there at present finds itself in a phase of strong growth. Even so, relatively little is empirically known about current on-line use.[281]

Early trends in Internet adoption

According to the ARD on-line survey from 1998, a panel of roughly one thousand on-line users aged over 14 showed the following patterns of usage: a very high proportion of all on-line users in 1997 were still only using Internet access during working or training hours at their place of work, school or university. At the time, some 60 percent of all people with on-line access tended not to use the Internet at home after work. Now almost two thirds of those asked are also using the Internet privately in their own home.

In Europe at least the Internet can still hardly compete with television use, but it is precisely here that figures from the United States are indicating a new trend. According to EITO figures, the average American citizen in 1995 was already spending almost half the time he would spend in front of the television at his computer. These figures admittedly include time spent working with the computer. Even so, the same EITO study also revealed that Web users are already watching 59 percent less television than the average viewer. There are even those who go as far as to predict that by the year 2005 the time spent in front of the TV screen will only be half of what is spent in front of the computer screen. Even though the total time spent watching television in Europe – in spite of certain target groups opting out – is still rising at a minimal rate, in the United States the total figure is already falling slightly.

Teletext on TV channels already provides content similar to some offers on the Internet

As an accompanying text for television, teletext is the most successful on-line medium in Germany. When an Internet user goes on the Net, however, it can be assumed that the time spent using it will be markedly higher than with teletext. The Internet must face comparison with the teletext as minimal competition in terms of how up-to-date and accessible it is in what it offers. What teletext offers for free rules out the provision of up-to-date news headlines as a potent sales argument for Internet access. The Internet's main use must thus lie in a different field from the provision of current information in brief.

As a consequence, what the Internet offers must be appreciably more attractive to users – both aesthetically and in terms of content – than teletext offers. When successful, the Internet then seems to exert a form of suction not noticeable to the same extent with teletext use. The reason is that along with the latest news headlines the Internet provides a whole host of other services from chats to extensive booking and ordering services, while in entertainment, particularly the erotic field, the offer goes much further than with the classical teletext. Nonetheless, with costs as they stand at the moment, the willingness to pay personally is still very low. The group of "payers" is markedly smaller than the

281 The following interpretation of empirical data is based on van Eimeren et. al. 1998, and EITO 1998

"Unlike the computer freak, the average consumer is looking for solutions that are easy to use and stable."

seven million or so German on-line users. In Germany at least, on-line does not equal Internet..

> *"Only 47 percent of German on-liners describe themselves as frequent Internet users. Even so, compared with the previous year a clear increase in the shift towards the 'network of networks' can be made out. In 1997 only 30 percent of German on-line users claimed that they frequently surfed on the Internet."* [282]

According to the same survey, on-line use among "on-liners" has recently undergone a huge increase and is starting to rival the use of other forms of mass media. On weekdays, for example, daily on-line use is around 76 minutes, while at the weekend it amounts to some 80 minutes. The most frequent user group is the 14 to 19 year-olds, who during the week spend a daily average of 91 minutes in the Internet and at the weekend as much as 121 minutes. Analogous to this is the lower on-line use among the over-fifties. These spend about 68 minutes surfing on weekdays and 66 minutes at the weekend. The competition this poses to the classical forms of mass media is undeniable. More than a third of those asked assert, for example, that their on-line use is at the expense of television.[283]

Internet content and presentation still geared towards PC users

If it is assumed that at present only between 15 and 25 percent of all Germans use a computer at home (the sources of data and bases for calculation are too variable for an exact figure to be reached here), then it follows that between 75 and 85 percent are in this way excluded from Internet use even in principle. This is not just on account of the technology available. Rather, it has to do with the fact that what is offered on the Internet is geared towards PC users, and a lot of these think and live differently from the majority of the population. Computer users are enthusiasts. They spend large sums of money on their favourite gadget, and with some of them the computer is already enjoying even greater favour than the motorcar, which always used to hold down the number one spot in earlier studies on the subject.

More ease of use for the mainstream customer

Beneath the graphics-oriented Windows surface, however, the remaining German citizens – and that means the majority of the population – still find computers too complicated and too liable to go wrong. Confusing access software (browsers) for the Internet with complicated log-ins and plug-ins make access to the Net more difficult. In addition to this, recurrent new generations of software and appliances as well as the continuous and rapid fall in prices have more than unsettled the average consumer. Unlike the computer freak, the average consumer is looking for solutions that are easy to use and stable, and it is this that the computer industry has hitherto been unable to supply. Set-top boxes for digital television have up to now found little resonance either, since people have suspected that these boxes will end up raising prices through the introduction of pay-TV. The opportunities for exploiting their inherent interactivity have therefore scarcely been used so far.

282 Ibid.
283 Ibid.

"The computer will leave the desktop and move from place to place throughout the home."

Towards the Volkscomputer

As has been said above, new applications for the individual household need not primarily prove to be useful, good value or efficient. Rather, they should be bound up with the special private qualities of domestic life in an agreeable and if possible unobtrusive manner. Even when new technologies can be integrated into a home with relatively little disruption of the private dynamics of the place (for example by making the appropriate structural alterations to link the flat up to the cable network), they are still viewed with suspicion and frequently rejected.

The "Volkscomputer" – a metaphor and a challenge

It is not so much new equipment that is called for, with the additional ordeal of new sets of instructions. Instead, it should be possible for the household to integrate the new products and services (which they are willing to use and thus also pay for) into their own comfortable, individually structured entertainment and information lifestyles at no great additional cost. Given current strivings for a user-friendly environment for new interactive services, the metaphor of the "people's" or "Volks" computer thus appears so plausible and essential that the German variant of the concept has already gone into the English vocabulary, bringing back memories of the success enjoyed by the Volkswagen and forming a central challenge for the marketing of new products.

New interfaces, new terminals, new appliances

In the interplay of television set, telephone, stereo system and computer, a new balance will be attained. New interfaces, new terminals, new appliances and systems are called for here, many of which may well bear little resemblance to the interfaces familiar today. The computer will change from being an instrument for word-processing and calculation to an information and entertainment machine, while the telephone may well be transformed into an Internet phone or a videophone. All the players active in the market in principle agree that the Volkscomputer will have to be simpler and more resistant to interference than its precursor the PC. The computer will leave the desktop and move from place to place throughout the home. Microprocessors have already arrived on the scene in countless household appliances and will in future be better integrated and assimilated with one another in their use.

New dedicated terminals

Given the expansion of appliance functions, however, the number of "remote control" devices – as new mobile interfaces within the household – must not be allowed to increase any further. A decisive and particularly innovative dimension in the conception and effectuation of easy-to-use interfaces might well be the development, for example, of a dedicated kitchen terminal. The kitchen is a place where products are consumed, wrong decisions consciously experienced as such, and orders planned and implemented. How about setting up an attractive electronic shopping mall here, leaving the classical TV users in the household undisturbed? Experiments have already been carried out with microwave ovens equipped with touchscreens on the front surface from which e-mails can be sent and orders made. Even more important than the appliance's consumer-friendly design, however, is a comprehensive and attractive communications and service

"Set-top-boxes are a metaphor for different speeds of innovation in different industries."

concept that can be followed and then also financed as necessary by the potential customer.

The integrated media box: the influence of appliance hardware on the acceptance of new systems

Set-top-boxes: a necessary backwards step

In the centre of the discussion concerning the appropriate appliance hardware for the private household are so-called "box-solutions." The box is a metaphor for the divergence in innovation rates between the two converging branches of consumer electronics and the data-processing industry. Boxes mediate between the "fast" computer industry characterised by short innovation cycles on the one hand and the "slower" consumer electronics branch, which suffers from the longer use phases for replacement investments in the private household. Situated between the two main players with their distinct innovation rates is the telecommunications industry, which has to adapt to the different innovation cycles with hybrid accessory solutions. In recent years, however, the telecommunications industry has changed from being a passive "standardiser" and "mediator" to an active market player that is learning how to build up and develop its own networks with a specific market strategy in mind.

New devices and the imperfection of technological vision

The use of accessory boxes on the television set is actually a return to the days when a special device functioning as a converter was put on top of the TV set for receiving additional programmes. Boxes have always made an appearance when older generations of appliances have not yet been adapted to new developments or when a new technological standard is still under discussion. This is certainly the case with digital TV, although the chances are that future generations of TV sets will be adapted to receive digital programmes even without a set-top box having to be connected. Behind the physical boxes themselves there are ever more complex systems at work, where the manner of signal transmission and its storage and modulation play a decisive role. Even so, it must also be possible for these technological visions to be put into real working devices. Boxes do indeed prove suitable in this respect, since they have "features" that can be fixed as their applications are being developed. What would be desirable, of course, is an integrated box for immediate use, which as a recognised standard would additionally have to function cross-platform and be as stable as possible, easy to operate and good value. Yet an integrated box of this kind as yet remains a thing of the future.

Plurality of choices: an example for versioning of hardware

It makes sense to divide Internet TV box solutions into four categories, although it can be assumed that in the long run all boxes will be able to be upgraded by modules and scaled appropriately to the applications desired of them.

1. Under $ 150 (very low end). For this price the customer will probably within the next few years get a simple browser adaptor for using the Internet on the television together with a modem and keyboard. What remains open to question is whether a card reader can be integrated as well for this price.

2. Under $ 250 (low end). Along with the basic very-low-end equipment, appliances of this sort might already include a hard disk solution.
3. Under $ 500 (medium level). These appliances will resemble today's PCs and be equipped with a DVD or CD-ROM reader as well as a hard disk or powerful memory chips.
4. Over $ 500 (high end). In the long run there will be virtually no upper limit to the degree of comfort and the modular extension of systems. Accessories and extensions will use a variety of interfaces to facilitate combinations of appliances with a whole range of capacities.

The "Universal Media Box"

The universal media box serves as a metaphor for an appliance that succeeds in combining the interactivity of the computer with the ease of operation of the television. While the TV set is maintained as a visual interface, the media box permits communication within the system to be two-way. The media box will in the long run be conceived in platform-independent terms and thus end up tied neither to the TV set nor the PC as interaction surfaces. It should also be possible to connect it up with a printer, as well as leaving open other dedicated interface configurations such as the Internet telephone, the mobile infoman or a "kitchen terminal." In this context television will just be one service among many others.

Internet TV

Internet TV boxes are targeting a new and different audience from the television market, even if these customers are for the time being only being offered what has been developed in the Internet for PC owners. In the United States it has long since been known that Internet TV users access different pages in the Internet and use the new technologies against a different background of applications. For them the Web is a complement to classical entertainment offers, and for this reason they want a connection that is as easy to operate as cable television. The new appliances are even expected to help them make better use of cable television. And the market for this sort of provisional solution is in the longer run possibly even greater than the classical PC market. Set-tops give chip and computer manufacturers the opportunity to reach considerably more households than was previously the case. While the number of PC households in the United States is just slightly over 40 million and growing only moderately, cable TV firms have already got more than 65 million households as customers – and with their infrastructure they could rapidly connect another 25 million to their network.[284]

It is not the technology that is impelling the development of Internet TV forward, however, but more the fact that behind these appliances are other applications and other markets, ranging from the manufacture of operating systems via applications software and on to video and computer games. It could further give a significant boost to "e-commerce."

In this way it is perfectly feasible that Internet TV boxes might function as a sort of catalyst for computer laypeople who later then move on to more powerful

284 See anon 1998b, p. 16

computers. These boxes actually function rather like computers do, and the similarities are genuine. Like network computers they could also be used to download computer-type software direct from the Net. Neither in the mass market nor at the high end is much effort being made to find more powerful provisional solutions that are acceptable and plausible for the coupling of the TV set and the PC.

To this end either a computer is installed into the television set or TV tuner cards built that can be inserted into the PC. Even though personal computers are certainly more powerful and more expensive than the afore-mentioned boxes, they will continue to have one essential disadvantage for a number of years to come: they are too complicated and too unstable for the normal user. This may pose particular problems when computer-type error messages meaning little to a TV viewer appear on the screen.[285] The drawbacks of the classical computer will thus persist even if the computer is incorporated in unchanged form into the TV set. At more than a thousand dollars, the prices of such appliances would be rather high for the majority of consumers.

From the Internet set-top box to integrated Internet TV offers – stages in opening the market to the next 20 percent

Internet TV promises easier access to the Internet than what can be offered by the computer, complicated and susceptible to interference as this is. For the first time, the TV set will be converted into an interactive information and entertainment machine that can also be used for carrying out transactions such as bookings or credit transfers. The mistake made by many Internet TV strategists in developing their business models is to assume that just because a technology is convincing all households will be provided with it in a matter of a few years. This is nonsense and is given the lie by the distribution curves of countless other new media and technologies.

Dissemination will not be immediate

For quite some time now, the triumphal advance into private households enjoyed by that wonderfully simple and popular technology the fax has been petering out. The equally easy-to-use answering machine will not be present in a hundred percent of all households for the next few years either. Internet TV is regarded as paving the way for a new growth market in the telecommunications branch in which computers are to be "pushed" into households but in a new guise. It is accordingly classified by Microsoft as among the most important strategic products. The main argument for this is that in the USA almost 80 percent of all households and in Germany for example more than 90 percent of households are not yet participating in what is happening on-line.

A rough idea of the chances of Internet TV spreading in Europe is provided by the figure of some 300,000 WebTV sets sold at short notice in the United States. These sets found a place in a large number of American households after just one year of relatively run-of-the-mill marketing. This is certainly an excellent

285 See anon 1998b, p. 16

figure insofar as it can be taken as showing the level of acceptance among the normal population, but what a pitifully small total in the context of a significant spread of Internet use among the total population! Innovative Internet TV offers – intended as a transition stage for new technologies still to come – could contribute to the fifth of the population already using computers in Europe being joined by another fifth of the population looking to find a simpler and cheaper way of enjoying new interactive services.

New browsers – a chance for European companies

In almost all European countries (excluding Scandinavia) there are already clear signs emerging that having been been bypassed during the introduction of computers into private households and schools, European companies may well once again be passing up their opportunities, this time in the introduction and application of popular Internet offers for a mass market.

WebTV by Microsoft

A non-European giant is already making massive incursions into the interactive TV business in the form of Microsoft's WebTV, even if the huge investments are coming up against a strong headwind. The market for operating systems seems to be slowly opening up again, and in Europe in particular the interfaces of the future for Internet TV have not yet been established. If European and especially German companies fail to respond within a year, however, then the inevitable market leadership of Microsoft in this field can be taken as given in Europe from the end of 1999 on at the latest.

German manufacturers are now starting to wake up. The Grundig box is being sold as the coming market leader in Germany, and 60,000 are already said to have been sold. Yet without a corresponding hard disk memory the Grundig box has so far seemed poorly suited for providing an integrated and convenient offer. Even Grundig are now beginning to think beyond marketing a single appliance, however, and together with Primus Online are planning to offer the box in a "strategic Internet TV alliance" as part of an integrated product.[286]

Multifunctionality

The defining characteristic that Internet TV offers is that of multifunctionality. The more functions there are built into a system, the better value the system appears as a whole. In addition, the prices for boxes are expected to undergo a further drastic decline within the next two years. With Internet TV appliances of this sort it is possible both to make use of Internet services and to receive and send e-mails – and this at a markedly lower price than when using a PC.

Internet TV is a comprehensive attempt to combine the television set with the Internet both technically and in terms of content. This attempt may involve various routes. Interactive TV based on set-top boxes has up until now not proved to be genuinely interactive. This may well change in the next few years thanks to new complementary offers coming from Internet TV. The concept of WebTV, for example, is much more than a purely technical set-top box solution and contains a broader concept for the preparation of Internet content. Its aim is to make

286 See anon 1998i

Internet content simple to operate on the TV set and thus also easily accessible. As well as this, Internet TV can also contribute to both an expansion and enrichment of the programme content offered by TV stations. Television stations are prepared for this development, and are already building up their editorial offices for teletext with the Internet in mind.

Internet access via TV channels can be attractive for PC users, if they are fast and cheap

Not everybody, however, is waiting expectantly for the new Internet TV offers. These will first of all have to win through on the market. Of the total population, at least the 20 to 30 percent of experienced computer users have by now learnt how to live with their PCs and the connection on-line. These people are more or less satisfied with the applications and the technology and are hoping simply for a more rapid and a cheaper connection on-line in order to be able to carry out transactions such as bookings or orders, exchange e-mails, take part in chatgroups, newsgroups and multi-user dungeons, or merely roam about as an Internet surfer. The capacity of new modems to provide even greater speed is limited. In future it is the network for the new data communication that will have to be expanded or its capacity increased.

New, reasonably low-cost Internet boxes that are independent of platform and provider – produced for example by firms such as Grundig or Pios – make it possible to use the Internet on the television set at home on the basis of a plug & play solution. As a rule these are appreciably easier to connect to the Net than a computer and their software furthermore makes them less liable to crash than a normal PC. As will be shown below, however, simply buying a new box does not on its own guarantee that Internet use will automatically be made quicker and cheaper.

Speeding up the Internet

To reduce the time spent on-line during Internet use, rapidly accessible caches (proxies or mirror servers) will be brought into play to download frequently used content from the Internet and so shorten retrieval times. If such servers are selectively made available in the cable networks of network operators or the housing industry, it could even become possible to call up Internet content free of charge without having to resort to the costly telephone network. The electronic "storage heater" – which allows content to be downloaded from the Net at cheap rates and thus made accessible to large numbers of users – seems to have become a real possibility.

Downloaded or edited Internet content will also be accommodated more and more frequently like teletext in the form of additional signals to the TV screen (as, for example, with Intercast or BOT technologies). These can then be appropriately picked out by a set-top box or modem and reproduced for free on the TV or PC. This will admittedly lead to new copyright problems, since not every content that is "fished" out of the Net can be passed on just like that. The more successful the Internet becomes, the more expensive and carefully protected its content will be.

In future, numerous new Internet TV suppliers such as WebTV[287] will appear, providing the private household market with a whole range of concepts for use and offers to complement the classical television.

WebTV Networks – the integrated box solution for the consumer?

Founded in 1995, WebTV Networks is based in Menlo Park, California. In August 1997 it was bought by Microsoft for 425 million US dollars. WebTV provides two services: WebTV and WebTV Plus. It currently has licensing arrangements with the appliance manufacturers Mitsubishi, Sony and Philips. WebTV Networks cooperate with some 100 content providers in the United States and had approximately 300,000 customers already in March 1998.

In conjunction with television sets WebTV boxes permit surfing in the Internet, e-mail, homebanking and a great deal more. The system was conceived as "enhanced TV" for users without any experience in computers or the Internet. When switched on, the first thing it does is automatically dial the selected Internet service provider, which supplies a service optimised for WebTV at a cost of some 20 dollars.

This involves not only boxes but also proxies specially tailor-made in accordance with the pages called up and the information needed by the consumer in question. WebTV is more than the simple adaptation of Internet content for the television by means of a box and is thus superior to most purely technical Internet TV box solutions.

With the successor system WebTV-Plus it is further possible to pick out Internet data from the TV signal. The system recognizes Web pages in the TV signal just transmitted and either announces these or stores them for later use on the hard disk. These "TV crossover links" make it possible, for example, to transmit additional information about the TV programme being shown at any particular moment. By means of an in-built TV-tuner, TV programmes can continue being watched even during Internet use (thanks to picture insertion techniques). Navigation between the individual WWW-pages is carried out by remote control or alternatively through a wireless keyboard.

The main criticism levelled at WebTV is that it has so far only allowed proprietary access, thus failing to support Internet standards such as Java or Real Player. But this has also been its strength. WebTV is a closed concept in its own right, and as such looks very useful. Without the special provider, however, the box is worth-less, and this makes it a decidedly risky investment.[288] For in general one of the decisive criteria for the purchase of an Internet TV box is supposed to be how open it is to various providers. WebTV too thus seems to be leaving a door open to rival providers.

287 WebTV Networks has been present on the German market too since early 1998 as a cooperation between Microsoft and Deutsche Telekom.

288 See Haas 1997, p. 24ff.

In view of the euphoria sparked off by Internet TV based on narrowband solutions, the high-end solutions have tended to be completely disregarded, although here too new developments have taken place. Under different circumstances, perhaps an interactive service based on broadband networks will prove capable of surviving after all. After WebTV, @home in the USA is likewise being treated as a shooting star that apparently offers a convincing technical performance.

Internet TV enhancing traditional TV

Internet TV is no new television, therefore, but it does enlarge and complement the classical electronic offers by providing new interactive possibilities and additional information about programmes. The moving TV picture will have explanatory data, diagrams or images supplied by Internet wrapped round it. Within a matter of years, video clips and sound recordings will likewise enhance programmes in this way, for example allowing the user to download a digital CD into a private stereo system at a small cost even while the music programme itself is still being watched.

The "Infoman" – extensions of the Internet

The struggle to benefit from the private household's successful entrance into the on-line world has already begun. Yet there are still a lot of questions to be answered, for example concerning the safeguarding of consumer information or the binding nature of expressions of will in the Net (for example in the form of an electronic signature for purchase or payment transactions). Competition for Internet TV is also emerging from the field of telephony. Modern telephones, whether fixed or mobile, will also be equipped with a display for Internet pages. The Internetphone and the Infoman[289] are extensions of Internet use that go beyond the TV set and the PC.

The market for the new Internet TV continues to be as complicated as ever and remains uncertain in terms of content, technology and costs. A sudden, short-term household revolution cannot therefore be expected from Internet TV. The transition to new forms of interactivity in private homes – to be looked at in more detail below – will instead be slower and more fluid in nature.

The development of the computer memory as an integrative link between hardware and software

Electronic storage devices will be central for broadband development

The electronic memory will in future come to play a pivotal role in private households. It is essential for making network traffic both cheaper and quicker through modifications in the on-line/off-line relationship. Given the artificially high telecommunications charges in Europe, the electronic cache (or "storage heater") should prove a vital business development. Cheap rates, for example at night outside peak traffic times, should be used for on-line traffic in order to carry out the expensive business of downloading data from the Net. But this function is only necessary if public or private networks have to be used for on-line traffic. "As much off-line as possible and as much on-line as necessary" should be the principle guiding private households using the Net if these networks are to be economically run.

289 A trademark registered for a kind of "walkman" device presenting information from data broadcasts.

Physical data carriers such as the floppy disk are a preliminary form of electronic memory and as a means of transporting data from appliance to appliance they remain as popular as ever. The CD-ROM has for the time being become the best-value and interactively most sophisticated means of transporting digital data. Mediated through the computer, CD-ROM readers will thus come to be included in growing measure among household equipment. The CD-ROM was developed first and foremost for the distribution of computer programs, but then became the most important carrier for electronic books and magazines as well as other multimedia applications. It is now also used more than averagely intensively by users of game consoles.

The storage capacity of the hard disks built into computers is growing by the year. Hard disks with several gigabytes are now taken for granted by private computer users. Hard disks prove ideally suited for quickly storing data that come up in on-line traffic and need to be saved in the short or medium term. Their drawback is the relatively high cost of doing so by comparison with the CD-ROM and the fact that they need to be permanently installed or integrated into the system. For hard disks to be used as a local saving device for data from central servers, they would regularly require a complicated process of steering and modification. In addition to this, when in continuous operation their mechanical parts (including the ventilator) give rise to an annoying background noise in the home. For these reasons, short and medium-term data-storage will in future be taken over by electronic components in the form of memory chips or cards.

The type of content to be stored is predictable

For certain applications the hard disk can be fitted with a cache that makes it possible to use Internet pages in particular – provided they do not change – over longer periods of time. While a cache is located on a local PC, providers also use proxies as a way of temporarily storing frequently used Internet pages. The browsers "remember" these temporary stores and so do not need to retrieve the data from the Net again. Proxies thus make dealings with the Internet considerably easier. WebTV, for example, claims that simply on the basis of statistics it is able to select and then keep on its own server more than 70 (!) percent of the pages needed and clicked by its users.

Electronic memories for the private household are also available as end appliances in their own right. A cache for receiving e-mails and faxes (while the PC is turned off) has already been developed in Berlin by the firm Tixi.Com GmbH. This makes it possible to receive e-mails and faxes all round the clock even when the computer is not running, and is of especial interest to those who make heavy use of the e-mail service. If the sender and the recipient have such a box, then communication costs are substantially reduced since no new charges are made for collecting the mail. Even so, the box can at regular intervals call the provider to pick up any e-mails that may have arrived there.

Intelligently selected memory management that is constantly being optimised in accordance with consumer interests is a decisive condition for the development of new media in the Internet.

Pull and push strategies – a reversed model for background storage

There are two fundamentally different strategies for delivering information to users. Either the users themselves ask for the data they require, a delivery model known as "pull", or the data is sent to users unsolicited (although an original pre-choice may have been involved), which is known as "push". Various systems will be tried out in the next few years in order to achieve a sensible interaction of push and pull, and all of these are in need of editorial support. With pull technology too it is possible to gain more profitable combinations through the use of electronic memory. Pages from the Internet can be stored in proxies, speeding up the process of retrieval for the individual user. With PointCast, users can selectively target the news or data they want to download on the basis of a menu choice in their own PC. By its own account this free service, which is to be financed by advertising inserted into the screen, already has 1.8 million registered users.[290]

Even though the asynchronous pushing and pulling of data has been exhaustively discussed in specialist literature, simple user interfaces that can be easily dealt with by computer novices have still to make an appearance. Yet the goal of providing news that is automatically geared to the user's exact personal interests no longer remains a pipe dream. Integration and convergence in this field really is making itself felt by the individual recipient and consumer. The Internet is in this way generating new forms of mass media.

Online memory helps smoothing network usage

On-line memory in the end appliance – whether in the form of internal or external hard disks in PC or box, memory cards without mechanical parts, or other solutions still in the making – facilitates the use of cheap transmission rates outside peak periods (which as a rule tend to coincide with those times spent working at one's place of work). As such they function as a sort of cheap storage heater for entertainment and information. It is a problem, therefore, that some boxes come without any such cache. In these cases, Internet use remains very expensive, since the data constantly have to be asked for on-line. But here too it is not the appliance but the architecture of its use that is decisive. It would be perfectly feasible for a box without a cache to be directly connected to the local server for house, estate or district, allowing a regional offer to be called up almost as quickly as if it were stored on the local hard disk.

290 See anon 1998

5.2.3 Financing and Implementing the Integrated Offer

TV will be more expensive than today

The private household budget for telecommunications services in Europe is unlikely to grow significantly in the coming few years. Up to now, for most consumers the expectation has been that electronic media should be delivered more or less free of charge. In relation to its comparatively high use-level per month, the television is exceptionally cheap and will remain so for the majority of the population in spite of the aggressive advertising of pay-TV suppliers. The public broadcasting service and "free TV" are in most European countries at present far from threatened with extinction.

In Germany teletext is still the most successful interactive data communication system for most of the German people. In 1997 teletext reached at least 3.36 million households every day or 4.22 million people from the age of 14 upwards. Yet this is still only 10.1 percent of all households per day or 6.8 percent of individuals over 14. Just 64 seconds per day and household or 45 seconds per person were devoted to reading teletext pages. Teletext provides the "average citizen" with a point of comparison with Internet use.

Revenues for new content will be difficult to generate

What emerges is that the current need for electronically retrievable information – even when delivered free of charge – is still best regarded as limited in extent. Only people in search of noticeably more information "in depth" will tend to fall back on the Internet. For this reason alone the German public can in the long run hardly be expected to spend more time and money on getting into the Internet and browsing through Web pages than on buying the daily paper. It can be assumed that monthly expenditure on the "electronic newspaper" will lie somewhere between $ 8 and 14 at the most. In the average household little more than $ 4 will be for this service alone, as it comes coupled in a bouquet with other services. This revenue alone would never allow for special billing nor the appropriate advertising. It is thus hardly likely to be possible to finance a new interactive system on the income from an information service.

Platform and content will have to match

A wisely chosen marketing strategy will not, therefore, sell the Internet's information offer alone. In an open, interactive communications market in good regional working order it is not primarily through the charges paid by individual consumers that money is earned in the long run. It is more the provision of platforms for advertising, banking and insurance activities as well as retail, wholesale and mail-order trade that brings in new revenue and takes care of the marketing for such a system as a whole. The most important condition for this in a highly populated urban area, however, is a market opening with an estimated 300,000 to 400,000 connected households at the least. 150,000 to 200,000 of these users may come from the ranks of PC-users who then proceed to get connected to the Net. For the other 150,000 to 200,000 households it would be necessary to find appropriate access to the system through a Volkscomputer or a system like Internet TV.

The Internet does not as yet know any really self-contained submarkets, even though these have long since existed in the form of user-links and lifestyle orientations. Not every "Intershop" gets noticed by all network participants or forms a real market for them. As access to Net shopping offers can vary greatly and – in spite of global accessibility – many a sales offer remains forever hidden from many an Internet user, virtual metropolises, or sites of high usage-density, will develop within the Net. "Communities" of this sort have already started clustering around the search engines and the development of shopping malls is particularly noteworthy in this context.

Minimum household penetration as precondition for success

If it is assumed that (as described above) at most 20 to 30 percent of all households will actually be won over by Internet TV offers, then a successful market opening will require a catchment area of at least 700,000 households if 200,000 real users are indeed (with optimal acceptance) to be reached by this technology.[291] This gives some idea of how difficult it will be to develop offers that go beyond the metropolitan boundaries while relying on regional and local servers.

Assuming that the average connected household's equipment in a few years will cost a maximum of $ 180, then the investment in technology prior to a market opening alone will come to some $ 40 million. Another $ 350 per home can then be expected to go on getting connected up to the server and the cable network. For a total of at least 200,000 homes, this adds up to a further $ 70 million. In order to attain a basic volume of this order in any single region, a concerted effort will be necessary on the part of the various financiers. Much of the responsibility here lies in particular with the banks, especially since these will be important beneficiaries of the homeshopping and homebanking to come.

Direct or indirect charging

What remains uncertain is whether the costs of connecting a household should be charged directly or indirectly. The aim of achieving blanket levels of usage in order to enjoy positive network effects can probably only be realised if the individual household is not made to pay for the network infrastructure until convinced of its usefulness by the benefits of the system as a whole. For building up the logistics of the retail supply system, a high connection-density is extremely desirable.

The possible benefits of cheaper shopping in the Net are unfortunately not always obvious to users prior to their first experiences with the system. The psychological dilemma is thus that a shopper might fail to recognise that a purchase has been made more cheaply even though this would more than compensate for the costs incurred in getting connected. Marketing is necessary here in order to make its value suitably clear to potential users.

The financing of the appliances and infrastructure could also take place within the context of modernisation measures undertaken by the housing industry. Appropriate concepts for the deployment of set-top boxes as tools for facility

291 Such estimates are neither unusual nor implausible. Even such a well-established and still relatively popular service as cable TV continues to have considerable problems with acceptance. Although in "cable-dense" Germany 66.6 percent of all households in 1996 were directly connectible to the cable network, only 44.6 percent, or two-thirds of connectible households, had actually taken advantage of this service.

management have already been developed to this end. If through these, for example, individual household connections to a powerful server of the housing industry were to be guaranteed and jointly financed, then the companies with an interest in the transactions could provide the information, entertainment or communications appliance free of charge. The household would then just have the transactional costs and the costs of the content bouquets along with the aforementioned shared modernisation costs.

Who will pay for infrastructure?

In this context, the role to be played by property companies, private real estate firms or other property owners and landlords is interesting. These have the chance to make a suitable initial outlay in order, for example, to supply whole estates with networks and servers "from one hand" analogous to the "local area networks" of firms and authorities. Although the return on investment for the property companies concerned is anything but certain given the conditions currently prevailing in Europe, the opportunity of providing a comprehensive technical service from one hand is nonetheless enticing. Many of the individual expenses faced by single households – for example for the answering machine or fax machine in the field of telephony, or equally for certain servers or even set-top boxes – could in this way be significantly reduced to the tenants' advantage. This would require a functioning, centrally run communications system known as a "campus solution," though these in turn entail further problems of their own.

Multimedia in the context of facility management

The fact that as owners or co-owners of technical installations the property companies are to be among those duly participating in the value added created through the provision of new services is something disputed neither by the state nor private industry. The topic of negotiation here tends to be the level of the one-off payment per apartment for the purchase of usage rights in the form of permission contracts (i.e. the right to serve the single tenant with telephony, TV, Internet access etc.) or the purchase price for such networks when sold to third parties. In the context of such negotiations for usage rights in the international field, prices range up to and beyond $ 500 per apartment, yet this applies to a different context in which the cable can be marketed much more lucratively. In Germany the copper coaxial cable remains a loss-maker for Deutsche Telekom, and the planned increases in prices – which would have still meant a price lying below the cost threshold – have been politically out of the question. Given the financial difficulties of laying copper coaxial cable from the point of view of the technical provider, negotiations for usage rights are difficult to carry through profitably. In the German market, where the housing industry is relatively poorly organised, it seems that amounts of between $ 80 and $ 140 per apartment are the general rule. Moreover, a number of property companies that in the past clinched long-terms deals with Deutsche Telekom at give-away prices are bound to be interested in these prices (which are substantially below international levels) remaining unchanged for as long as possible. In calculating the value of usage rights it is also interesting to ask what influence the housing industry has on the choices and customer loyalty of its tenants, and what factors this influence

depends on. Like the network operator firms, the property companies are after all also being promised cable fees. According to a statement issued by DF1 at a public event in Leipzig, these could amount to some $ 0.01 per channel and apartment. In addition to net charges of $ 80 per year, quality bonuses of 2% per month would also be paid.

Competition for bundled access and bundled services

Other branches too are seeking to win over the private household and to the extent that adequate transactional profits can be made here they will pay accordingly for access to the individual household. As has been shown earlier, telecommunications will revolutionise customer relations. In Germany, for example, the opening of the telecommunications market has already led to established rules in the field of comparative advertising being waived wholesale prior to the general opening of the market through the European Union. And it is precisely the capacity to draw immediate comparisons between products being offered that most specifically characterises an interactive system. Comparative advertising is now permitted in the telecommunications market.

The development and financing of integrated applications, software, "content", and services

"Paperless offices", "virtual companies" and the private telecommunicative "cocoon" that users will be spinning around themselves along the lines "my Internet TV is my castle" are visions that have little to do with the practical aspirations of ordinary citizens, employees or tenants living today. Even so, prognoses relating to the distant future do help provide us with a degree of orientation as we go along.

The crucial role of stable browsers

The spread of browser software is of crucial significance for the provision of additional electronic services. While the market for operating systems seems still to be relatively well ordered (even though the change in platforms could well produce another tabula rasa giving rival suppliers completely new chances of success), the browser market for computers – in spite of having only a few suppliers – is characterised by the immense diversity of program "versions".

Developers can thus never be sure that the products they have designed will actually reach consumers in the form in which they conceived it. It is precisely here that the mass market demands greater standardisation, however, and this possibly represents the long-term opportunity to dominate the market of the future, above all for proprietary Internet providers, digital TV providers or the providers of new Internet TV solutions. As yet the market for Internet TV browsers is even less transparent than that for PC browsers. Even though the systems for reading HTML pages are generally open, as a rule each end appliance has its own proprietary product.

Another important area relates to questions of copyright, where considerable changes could come about in the very near future. Up to now it has been relatively unproblematic to create links between pages. This could change very soon. There

"The assumption that Internet content will for ever remain "free for all" is a misleading one."

are already suppliers that have successfully gone to court over the linking of their product. Most suppliers continue to be interested first and foremost in the greatest possible dissemination of their product. But once Internet content is used to earn money, then it could well come to pass that suppliers start defending themselves against the joint use of their products. This applies in particular to content that demands a substantial input of editorial effort. The assumption that Internet content will for ever remain "free for all" is a misleading one. In calculations relating to future services and business plans beyond the year 2000 it is therefore absolutely essential to take into account the appropriate "licence costs" or royalty charges payable to GEMA or to other organisations and enterprises.

Copyright ... and how about links?

Many providers of new services could in the coming years go drastically wrong in their calculations if they continue to take it for granted that the varied and interesting offer available today on the Internet will in the long run remain free of charge and that the current situation will persist unchanged. WebTV, for example, has already had to contend with problems of this sort in the United States. An interesting question is whether temporary storage in a proxy or other server and changes in the aesthetics of the page-layout by a proprietary browser constitute an intrusion upon the author's rights.

Yet how is a regional and local information or entertainment service to take shape if not primarily through adopting the material available in the Net? Beyond the classical mass communication offers and their double or triple use, there at present seems to be no scope for financing further new editorial services and above all for professionally edited electronic all-in services comparable with a traditional local newspaper. This may be easier in the case of simple freesheets, since these gain much of their revenue in a different manner and adverts are excellently suited to being stored and universally and flexibly called up in electronic databases and networks. The daily newspapers thus also frequently provide a well-structured and up-to-date small ad market as an add-on to their service in the Internet. Regional and local content for city and citizen information systems will for years to come continue to be available only via services provided by interested sponsors and PR organizations. The willingness to pay for this is restricted to expensive "high-end" special offers targeted at small circles of specialists. This of course ignores the universal and individual access to erotic content.

Not much room for revenues for new audiovisual content

The Internet is an ideal and open electronic marketplace that also offers its users products lying outside their own networks. As was pointed out above, the choices made by households in relation to media and technology usage are seldom based on the pure logic of rationalising their purchases or orders in terms of time or money. Shopping remains an adventure and it will continue to be frequently done outside the individual household. This clearly does not apply insofar as new Internet services make it possible to avoid irksome chores such as going to the bank to fill in transfer forms or dragging crates of bottles through the

Online shopping attractive for repetitive buying of standard items

streets. Yet it seems that the spending power enjoyed by many households nonetheless remains insufficient for such luxury offers, not allowing them to recover the still high transactional and logistical costs quickly enough through lower prices and the gains of rationalisation in the purchase of products.

Delivery service still too expensive

It remains to be seen how the logistics of the retail supply system will be built up. At present, all the transactions – from ordering the goods to having them delivered to one's home – are excessively costly. If a $ 5 delivery charge is compared to an average shopping bill of around $ 25, then it becomes clear that a system of this sort will only be worthwhile in the context of a more advanced delivery culture incorporating a high percentage of households.

High Internet density is not in itself an automatic guarantee that on-line shopping will be accepted across the board. Be that as it may, experts estimate that even in Germany some 700,000 people use the WorldWideWeb to do their shopping. It is nonetheless possible to predict the limits to the turnover generated by shopping in the Net by looking at the teleshopping offers available on cable television. Here, in spite of the excellent opportunities for a "broadband" presentation of products, Home Order Television has calculated that for every 30,000 units of accommodation turnover of no more than $ 250,000 per year is generated.

It is now beyond dispute that the sale of books and CDs through the Internet functions well and that considerable revenue is also produced in the field of software distribution in the Net. What remains open to question, however, is whether the total amount spent by an average European household on such products is sufficient to ensure that the necessary infrastructure pays for itself. Doubts also persist regarding who should furnish the initial outlay that is required: the individual user, the household, the supplier, the beneficiary of orders or transactions, the network operator or the service provider?

Changing structure of goods and services that will be acquired online

Even if it is cheaper going shopping in the Internet, it remains uncertain whether the gain from this rationalisation within the individual household is perceived as being relevant to an investment decision. Apart from making transactions easier, it is the supply of "multimedia" content that seems to be what is qualitatively new about the multimedia market. Yet so far little money has been spent by individual households on multimedia content in the Net. More lucrative has been the use of the new network for the distribution of familiar goods. This development will become all the more interesting, however, when the products to be marketed no longer depend on physical "carriers" such as records, CDs or video cassettes but can be transmitted and sold directly in the Net without any additional logistics involved (as with book delivery, for example).

To this extent, the retrieval of audio-clips is a new market worthy of scrupulous attention, a booming business (as the 1998 Cebit Home fair had us believe). The electronic sound-carrier now seems to have found its public – even if the retrieval of individual files continues to be pretty expensive once all the

additional costs have been taken into consideration. Given the financial hurdles that would be entailed in the field, it makes no sense providing a more comprehensive audiovisual entertainment offer in the Internet – capable of competing with interactive cable or satellite TV – until well into the next decade. The prices of licences for entertainment products such as feature films have been rocketing. Even with access to powerful transmission lines and the appropriate servers, a medium-sized supplier in the Internet will in the long run be unlikely to take the financial pace.

New audiovisual offers unlikely on a regional scale

Even if local suppliers have the clear advantage of being "in on things" in the regional context, a comprehensive regional programme will likewise be virtually impossible to finance or will always tend to appear "inferior" in the light of the production expenses borne by stations for film and TV productions. In public broadcasting services, costs of $ 1,800 to $ 2,300 per minute of programme[292] are taken as the norm for the fields of family and information programmes, which are similar to those that would have to be developed for commercial regional productions. Even if the costs per minute in commercial TV could be more than halved by clever management, this would still leave little scope for productions of their own and the corresponding return on investment.

Assuming that $2.50 could be raised per household and month for an attractive regional offer and that through cooperations between suppliers the targeted minimum potential of 300,000 households for the market opening could be attained, then – provided the money was not used for any other purpose – it would be possible to provide not much more than 600 minutes or a total of 10 hours of programmes per month at the level of commercial regional television.

For local and regional offers, television requires substantially greater viewing figures for a market opening than multimedia offers with few moving pictures:

> *"A survey carried out for North Rhein-Westphalia by Rinke Treuhand GmbH starts from the assumption that a population of 1.5 million is sufficient in order to be economically successful. This claim should be taken as allowing for considerable regional variations. Cologne and the surrounding area are judged to be viable, while Aachen, Bielefeld and Düsseldorf on the other hand are not. These calculations are based on local advertising expenditure and the market share to be reached in each case. Estimated advertising revenue of 13 million DM (1995) to 22 million DM (2000) is here assumed, as well as a 5% market share and a nationwide marketing combine in the production of features, the so-called 'cover' programme"*[293]

No supplier is likely to have such favourable conditions for its own local or regional programme all to itself. Not even a conurbation like Berlin Brandenburg, the most concentrated regional market in Germany, seems to have the necessary scope for a viable regional television service. Furthermore, the credibility of a monopolistic regional supplier would be more than dubious from the user's point of view irrespective of who actually finances the service. Quite apart from the lack

292 See Clevé 1998
293 See Clevé 1997, p. 74

of audience acceptance that can certainly be expected for a pure "landlord" TV, for example, such plans for comprehensive services from the housing industry would fail simply on account of the cost and refinancing factors mentioned above. Time and again new suppliers in regional and local television have financially overstretched themselves. The on-demand market for services of this kind will thus be even more difficult to establish.

Targeted offers need high-income customers

It is only really in Munich, if at all in Germany, that a reasonably viable conurbation-based TV has so far had any success, although here too it is a comprehensive regional programme that is being marketed and not a genuinely local one. The yearly budget of some $ 20 million makes it patently clear that even a number of small or medium-sized suppliers would be biting off more than they could chew with such a programme initiative. There is hardly a supplier outside the classical media companies acting on a national or European scale that would be capable of developing such a business field on their own. Newcomers from outside the branch further lack the commercial, methodological and aesthetic qualifications essential to the media field in order to develop a market of this sort on a local level anywhere near lucratively.

Audiovisual products from the Internet or an Internet-like on-line network for the private household will for the above reasons consequently be of little relevance for regional offers and for the time being continue to lag a long way behind text or graphics-oriented offers in terms of turnover volume. An exception here could be small, well-made inserts or trailers. For reasons of cost, the multimedia designers of smaller regional suppliers will have to fall back on other forms of multimedia representation than those of classical audiovision, with a final product that is print-based rather than TV-based. Costing prices per Internet page have to be fixed here, and these can vary greatly depending on the extent to which the initial company expenses can be met by the regional suppliers (from cornershops to department stores) themselves.

The "one-stop-shop" – integrated solutions for the private household

Multimedia one-stop-shopping expected by consumers

The integrated telecommunications supplier of the future need be neither a classical network or service provider nor the owner of flats or estates. It might equally well be a bank, a chain of department stores or quite distinct companies from a different branch but with direct access to the private household (for example book clubs or insurance companies). Given their mobility within telematic networks, telecommunications suppliers will not allow themselves to be tied down within narrow limits. Even so, a universal telematic general store entails a whole new set of financing problems and investment risks of its own.

Deutsche Telekom itself is the best example of such a general store. With its "mega" supply of goods and services, it on the one hand demonstrates the opportunities that exist for expanding value added possibilities beyond one's core activity, while on the other hand also drawing attention to the limits of a diffuse and dispersed general offer when it lacks a distinct profile.

Temptation for telecommunication companies to engage in providing content

Private customers look for an enterprise that can provide telecommunicative services that are reliable and comprehensive. This only becomes problematic when the quality is bought at the cost of the economic strength of the supplier, who ends up only being able to provide goods and services that are manifestly overpriced. Big businesses are prone to offer everything to everyone. With their easy access to customers and the corresponding customer links, telecommunications suppliers may well in future find themselves seduced into offering other supplementary services.

Beyond dispute is that the front door continues to be the main entrance to the most important purchase decisions. It is for this reason that the battle is now raging for the electronic entrance. Once this has been opened, the type of media access and the technology used to provide a product fade into the background. The aim is to earn money through all channels that private consumers can be electronically reached by.

The chairman of Bertelsmann, Thomas Middelhoff, has announced that the company's media products are in future no longer to be marketed by media type but rather thematically. The underlying supposition here is that the important thing for the consumer is the content and not the medium of distribution involved. The general assumption is that the developments sketched above will lead to a whole new selection procedure.

Integrated marketing is a double-edged sword

It is common knowledge that integrated marketing is a double-edged sword from the point of view of the private customer. Comprehensive service often comes coupled with the misuse of market power. The value added process in media and communications technology and its commercial translation to the market is here hardly any different from that in other industries. The type and number of possible scenarios leading to the expansion of the communications infrastructure in private households is diverse, to put it mildly, and full of contradictions. This should be taken into account in the planning of business models before they inevitably founder in the market through inattention. As far as they concern the private household, value chains have become more varied and complex, containing new links and intermediaries.

On balance, the technical problems associated with integration and convergence can clearly be seen to be insignificant in comparison with social and economic problems. Technical feasibility has long since ceased to be an automatic guarantee of market success in the fields of communications, entertainment or information. The integration of private households into the new network economy needs considerable forethought and planning. The above facets of an integrated view of a new private household economy are anything but complete. They are intended to serve as the foundations and guidelines necessary for an adequate development model.

5.3 The Internet Economy and Regulation

5.3.1 Political Implications of the Internet Economy: Why regulate, after all?

The analysis thus far has concentrated on the (new) rules of the Internet economy, the strategic implications these rules may have for business enterprises, and the connections that may arise with possible and probable changes in user-behaviour and patterns of usage. The question of regulation has as a result played but a bit part so far. It can clearly not be ignored, however, that not only has the media and communications market up to now been substantially determined by government regulatory measures, but above all the developments of the future depend on the systematic political measures (limitations and stimulations) that are yet to take effect. Market developments will be crucially influenced by whether basic conditions are actually imposed in particular areas, by the aims underlying any such policy and the ways in which these aims are to be achieved, and by the institutional framework in which the government regulations are embedded.

Economic development needs an appropriate regulatory framework

On a technological level, digital technology has meant that convergence is already a fact of life, in turn providing the basis for further technological progress and the economic advances this permits. On an economic level, convergence presents the firms taking part (as well as those affected only indirectly) with great challenges. Having previously operated in separate markets, considerable adjustments to the new conditions are necessary if chances are to be taken and hopes realised on the road to an information society. On a political level, efforts are thus being made to find an appropriate regulatory framework that permits, and possibly even promotes, a more rapid development of supply and demand in Europe at least, while at the same time also keeping the public interest – itself in need of rethinking and possible redefinition – firmly in view. There are two good examples of this discussion process stemming from Europe, firstly the British "Consultation Paper" on the regulation of communications in the Information Age,[294] and secondly the reports from the German Enquete-Kommission on "The Future of the Media in Economy and Society; Germany's Route to the Information Society."[295] The discussions focus on regulation content (priorities), regulation styles (sovereign, arbitrational, self-regulating) and regulation competences (vertical and horizontal).

The line of argument currently prevalent in Europe can be summarized in such notions as confidence in the market and competition, rigorous deregulation and simplification in the application of the law, and the priority of self-regulation in the face of outside regulation. Within the EU the principle of subsidiarity holds sway. According to this as much as possible is to be regulated on a national level and only as much as is necessary on the common European level. Much of this is right, but the prevailing consensus is only in part the result of mutually corroborative analyses and a common will to action. In part it is also the expression of a lack

294 vdti 1998
295 Enquete-Kommission 1998

of appropriate concepts in the face of the undeniable failure of traditional government politics. The debate about (de-/re-)regulation has all too often assumed the character of an argument between distinct doctrines: "government failure" and "market failure" are not analysed in a sober fashion but bandied about as a way of apportioning blame. On the one hand, sweeping (but understandable) denunciations of excessive regulation impede a clear-headed assessment of the social and economic role regulation undoubtedly plays. On the other hand, an exaggerated (but equally understandable) defence of the social virtues of regulation does not get us very far when what is at issue is what needs to be regulated and how this can be achieved most "adequately". The current situation is characterised not only in Germany by a "paradoxical situation of asymmetrical deregulation": "while the network of potentially dysfunctional regulations for specific areas continues to proliferate, in Germany some of the cornerstones essential to the requisite government objectives have been thrown into question, and the government guidelines necessary for company investment policies have been delayed to such an extent that one must even speak of a regulations gap."[296]

Predominant rhetoric on deregulation shows a severe lack of appropriate concepts

The two traditional patterns of argumentation mentioned above start from the assumption that it is either the state (through over- or misregulation) or the companies (through under- or bad investment and/or monopolistic conduct) that have contributed to development problems in various areas. The Internet economy suggests a different interpretation. Both the government's convergence and deregulation policies and the market development strategies devised by companies are based (in part) on old economic models, and both thus fall short of the mark. Newer interpretations of the Internet economy allow a more realistic view of things to be taken from which possible guidelines for action can be derived not only for the companies but also for the political bodies concerned. This applies both to the restrictive regulation of company behaviour deemed unacceptable (i.e. illicit competitive behaviour and/or conduct that is harmful to the consumer) and to positive regulation in the form of government guidelines for developments considered acceptable and the political, legal and/or financial measures to promote them.

Government failure and market failure

The most important conceptional change in regulation policies lies in the field of competition. There are two reasons for this. Firstly, it can be assumed that convergence will result in regulations specific to traditional sectors having to be given up and that as a consequence more universal competition policies will come to assume increasing significance. Secondly, the nature of market dominance – essential to an analysis of laws regulating market competition – will have to be redefined, with greater weight placed on dynamic market perspectives as opposed to static market models. The significance now attached to standards and the special revenue models now in operation (such as "follow the free") make rapidly developing monopolies almost a defining characteristic of the Internet economy. At the same time, the short innovation and product-cycles and the incalculability of user and buyer behaviour mean that such monopolies are under considerable

Competition policy – two reasons for growing significance

296 Zerdick 1995, p. 73

structural threat and thus in principle less enduring than the classical oligopolies of traditional economic models.

Market structures become more transient

The overlaps and intersections between cooperation and competition characteristic of "business webs" make it increasingly complicated to develop a concept for new competition policies. Yet they are an integral aspect of tomorrow's reality and it is imperative that they should be part of any conceptual consideration of the field. They also hamper any attempts at market analysis conducted for the use of anti-trust commissions. Their theoretical and empirical foundations accordingly require further examination, with attention paid both to the extension and integration of value chains in the "value added communities" described above and to the various globalising components of (actual or potential) competition. Market structures can be expected to become more transient and thus less suited as a basis for laws regulating market competition. Market conduct and market results will gain in significance as criteria in the attempt to limit unacceptable business strategies (alleged or actual).

European governance of TV

Television programmes here in Europe tend to be evaluated much more highly than in the United States in terms of quality, while many technical and economic developments in media and communications are felt to be unsatisfactory. Selective measures to rectify these deficiencies have thus become a key aspect of government policies, be this at the level of the EU, the member state, or in Germany for example at the level of Land or municipality. The integration of the value chains in the Internet economy means not only that companies become dependent upon one another for success but also that government support must make a point of targeting such interrelations in order to have the appropriate effect.

Five types of deficiencies

The problems faced by businesses in crucial sectors can be summed up in terms of five typical deficiencies either on the supply or the demand side: deficiencies of standardisation, of cooperation, of financing, of acceptance and of willingness to pay. A market is felt to be flawed enough to warrant political intervention if – though regarded as innovative or important – it nonetheless fails to develop or develops only haltingly or markedly more slowly than in comparable countries. (This constitutes "market failure"). The signs of structural market failure – failure that cannot be remedied by the market participants alone – lead to demands for new strategies and forms of "structural coupling" between market and politics (in turn anchored in a legal base), i.e. for new regulations. The essential elements here are easy to define. Credible objectives are required, as well as moderation and coordination in the rapid development and effective implementation of standards and basic conditions. There must be strict limitations in important questions of principle (for example through regulations on certain contract clauses in the case of companies dominating the market, or through the establishment of clear rules aimed at consumer protection – such as the possibility of contacting the customer service by phone within 90 seconds – instead of targets formulated in purely general terms). Effective deregulation is

necessary to provide markets with swift relief from excessive regulation and to focus government institutions on essential questions of principle. At the same time, flexible re-regulation is required to do justice to technical and economic developments without harmful delays.

Action needed – quickly, and effectively

Given these objectives, we are faced in Europe not only with a partially flawed market but also with a (more than partial) failure on the part of the government. Fortunately, however, we do have concrete starting points for defining the core areas for government regulation and its institutional framework. From the viewpoint of the Internet economy, the most important thing is to translate them into action as quickly and as effectively as possible.

5.3.2 Realms of Governance: What remains to be regulated?

Regulation of the future – more abstract, more concrete, more experimental

The globalisation and the convergence of the media and communications markets will result in an increase rather than a decrease in the demands made on the level(s) of regulation. In order to prevent these demands from becoming excessive in the long run, regulation must simultaneously become more abstract, more concrete and more experimental. This perhaps sounds like a contradiction, but it is not. Regulation becomes more abstract when a conscious effort is made to dispense with rulings pertaining to a large number of individual aspects, and instead of this the focus shifts to key fields and fundamental goals. Regulation becomes more concrete when these goals are not expressed in terms of general wishes but clear and solid guidelines. Regulation must become more experimental because neither the development of new markets nor the effectiveness of new regulatory approaches can be sufficiently predicted in advance.

> *"As new services become more significant, changes to regulatory aims and methods will be needed. Making changes early in the process might help to ensure a regime appropriate to the digital future at an early stage. It could give (our country) 'first mover' advantages. However, it is impossible to be sure how the markets will develop. We risk developing a system of regulation around a prediction of what the digital world will look like which, if it turns out to be mistaken, will leave the regulatory system obsolete."*[297]

Yet even with this emphasis on the experimental nature of regulation, it should not be forgotten that excessively rapid (or simply uncertain) changes to the rules can be problematic for investing companies (and thus also for market development).

Withdrawal from micro-regulation, and stronger emphasis on issues of principle

Technological possibilities and market conditions in the digital world change too quickly for regulations to be able to keep up with them, let alone get ahead. The impression is one of perpetual political failure. One possible way out of this situation is to retreat from this intensity of regulation (i.e. deregulation), at the same time reinforcing a "strong" and effective regulatory framework as protection against infringements in fundamental matters, e.g. going beyond preannounced

297 dti 1998, p. 5

intervention thresholds. Strong regulation of this kind is indispensable for European and national fair trading law, questions of copyright, the protection of children and young people, data protection and consumer protection. This is not the place for entering into details, but we do at least want to name the main systematically deducible fields for action that have emerged from discussions as being grounded in consent and promising for the future.[298]

British "Consultation Paper" defines core public policy objectives

In July 1998 two British government departments presented a joint programme – "Regulating communications: approaching convergence in the Information Age"[299] – that is of particular interest in the context of regulations that in many areas are traditionally considered exemplary. Three aspects are especially noteworthy. Firstly the programme is published only as a "consultation paper," and, in spite of the outstanding analysis it presents, its reflections on future regulation are couched in the form of questions put forward for public discussion. Secondly the document includes telecommunications and expressly covers not only commercial but also public broadcasting, with reflections on the regulation structures that could unite all these fields. Thirdly the core areas of government regulation ("public policy objectives") are defined in terms of six general economic fields and five demands on media content that could apply equally well to other countries and the EU in general:[300]

Economic and social objectives

- access
- universal service
- consumer choice
- achieving competitiveness
- promoting investment
- fostering competition

For the economic and social objectives, the fundamental principle is that "the Government will regulate, where necessary, in the consumer interest, but regulation will be no more than is required to achieve the necessary protection. Wherever possible, the Government will support and encourage the operation of the market to encourage investment and the development of new services."[301] Universal service and consumer choice are concepts that have long since been the basis for regulation in the USA but are now increasingly coming to be so in Europe too. Yet experience in the USA makes it clear how difficult it can be to put these relatively simple and straightforward goals into practice when they have to be coordinated with one another and confronted with the imperfections of the real market.[302] The concept of "universal service" means that the first thing to be decided is which standard should apply for a product such as the telephone service.[303] It then has to be established which supplier(s) are to be given the responsibility for "universal service." And finally it must be worked out how high the resulting costs will be, these possibly being met by payments from third parties.[304]

298 This is illustrated by numerous examples of new regulatory mechanisms. An acknowledgement and analysis of these warrant an investigation of their own, however.
299 dti 1998
300 dti 1998, p. 13ff.
301 dti 1998, p. 13
302 See also Economides 1999
303 In our conversations Spencer Reiss coined the term "universal crummy service": this is what applies when (cross-)subsidies are used to cover only a minimum standard for everybody, while additional functions are offered at market prices.
304 The hypothetical cost models necessary for this have hundreds of variables – and yet are hotly disputed.

Objectives for content-based services[305]

- plurality of voice
- impartiality
- diversity of content
- high quality of content
- taste and decency

Special regulatory issues for broadcasting

In determining these demands on content it is assumed from the outset that in the field of broadcasting both positive regulations (which define certain aspects of content considered desirable) and negative regulations (which exclude content considered unacceptable) will continue to be necessary as long as (1) broadcasting has a universal reach, and (2) is freely accessible to all; (3) audiovisual programmes are (politically and culturally) particularly powerful; (4) a high level of institutional control is necessary (given shortage of frequency) or possible (given concentration).[306] All these factors are independent of the form of distribution (analogue or digital; cable or terrestrial/satellite). Distribution through the Internet likewise fulfils the fourth criterion, if control of the market and public opinion is taken to replace shortage of frequency. However, there are real limits to the likely effectiveness of content regulation in the new media age.

Suggestions for Germany

For Germany, however, closer thought on the matter seems to be called for if the level of the British (and European) discussion is to be reached. What is required in general is for regulation and financing to be coordinated in such a way as to ensure that state subsidies and/or finance guarantees (as with radio or TV licence fees) can be explicitly combined with clearly defined objectives. Thought should be given to the possibility of bundling together competences for the entire dual broadcasting system and for telecommunications as regards both content and institutions. The discussions on the matter alone could achieve a great deal. For broadcasting as a whole, it makes sense to impose clear (and binding) deadlines for the conversion to digital technology, and for the public broadcasting service a binding function assignment should be fixed and in conjunction with this a commitment to financing exclusively by radio and television licence fees. Fixing the function assignment could also include conferring a pioneering function for digital ranges (DVB and DAB) on broadcasting corporations financed from public funds. Even a role as an Internet access provider – along the lines of a basic supplier – seems feasible within the context of a newly defined function assignment. The aim is not so much to occupy a market as to open it up and develop it. In this way public broadcasting will quickly be able to open up socially beneficial new fields not only for itself but for private companies as well.[307]

Two further areas that can be ascribed to government regulation appear to be of particular significance for the continued development of the media and communications sector: taxation, which can (or should) take a form that either boosts, impedes or is neutral towards the development of the market, sector or

305 dti 1998, p. 13ff.
306 dti 1998, p. 13ff.
307 A good example of this strategy in the field of TV technology is NHK in Japan.

"The US policy framework ... makes rational people less willing to trust their safety on the Internet."

product in question; and the creation and/or maintenance of confidence in the transmission process in terms of data security and data protection.

Data security and privacy

In the sphere of data security and data protection, Europe is at a clear advantage compared with the United States. This could well contribute to the economic development of the markets too, especially once the phase of experimentation on the part of technophiles has been and gone, and the transition to daily use by the majority of the population has taken place.

In the USA the use of reliable coding procedures is limited for reasons to do with combating crime. It is thus accepted that firms and private persons cannot always achieve the data security they might regard as necessary and for this reason do without using the networks on such occasions, in turn leading to a failure to exploit the full potential of network use. In Europe similar considerations produced an outcome in favour of greater coding and thus greater data security. In Germany it is even assumed that a specific competitive edge is thereby produced for software companies who specialise in coding systems.

Lack of privacy protection can be an impediment to commercial use of the Internet

A part of possible Internet use is thus obstructed because users are afraid of their data being misused by third parties. The use of "cookies" to gain automatic access to figures for Internet use and the reuse of these in the context of so-called "data-mining" generate additional problems of unwanted data use. These range from the transmission of advertisements to the uncontrolled divulgence of data to other companies, and on to the fraudulent use of credit card and bank information for criminal purposes. It is obvious that this level of insecurity is a hindrance to developments and thus that data protection (quite apart from its immediate objectives) can also make a positive contribution to market development (confidence-building as a bridge to acceptance). The US government wants to leave both data protection and consumer protection to the forces of self-regulation.[308] A critical voice from the Silicon Valley comments on the resultant competitive disadvantage for the United States:

> *"Contrary to the available evidence, the administration still maintains that private businesses can somehow regulate themselves when it comes to protecting people's privacy. Not only is this visibly untrue, but laissez-faire simply won't cut it with the European Union, where governments have very different attitudes. The EU has adopted tough rules designed to prevent the misuse of personal information collected in databases, and U.S. bleatings about self-regulation are falling on appropriately skeptical ears there. The latest policy framework also, as in the past, all but ignored encryption, the scrambling of data to keep it from prying eyes. Once again, the administration persists in a policy – making life difficult for Americans who want to use or sell strong encryption – that has two negative effects. It makes U.S. companies less competitive, and it makes rational people less willing to trust their safety on the Internet. These policies don't help the Net's growth; they impede it."*[309]

308 See The White House, Office of the Press Secretary: Memorandum for the Heads of Executive Departments and Agencies, Subject: Successes and Further Work on Electronic Commerce. November 30, 1998; www.oma.eop.gov.us/1998/11/30

309 Dan Gillmor in: San Jose Mercury News, December 1, 1998; www.mercurycenter.com/columnists/Gillmor

Taxation

The question of taxation as a means of state regulation for the media and communications sector is among the most interesting areas for further discussion. The current US government has set itself the target of promoting the Internet's development by – among other measures – dispensing with (additional) taxation, as well as imposing this promotion on an international level. Two fundamental positions can be distinguished here. The radical position represented by Nicholas Negroponte, whose text "Taxing Taxes" is printed at the end of this section,[310] starts from the assumption that immaterial information goods will end up completely replacing material ones. This proves decisive to the extent that it calls the physical and spatial basis of taxation into question. Making fun (justifiably enough) of the notion of a "bit tax" (a notion which is still under discussion), he comes to the conclusion that in the end real estate will be the only remaining basis for any form of taxation.

Negroponte: taxes just won't work on the Internet

The second – more realistic – position is one held not only by governments in Europe but also the federal states in the USA. It is summarised (and taken even further) by Gillmor:[311] to the extent that the Internet facilitates transactions related to material goods (and in the future too this will continue to be the vast majority of transactions), tax exemption for Internet trade represents an unfair advantage for the electronic mail order business over all other forms of trade and also an erosion of the tax basis of federal states and municipalities (to the detriment of poorer members of the community). The complexity of payment and distribution procedures that American discussions have focused on are taken by Gillmor as an opportunity for making fun of their supposed worries:

Gillmor: tax exemption of the Internet unnecessary and unfair

> *"Suppose Congress agreed that mail-order, telephone and Internet sales – they are identical for this purpose – should not keep their current loophole. Defenders of the exemption worry aloud about the complexities of doing away with it. They worry too much. Where to tax the item is the easiest part. ... Another hang-up is the complexity of figuring out state taxes, and then sending payments to the states. This sounds like a software problem to me – but not all that tough of a problem, especially for online merchants who, after all, are doing their business with computers."*[312]

The question then remains whether tax encouragements are wanted for business taking place in the Internet or partially facilitated by it, and if so, in what measure. If electronic trade entails lower transactional costs, then the tax payable is correspondingly lowered – and tax encouragements seem unnecessary. If selective (and probably temporary) tax encouragements are considered desirable to promote the development of this form of trade, then either just the VAT share of the Internet agent could be dropped (or reduced) or the traded product as a whole could be exempted from taxation. In any case, it is certainly feasible that medium-term tax measures (or the decision to do without them) will have a greater effect on development than other types of state regulation.

Tax incentives as a type of government regulation

310 Negroponte 1998b
311 Gillmor 1998
312 Gillmor 1998

"Maybe the information highway metaphors have gone to the heads of digitally homeless economists, who think they can assess value by something akin to counting cars."

Taxing Taxes

by Nicholas Negroponte, MIT Media Lab

The mind-set of taxes is rooted in concepts like atoms and place. With both of those more or less missing in cyberspace, the basics of taxation will have to change.

After discovering the basic principle of electromagnetic induction in 1831, Michael Faraday was asked by a skeptical politician what good might come of electricity. "Sir, I do not know what it is good for," Faraday replied. "But one thing I am quite certain – someday you will tax it." Little did he know how right he was, though more than a century would pass before the word *bits* existed.

The idea of taking a tax bite out of digital communications comes courtesy The Club of Rome, specifically Arthur Cordell and Ran Ide's 1994 report "The New Wealth of Nations."

More recently, redistributing the benefits of the information society has been championed by influential economist Luc Soete, director of the Maastricht Economic Research Institute on Innovation and Technology. Despite their repute, supporters of such a bit tax are clearly clueless about the workings of a digital world.

Tax bytes

A typical book contains about 10 million bits, which might take even a fast reader several hours to digest. By contrast, typical video – digital and compressed – burns through 10 million bits to produce less than four seconds of enjoyment. A bit consumption tax, in other words, makes no more sense than tariffing toys by the numbers of atoms. Maybe the information highway metaphors have gone to the heads of digitally homeless economists, who think they can assess value by something akin to counting cars.

Of course, collecting taxes can be tough enough without trying to assess something you can't see, especially when you don't know where it is going or coming from. This helps explain why the Clinton administration in late February reaffirmed its commitment to making cyberspace a global free-trade zone. The policy's purpose, the brainchild of White House senior adviser Ira Magaziner, is both economic stimulus and practicable fairness. So whether or not Congress has kept its promise to vote on the related Internet Tax Freedom Act by early spring, the legislation has the full force of careful deliberation – and historical inevitability – behind it. For these and other reasons, Europe abandoned the bit tax. But the idea still survived three and a half years of consideration, despite the growing awareness that bits by their very nature defy taxation.

The locus pocus of sales tax

Even so, the principled position taken by Clinton and Congress comes, in part, because making the Net a free-trade zone works for the US federal government. The treasury derives most of its revenues from personal and corporate income taxes. If the economy sees a boost from any form of free trade, the Feds will see a proportionate rise in their own intake. Simple arithmetic.

However, many countries and most states don't work that way. Instead, a sales tax is the means – often the principal means – of filling government coffers. Ohio governor George Voinovich, chair of the National Governors' Association, declared that the Internet Tax Freedom Act "represents the most significant challenge to state sovereignity that we've witnessed over the last 10 years". Both he and the act may be right.

The sales tax is also particularly popular among bureaucrats in developing nations, where collecting income tax is even harder because the poor make so little and the rich can avoid so much. Plus, the sales tax turns retailers into a nationwide web of tax collectors. And the tax is "fair" because it's based on what you spend versus what you earn.

Still, Voinovich and company would be smart to start looking elsewhere, because their receipts will plummet as we buy more and more online, especially if what we buy are bits.

The VAT vat

While the sales tax is fairly commonplace, the value-added tax is more or less unknown in the United States. Loosely speaking, it taxes the various stages of transforming raw material into a finished product, the last stage of value added being what you pay at the retail counter (and get back at the airport's VAT-refund counter).

This kind of tax makes even less sense in the world of bits.

Assume that bits are my stock in trade and I use Microsoft Word to refine my raw material: Should I pay a VAT for spellchecking each story? Should I pay a VAT to have it encrypted and another to have it decrypted, not to mention on each of the layers of value added by various editors? In fact, as a cheerful taxpayer, if I have to pay taxes on bits – at least those that make up words – I would be willing to pay a higher VAT for the fewest possible bits: just the right ones, please. That would be value added indeed.

Jurisdiction in jeopardy

But the most taxing aspect of cyberspace is not the ephemeral nature of bits, the marginal cost of zero to make more of them, or that there is no need

for warehouses to store them. It is our inability to say accurately where they are. If my server is in the British West Indies, are those the laws that apply to, say, my banking? The EU has implied that the answer is yes, while the US remains silent on the matter.

What happens if I log in from San Antonio, sell some of my bits to a person in France, and accept digital cash from Germany, which I deposit in Japan? Today, the government of Texas believes I should be paying state taxes, as the transaction would take place (at the start) over wires crossing its jurisdiction. Yikes. As we see, the mind-set of taxes is rooted in concepts like atoms and places. With both of those more or less missing, the basics of taxation will have to change. Taxes in the digital world do not neatly follow the analog laws of physics, which so conveniently require real energy, to move real things, over real borders, taxable at each stage along the way. Of course, even analog taxation without representation is no tea party.

Getting physical

Looking ahead, taxes will eventually become a voluntary process, with the possible exception of real estate – the one physical thing that does not move easily and has computable value. The US has a jump-start on the practice, in that 65 percent of local school funds come from real estate taxes – a practice Europeans consider odd and ill advised. But wait until that's all there is left to tax, when the rest of the things we buy and sell come from everywhere, anywhere, and nowhere.

Wired Magazine, May 1998, p. 210.
With permission from Nicholas Negroponte.

"One of the main tasks faced by regulatory policies of the future will be to make the jurisdiction simpler, more transparent and better integrated."

5.3.3 Structures and Institutions: Who should be responsible?

Established regulation – ineffective at best, counterproductive at worst

Today's competences for regulation are becoming increasingly unsuited to the continuous processes of convergence and integration in the media and communications markets. Established regulations are as a result proving at best ineffective and at worst counterproductive.

Germany leads in regulatory complexity

The decision-making bodies are too broadly dispersed and not sufficiently coordinated. It is too complicated and takes too long to implement what is decided upon. In many cases this makes it impossible to achieve what needs to be achieved. In Germany in particular, the regulatory framework for digital broadcasting is splintered both by sector and vertically: the European Union, Federal Government, the sixteen Länder; the regulatory authority for post and telecommunications, fifteen Landesmedienanstalten (regional licensing authorities), the KEK (the Commission of Enquiry into Media Concentration), the Supervisory Committee for the Public Broadcasting Service, the KEF (the Commission of Enquiry into the Financial Requirements of the Public Broadcasting Service). This splintering has led to a lack of strategic orientation in the market participants and if anything has put a brake on market development.

One of the main tasks faced by regulatory policies of the future will be to make the jurisdiction simpler, more transparent and better integrated.

Convergence leads to regulatory overlap

The formulation of these objectives is based on one assumption. It can also lead to a natural confusion. And it permits a number of particularly constructive conclusions to be drawn as well. The assumption is that the technological and economic convergence between the various branches of the media and communications industries is leading to increasing overlap between traditional regulatory tasks and thus to more and more conflicts of aims – and that these conflicts of aims can be resolved by a reduction in the intensity of regulation and by cooperation (or amalgamation) between the institutions involved. In this context, Herbert Ungerer has made the important point that ultimately it is not just media and telecommunications that are converging but also telecommunications and financial services or telecommunications and trade.[313] The overlap between regulation fields and institutions may thus come to affect an even greater problem area.

WorldWideWeb, worldwide regulation – an attractive proposition, but thoroughly misleading

This assumption can give rise to a natural confusion. If branches are converging and regulation fields overlapping, it may at first glance seem to make sense to have regulation covering the whole area implemented by a single institution – no more inconsistency or conflicts of aims, just a maximum of efficiency. Fortunately, this train of thought quickly leads ad absurdum. As the convergence of sectors in our field goes hand in hand with the globalisation of markets, unified regulation of this sort would have to be anchored in a single institution worldwide – not a nice thought even for the most hardened supporters of regulatory

313 Quoted in Winsbury 1998

simplicity and clarity, given our experience with international decision-making and enforcement processes at the United Nations.

Extended scope of international treaties

The patent impossibility of unified regulation does however give scope for another option that is successfully applied in other fields (such as company assessment and the reduction of trade restrictions through the WTO): the existing institutions continue to exist, keeping more or less the same scope for various individual rulings while coming to joint agreements on unified regulations in questions regarded as particularly important. This concept of "harmonisation" seems to make sense in principle, but it can prove dysfunctional in the media and telecommunications sector, where the process of development is central:

> *"Ironically, for a world full of conflict telecommunications are probably the one sector that historically has been cursed with an excess of policy collaboration and with a compulsion to protect. But today's priorities are not the international solving of every problem, but the creation of opportunities for the new information age. Where problems emerge, they can be dealt with at the time. ... It might be different tomorrow. But today, the world of telecommunications needs more policy experimentation and less harmonization."*[314]

Different issues on the national and on the EU levels

These considerations only apply on an international level, however. A different structural analysis is required for the situation within the individual countries as well as for the EU as it grows together as a unified market.

The US model

It continues to be taken for granted that the structures in the United States (having grown naturally and without any explicit intention) are particularly suited to the process of convergence taking place at present. The Federal Communication Commission (FCC) has been responsible for telecommunications and broadcasting since its inception. This horizontal integration clearly makes it easier to resolve such questions as relate, for example, to the spread of broadcasting in the Internet. The vertical separation of powers and responsibilities between the FCC and the state-run "Public Utility Commissions" was changed in 1996, granting increased competence to the FCC. But this process is by no means over yet, as became apparent in California in December 1998, when the firm PDO tried to offer cheap Internet connections on telephone lines belonging to Pacific Bell (still a market leader in local telephony). Pacific Bell rejected this, and appeals to the Administrative Court found it incapable of reaching a verdict, since the matter did not fall (or no longer fell) within the competence of the federal state. PDO then appealed to the California Public Utility Commission, requesting it not to endorse this interpretation of the law and to oblige Pacific Bell to make their lines available. As this report goes to print, the issue is still undecided.[315] Within the USA the FCC model comes in for some pretty heavy criticism from various quarters. Yet this criticism is concerned less with fundamental questions of competence than with internal developments felt to be undemocratic or overbureaucratised. From a European perspective, how-

---

314 Noam/Singhal 1996, p. 780
315 Healey, John: Fight over phone lines, In: San Jose Mercury News, December 16, 1998; www.mercurycenter.com/highspeed121798

ever, two aspects of the FCC model remain of interest: firstly the above-mentioned horizontal and vertical bundling of competences, and secondly its own understanding of its role less as a sovereign regulatory institution than as working in the public interest (and with public interests in mind) as a stimulus and mediator for the economy.

Models for the European Union

The discussion within the EU was initiated on a broad front in late 1997 with the "Green Paper on the convergence of the telecommunications, media and information technology sectors."[316] As regards institutions, three options were put forward for discussion: (1) building on the existing regulation structures; (2) developing a separate regulation model for the new services, to exist parallel to the legal regulations for telecommunications and broadcasting; and (3) gradually introducing a new regulation model that embraces numerous existing and new services.[317] This third variant was clearly the one that involved the greatest changes and thus also the greatest risks, while the second – given the conditions of convergence described in detail above – was visibly dysfunctional. It came as no surprise, therefore, that the first variant – conservative in institutional terms but at the same time open in terms of content – was the one to gain by far the greatest support in the ensuing discussions.[318]

A reasonable suggestion by Rex Winsbury: an independent pan-European regulator

This does not alter the fact, however, that technological progress is advancing apace, the European market is continuing to grow together, and the convergence of problems will thus keep up the pressure for a reorganisation of regulations. One of the reasons that the ideas of Rex Winsbury[319] are so stimulating for the debate still to come is that they envisage an institution independent not only of the individual member state governments but also the European Commission.

Why a pan-European regulator for telecoms is only a matter of time

by Rex Winsbury, InterMedia

The question of whether there should be a pan-European regulator for telecommunications is now on the political agenda of the EU. (…) New legislation (if any is needed) could be in place by about the year 2002, when other telecoms reforms (if any are needed) would also come into force. The year 2002 is also the date at which a major extension of the EU could take place, to bring in a large number of new member states. It is also the date at which UMTS (Universal Mobile Telephony Service) could make its debut in Europe. These three events in 2002 are not unconnected when considering the need for a Euro-regulator.

The cautious approach

EC Commissioner Martin Bangemann (…) recently said, with uncharacteristic caution, that the issue of a pan-European regulator was still too politi-

316 European Commission, DG XIII/A4 (ed.) 1997
317 European Commission, DG XIII/A4 (ed.) 1997, p. 41-42
318 European Commission, DG X (ed.) 1998, p. 11
319 Winsbury 1998, excerpts of which are printed below.

cally sensitive, and that he preferred to start with more cooperation between national regulators. Many countries have after all only recently acquired a national regulator, and argue that it is too soon to launch yet another upheaval. Even the UK government, so often a European trend-setter in all these areas … does not intend to tear up the present regulatory structure in the UK, with its distinction between telecoms and television – not yet, anyway. But the grounds for this decision seem to be more of a prudent refusal to anticipate a still-unclear future by premature change, rather than any affection for existing structures. …

For (the UK and) Europe as a whole, the question is, what pressures could cause this cautious political approach to be abandoned, how forceful are these pressures, and when could or should change take place?

Where should power lie?

The question of a Euro-regulator is not a new one, and has often been phrased as ‚should there be a Euro-FCC?‘, equivalent to the US system of telecoms governance through the Federal Communications Commission. There is a certain irony in posing the question this way. For, to that hypothetical visitor from Mars, it would surely be perfectly obvious that the European Union does already have a Euro-FCC. It is called the European Commission (EC). The EC has wielded enormous power and influence, from grand theory right down to fine detail, over both the structure and the economics of European telecommunications. …

The original refusal of the governments of the EU member states to create a pan-European regulator was based on the principles of the Maastricht Treaty, notably the doctrine of ‚subsidiarity‘, which ordained that if a thing could be done at national level, it should be done at national level. Only if it could not be done at that level, should it be moved to supra-national EU level. And the right to regulate was something that governments were reluctant to give up, for both political and revenue reasons.

So it is now clear that, at the political level, creation of a pan-European regulator would demand not one, but two surrenders of power if that regulator is to be no mere figurehead, but an effective force. One surrender of power is by the governments of EU member states. The other is by the European Commission itself. And that might be equally difficult – and desirable. …

Pressures for change

Where would the pressure come from to drive these changes and create an ERA (European Regulatory Authority)? Not from one source, but from several, but all pointing the same way.

a. Pan-European licences – For example, there is that long-held ideal of a pan-European telecommunications licence, for a particular type of service or activity. This was an idea built into the original concept of a liberalised telecommunications regime in Europe, to make it easier for companies to compete anywhere withµin the EU, and to bring telecommunications into line with the ideal of the ‚single market'. … But … governments resisted it. So it is still the case that operators have to apply for 15 separate licences. … This is surely a nonsense … what sort of a single market is that? Perpetuation of 15 separate regulators would itself institutionalise the fragmentation of the European market.

b. 15 regulators, one system? – The present approach is to ensure that regulators from the different countries meet frequently enough to create a common position and sort out disparities. The EC is also trying to ensure that all 15 regulatory bodies have similar (and good enough) levels of staffing and expertise. – Fine. But if they are all doing the same thing with the same resources, why have 15 separate bodies? One body would be cheaper, and more efficient, and better able to argue trade issues with the USA and other global protagonists … – the voice of Europe.

c. UMTS and frequency management – UMTS will be an important gateway technology into mobile multi-media services. But it could … become a mess if frequencies are allocated to it by a patchwork of national regulatory authorities, if charges, fees, any auction procedures and the licensing system vary country by country, if spectrum allocation for individual services varies within allowed frequency bands, or if there are different rules about who can offer which new services.

d. Numbering – Numbering can be looked at in the same way. Numbers are a scarce resource, like frequencies, where a single market demands the same numbering for the same (or identical) service offered across boundaries. So should there be a trans-European 'numbering space', and who should administer it, if not a Euro-regulator?

e. Product life-cycles – With the commercial life-cycle of telecommunications products getting shorter and shorter, perhaps now measured in months rather than years, there should be one reference point for approvals and licences, not 15, simply to cut out delays that slice away profitability, and for Europe to stay competitive in global telecommunications markets. This problem is made worse by the fact that licence categories as well as fees and time taken to issue licences, are not always the same across the member states of Europe. Are such different licence regimes sustainable in a single market?

f. Relation to incumbent – Then again, take the still dominant position of the so-called incumbent operators in the main EU countries. There is

almost bound to be a close relationship between a national incumbent and a national regulator, sometimes over-friendly, sometimes over-hostile, but hardly arms-length. Industry pressure will surely grow for single rules for a single market, administered by a clearly independent body.

Which model?

The word from the European Commission is that the basic question now to be examined for telecoms regulation in Europe is this: for any given issue, what is the appropriate level for its resolution? Then, if the answer is at EU level, what is the best way to do it?

EC officials stress that the FCC in the USA is not the only model upon which a pan-European regulator could be founded. But it is hard to see how it could be very different. For one thing, the Euro-FCC would probably have to be politically answerable to the European Parliament (much as the FCC is answerable to the US Congress) in order to detach it from the European Commission.

For another, it would surely have to cover both telecommunications and broadcasting (to use an already anachronistic term). By the time any such Euro-FCC gets created, the process of ‚convergence' will have moved on much further, and it will make no sense to set up a new body based on old distinctions that no longer apply, and which would inhibit it from taking key decisions in the new all-digital environment.

Not whether, but when

The evidence cited here surely points to increasing pressure from industry for EU governments to recognise that the doctrine of subsidiarity is inimical to the growth of a genuine single market in telecoms in Europe. It is just a matter of time: but not too much time.

Excerpts from: InterMedia, the Journal of the International Institute of Communications, ISSN 0309 11 8X, October 1998, pp. 33-35.

Chapter 6

Facts and Figures – Appendix

The following appendix of tables (and figures derived from them) gives an owerview of the developments of European media and communication markets, and – to allow benchmarking-type comparisons – of those in the USA and Japan. As a rule, the most recent five years covered by consistent data have been included; for basic demographic and economic data we have included projections for future developments. This appendix thus supports our analysis in the preceding chapters.

- Section 6.1 shows the most relevant basic demographic and economic data which set the framework for media and communication markets. It includes prognostic values which in these highly aggregated dimensions are generally less susceptible to error than those for individual markets.
- Section 6.2 is dedicated to distribution infrastructures and private household access to these infrastructures and to services measured in household equipment.
- Section 6.3 covers important industry characteristics, i.e. research and development, production and supply.
- Section 6.4 deals with spending on information and telecommunication technologies as important categories of demand by the business sector.
- Section 6.5 covers advertising markets, where demand (for advertising) comes mainly from the business sector, but is targeted predominantly towards private household comsumption (first of advertisements, and later of products and services offered).
- Section 6.6 finally deals with (predominantly private household) demand in the most important media markets.

In order to ensure that this appendix is clearly structured for ease of use and that it will facilitate comparative interpretation also across different tables, the following basic principles have been followed as far as possible:

- The upper part of each table gives basic information in terms of absolute values. In the lower part you will find ratios between these values and others which will emphasise the most meaningful aspect of the basic values. These ratios can be, for instance, growth rates, per capita or per household figures, and they will be shown in tables or graphs.
- Figures are usually given for the five largest European countries individually (France, Germany, Italy, Spain, UK), for the total of all EU member countries,

and for Western Europe (additionally including EFTA member countries). Whenever possible, figures for the USA and for Japan have been used to allow comparisons.

- If and when up-to-date consistent data have been available, figures for 1993 through 1998 have been aimed for; in some cases, we have accepted a small number of missing elements. In some cases, however, the most up-to-date consistent data do not extend beyond 1997.
- All monetary figures are given in ECU at the 1997 exchange rate. On the one hand this ensures comparability of data between individual countries, on the other it avoids potential misinterpretations due to fluctuations of currency exchange rates. Finally, this procedure will make it easier to compare the data presented here with future data after the introduction of the EURO.
- All per capita and per household figures have been calculated based on the data in section 6.1. Thus there is always a consistent frame of reference, and readers will have relatively little additional trouble calculating other values by themselves.
- The most important aspects of any given table are commented on in the text at the margin of the page.

In principle, preference has been given to completeness, compatibility and reliability of information rather than using snapshots of individual data which would be without context and the validity of which would be hard or impossible to evaluate. It should be noticed that in spite of a growing pool of data for instance at the European Audiovisual Observatory, at Eurostat or at the Advertising Information Group, there is still no or only very insufficient information on some important media markets. An example for these insufficiencies is the whole market for magazines. Also, relatively little light has been shed on the whole area of company and business related information, which has high access barriers due to their competitive potential.

The selections made for this appendix may leave some questions unanswered and many open to further research; also in a few cases the consistency of data could not be ascertained completely. However, due to the strength of the basic principles applied, the following tables will be helpful as a clearly structured and reliable database.

Chapter 6

Facts & Figures – Overview of Content

Basic Population and Economic Data

There will be no significant change in total population within the next 15 years, but the share of the main advertising target group (15 to 49 years) will decrease in all countries. This will also cause changes for some very youth-focused markets, e.g. the music market.

6.1.1 Forecast: Resident Population

in 000s	1990	1993	1996	2002	2010
France	56,735	57,655	58,265	59,171	59,782
Germany	79,365	81,179	81,818	80,839	78,631
Italy	56,737	57,049	57,333	56,951	55,840
Spain	38,959	39,083	39,241	39,390	39,324
UK	57,465	58,191	58,697	59,037	58,925
EU5	289,261	293,157	295,355	295,388	292,502
EU15	364,430	369,715	372,666	373,243	370,195
Western Europe	375,468	381,281	384,367	385,085	382,052
USA	249,924	257,925	264,162	272,415	280,652
Japan	123,505	124,670	125,530	126,830	126,654

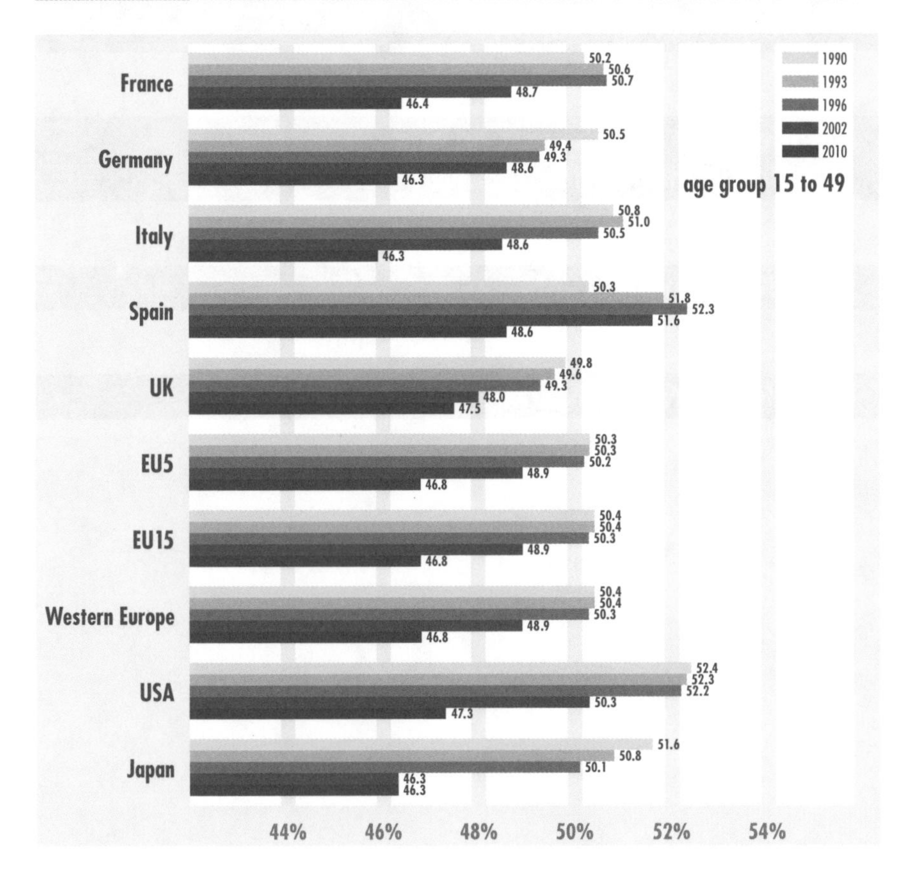

Sources

Prognos.
Resident population: projections include migrations.
Lower figure, age group 15 to 49: projections exclude migrations.

6.1.2 Forecast: Households

in 000s	1990	1993	1996	2002	2010
France	21,921	22,511	23,274	24,163	24,817
Germany	34,950	36,230	37,047	37,702	38,243
Italy	20,898	19,534	20,500	21,339	21,684
Spain	11,392	11,867	12,244	13,033	13,889
UK	22,720	23,647	24,482	24,656	24,997
EU5	111,881	113,789	117,547	120,893	123,630
EU15	140,568	143,698	148,293	152,657	156,134
Western Europe	145,212	148,594	153,492	158,173	161,891
USA	93,272	96,391	100,118	106,883	113,815
Japan	40,573	41,826	43,852	46,281	48,779

in %

single HH	1990	1993	1996	2002	2010
France	27.1	27.7	29.3	30.0	30.3
Germany	33.7	33.7	35.1	35.6	36.1
Italy	22.4	21.6	22.7	23.0	23.5
Spain	10.6	12.0	13.1	15.5	18.8
UK	25.5	28.2	28.3	28.6	28.9
EU5	26.3	27.0	28.1	28.7	29.3
EU15	26.3	27.2	28.3	28.8	29.5
Western Europe	26.5	27.5	28.6	29.1	29.8
USA	24.7	24.5	25.1	25.5	26.3
Japan	20.9	22.3	25.1	27.0	29.6

The total number of households will increase in each country, especially in Spain and the USA. Nevertheless, Italy, Spain and the USA will have low shares of single households in 2010.

Sources

Prognos AG.

Unemployment is the main economic and social problem in most European countries. Despite of the decrease in the last four years, Spain maintains the unfortunate leading position. Amongst other problems, unemployment includes the risk to be not able to participate in the media and communication future for significant parts of the society.

6.1.3 Unemployment rate

% of labour force	1993	1994	1995	1996	1997
France	11.7	12.3	11.6	12.3	12.5
Germany	8.9	9.6	9.4	10.4	11.5
Italy	10.2	11.3	12.0	12.1	12.3
Spain	22.7	24.2	22.9	22.2	20.8
UK	10.3	9.5	8.3	7.6	5.8
USA	6.8	6.1	5.6	5.4	4.9
Japan	2.5	2.9	3.1	3.4	3.4

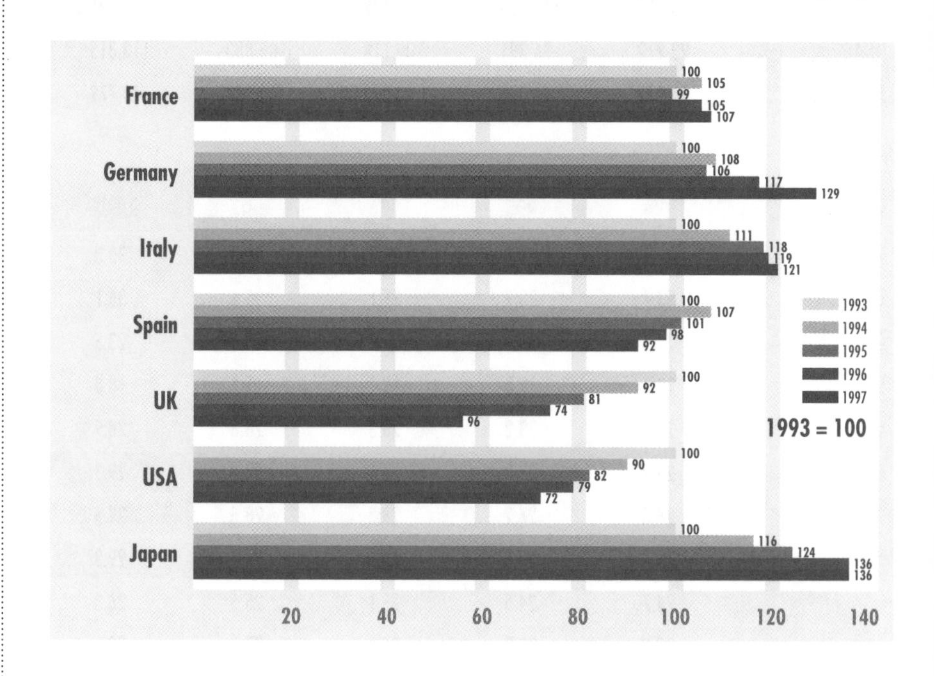

Sources

Advertising Information Group AIG.
Italy, Spain, UK and Japan: % of total labour force.
USA, Germany: % of civilian labour force.

6.1.4 Conversion Rates

1 ECU =	1993	1994	1995	1996	1997
France (FRF)	6.63	6.58	6.53	6.49	6.61
Germany (DEM)	1.94	1.92	1.87	1.91	1.96
Italy (ITL)	1,841.23	1,915.06	2,130.14	1,958.96	1,929.30
Spain (ESP)	149.12	158.92	163.00	160.75	165.89
UK (GBP)	0.78	0.78	0.83	0.81	0.69
USA (USD)	1.17	1.19	1.31	1.27	1.13
Japan (JPY)	130.15	121.32	123.01	138.08	137.08

1 USD =	1993	1994	1995	1996	1997
France (FRF)	5.66	5.55	4.99	5.12	5.84
Germany (DEM)	1.65	1.62	1.43	1.50	1.73
Italy (ITL)	1,572.00	1,621.40	1,628.90	1,542.90	1,703.10
Spanien (ESP)	127.20	133.96	124.69	126.66	146.66
UK (GBP)	0.67	0.65	0.63	0.64	0.61
USA (USD)	1.00	1.00	1.00	1.00	1.00
Japan (JPY)	111.20	102.21	94.06	108.78	120.99

Sources

EU, prognos, Advertising Association.
For all tables the original data in local currency have been converted into ECU figures using the 1997 conversion rate. Therefore some differences to other sources with current conversion rates are inevitable.

6.1.5 Forecast: Gross Domestic Product (GDP) at 1990 Prices

The start of the European Monetary Union EMU on January 1st 1999, which included the launch of the single currency, the Euro, is expected to make a significant contribution to a more than 2-%-GDP-growth p.a. within the next 15 years. The elimination of currency transactions will increase the competition in the single European market and stimulate
innovations.

in bill. ECU	1990	1993	1996	2002	2010
France	984	993	1,057	1,197	1,421
Germany	1,235	1,398	1,486	1,693	2,020
Italy	679	683	723	813	965
Spain	302	308	330	388	476
UK	794	790	859	996	1,205
EU5	3,995	4,171	4,455	5,087	6,087
EU15	5,007	5,201	5,578	6,386	7,654
Western Europe	5,290	5,490	5,883	6,730	8,060
USA	5,065	5,279	5,723	6,623	7,942
Japan	3,137	3,300	3,486	3,979	4,825

average growth p.a.	90/93	93/96	96/02	02/10	96/10
France	0.3	2.1	2.1	2.2	2.1
Germany	4.2	2.1	2.2	2.2	2.2
Italy	0.2	1.9	2.0	2.2	2.1
Spain	0.6	2.4	2.7	2.6	2.7
UK	-0.1	2.8	2.5	2.4	2.4
EU5	1.5	2.2	2.2	2.3	2.3
EU15	1.3	2.4	2.3	2.3	2.3
Western Europe	1.2	2.3	2.3	2.3	2.3
USA	1.4	2.7	2.5	2.3	2.4
Japan	1.7	1.8	2.2	2.4	2.3

Sources

prognos; original values at constant (1990) prices are converted into ECU at constant (1997) conversion rates.

6.1.6 Forecast: Private Consumption at 1990 Prices

in bill. ECU	1990	1993	1996	2002	2010
France	587	602	633	707	831
Germany	672	814	850	965	1,138
Italy	419	424	441	497	600
Spain	189	194	202	234	284
UK	499	500	539	628	766
EU5	2,366	2,534	2,665	3,032	3,618
EU15	2,947	3,136	3,310	3,769	4,498
Western Europe	3,100	3,294	3,475	3,951	4,710
USA	3,385	3,565	3,870	4,519	5,423
Japan	1,819	1,924	2,057	2,340	2,787

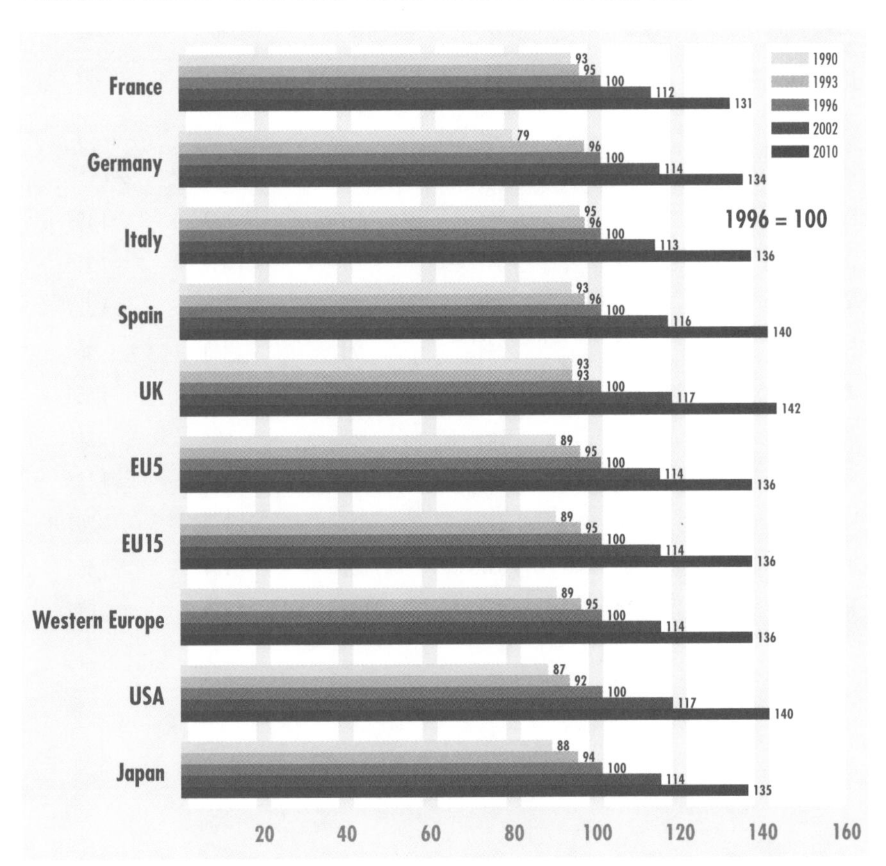

The highest increase of private consumption is expected for the UK, Spain and the USA.

Sources

prognos; original values at constant (1990) prices are converted into ECU at constant (1997) exchange rates.

6.1.7 Forecast: Expenditures for Education and Cultural Services at 1990 Prices

Expenditures for education and cultural services are part of private consumption. They include the media spendings, e.g. for pay-TV, and are therefore an indicator for the development of media markets. European Countries will not reach the high level of the per capita expenditures in Japan and the USA in the near future. The gap will even increase.

in bill. ECU	1990	1993	1996	2002	2010
France	44	46	50	58	72
Germany	68	85	83	86	95
Italy	39	40	45	55	74
Spain	13	13	15	17	22
UK	49	51	61	73	92
EU5	213	235	253	290	355
EU15	264	289	312	359	440
Western Europe	277	303	327	375	459
USA	340	397	423	540	720
Japan	224	244	262	312	389

in ECU per capita	1990	1993	1996	2002	2010
France	784	795	851	977	1,199
Germany	861	1,047	1,017	1,069	1,210
Italy	680	702	782	959	1,329
Spain	332	344	371	443	560
UK	850	879	1,041	1,243	1,566
EU5	737	803	858	981	1,215
EU15	725	783	838	961	1,187
Western Europe	738	795	851	973	1,201
USA	1,362	1,537	1,602	1,984	2,565
Japan	1,813	1,954	2,086	2,456	3,070

Sources

prognos; original values at constant (1990) prices are converted into ECU at constant (1997) exchange rates.

6.1.8 Forecast: Expenditures for Transport and Communication

in bill. ECU	1990	1993	1996	2002	2010
France	99	96	106	119	140
Germany	113	126	138	148	175
Italy	52	50	55	64	82
Spain	87	85	96	114	143
UK	30	29	31	36	45
EU5	381	387	425	482	585
EU15	465	472	519	592	720
Western Europe	484	490	539	615	747
USA	312	311	346	402	475
Japan	211	222	234	260	306

in ECU per capita	1990	1993	1996	2002	2010
France	1,748	1,671	1,817	2,015	2,338
Germany	1,425	1,558	1,681	1,831	2,225
Italy	908	879	957	1,130	1,470
Spain	2,239	2,169	2,440	2,886	3,636
UK	522	504	526	617	760
EU5	1,317	1,320	1,439	1,631	1,999
EU15	1,277	1,276	1,393	1,586	1,944
Western Europe	1,288	1,284	1,403	1,596	1,955
USA	1,249	1,204	1,311	1,475	1,694
Japan	1,707	1,779	1,865	2,048	2,413

Another media related part of private consumption are the expenditures for transport and communication, which include for example telephone rental or call charges. In contrast to the expenditures for education and culture the per capita level is higher in Europe than in the USA.

Sources

prognos; original values at constant (1990) prices are converted into ECU at constant (1997) exchange rates.

Distribution Infrastructure and Household Equipment

6.2.1 TV Households

Practically each European household has at least one TV set. Therefore the number of TV households will only increase with the number of total HH. Digital reception is only important in countries with successful digital Pay-TV: USA (DirecTV), France and Spain (Canal +).

in 000s	1993	1994	1995	1996	1997	1998
France	21,442	21,517	21,557	21,701	21,911	22,159
Germany	34,748	35,158	35,615	35,862	36,334	38,700
Italy	18,733	18,886	18,886	18,976	19,003	19,400
Spain	11,314	11,332	11,277	11,305	11,359	12,000
UK	22,174	22,382	23,388	23,635	23,973	24,239
EU5	108,411	109,275	110,723	111,479	112,580	116,498
EU15	136,085	137,205	138,992	140,143	141,580	145,798
Western Europe	140,688	141,880	143,729	144,932	146,323	150,651
USA	94,178	95,500	96,000	97,000	98,000	98,000
Japan	40,667	41,328	43,000	43,560	44,000	n.a.

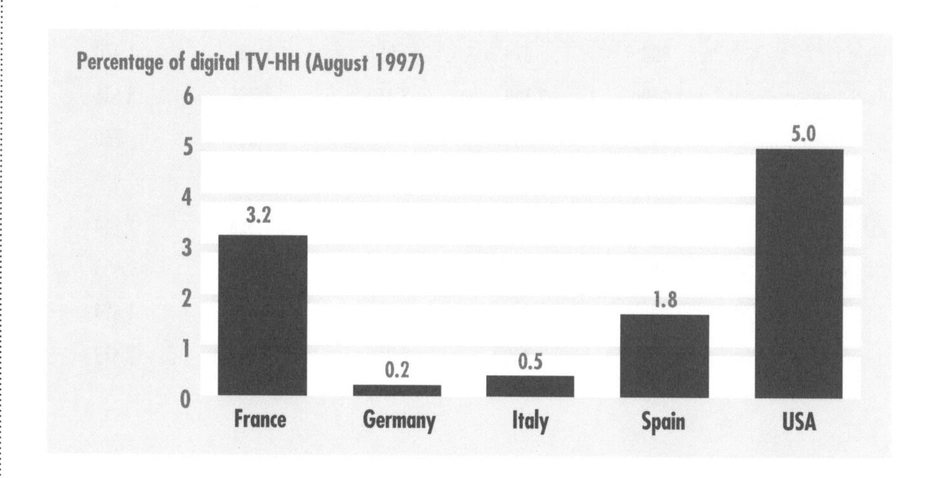

Sources

ITU, OBS, Screen Digest, prognos.

6.2.2 DTH Households

in 000s	1993	1994	1995	1996	1997
France	161	1,010	1,250	1,900	2,470
Germany	6,880	8,320	9,302	10,400	11,150
Italy	100	100	520	660	760
Spain	107	390	856	1,000	1,130
UK	3,123	3,390	3,700	4,000	4,310
EU5	10,371	13,210	15,628	17,960	19,820
EU15	12,150	15,697	18,869	21,873	24,380
Western Europe	12,452	16,092	19,335	22,383	25,090
USA	3,600	3,800	4,565	6,563	8,400
Japan	7,355	8,328	9,129	10,965	12,065

% of all HH

	1993	1994	1995	1996	1997
France	0.7	4.4	5.4	8.2	10.5
Germany	19.0	22.7	25.2	28.1	30.0
Italy	0.5	0.5	2.6	3.2	3.7
Spain	0.9	3.3	7.1	8.2	9.1
UK	13.2	14.1	15.1	16.3	17.6
EU5	9.1	11.4	13.4	15.3	16.8
EU15	8.5	10.8	12.8	14.7	16.4
Western Europe	8.4	10.7	12.7	14 6	16.3
USA	3.7	3.9	4.6	6.6	8.3
Japan	17.6	19.5	21.0	25.0	27.3

5 10 15 20 25 30

The share of DTH-HH in Germany is twice the European average. Only Japan reaches the same level. In Italy over-the-air reception is still dominating, but the number of DTH-HH is increasing.

Sources

ITU-Database 1997, Screen Digest 5/98, EAO99.

6.2.3 Cable-TV: HH Connected

In similarity with DTH reception levels Germany has the highest share of HH connected to cable systems. More than 75 % of German HH can receive at least 30 channels. This is more than the US average. The only European country with no registered cable systems is Italy.

in 000s	1993	1994	1995	1996	1997	1998
France	1,056	1,287	1,608	1,858	2,136	2,392
Germany	13,309	15,655	15,782	17,080	18,020	18,650
Italy	0	0	0	0	0	0
Spain	122	130	130	142	145	400
UK	652	783	1,056	1,327	1,900	2,374
EU5	15,139	17,855	18,576	20,407	22,201	23,816
EU15	29,194	32,166	33,428	35,125	37,156	39,421
Western Europe	31,760	34,862	36,275	38,038	40,124	42,989
USA	57,200	59,700	62,100	63,500	64,200	65,400
Japan	9,228	10,255	11,000	12,629	n.a.	n.a.

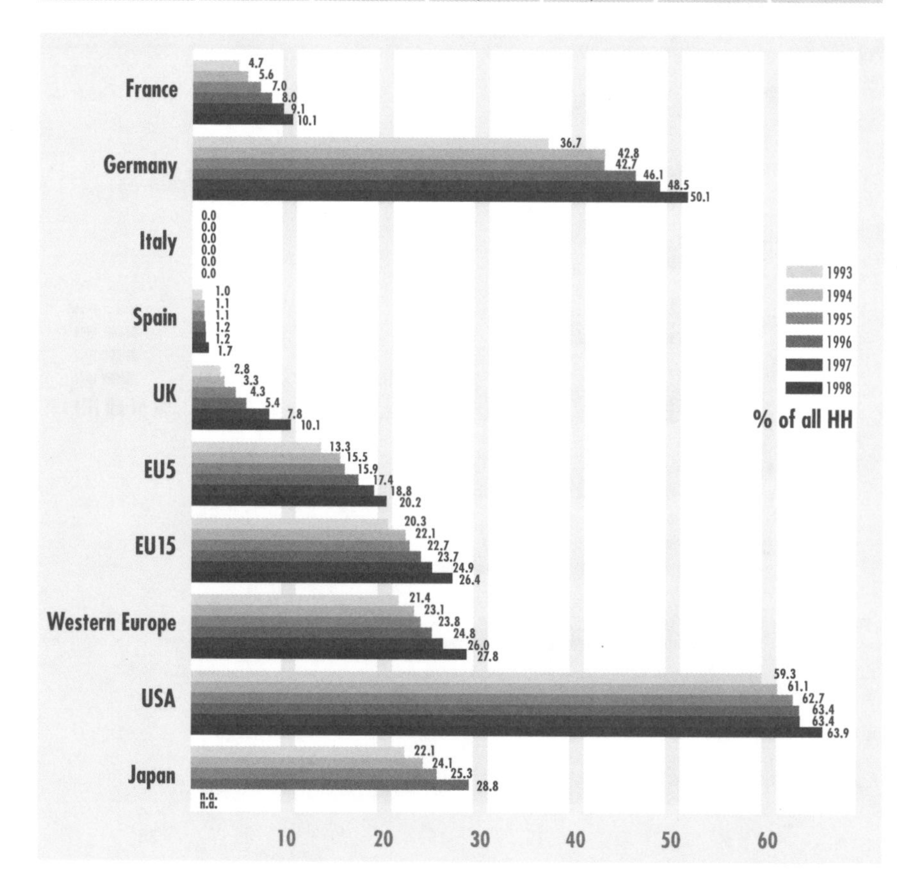

Sources

ITU, OBS; Prognos calculations.

6.2.4 Households Additionally Connectable to Cable

in 000s	1993	1994	1995	1996	1997	1998
France	3,630	3,996	4,194	4,394	4,496	4,608
Germany	12,412	12,535	14,556	14,306	14,180	14,250
Italy	0	0	0	0	53	50
Spain	778	790	810	813	855	600
UK	2,261	2,863	3,801	4,715	6,500	8,320
EU5	19,081	20,184	23,361	24,228	26,084	27,828
EU15	21,609	22,877	26,023	27,205	29,827	31,409
Western Europe	21,793	23,103	26,536	27,677	30,584	32,795
USA	33,400	31,900	30,600	30,200	30,000	29,700
Japan	n.a.	n.a.	n.a.	n.a.	n.a.	n.a.

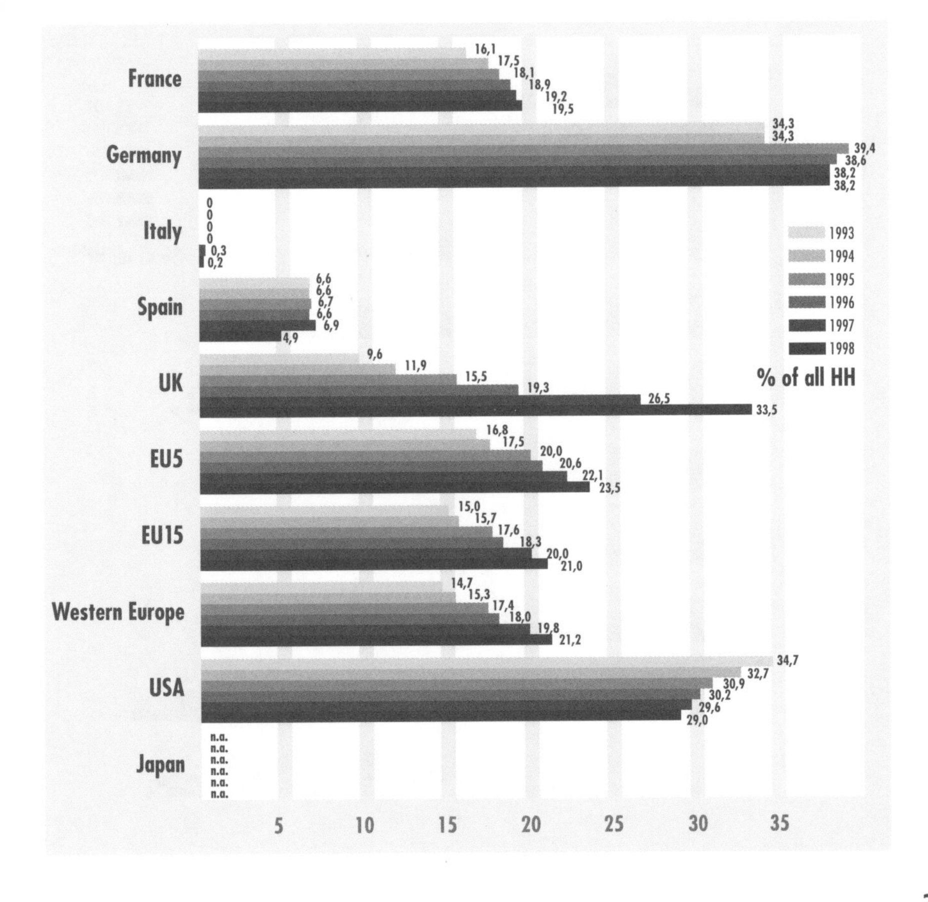

Although the number of additionally connectable HH was still increasing in the last five years, in Germany and the USA the extension of cable systems stopped. It is uncertain whether the homes passed will connect to cable in competition with DTH reception.

Sources

ITU, OBS; Prognos calculations.

6.2.5 VCR Households

The share of VCR-Households is still increasing. Until 1996 no European country had reached the saturation level of 70%, which had been generally expected. Considering the Japanese and US figures, this may be questionable anyhow. It is remarkable that the USA recorded the highest increase from the highest level in the last five years.

in 000s	1992	1993	1994	1995	1996	1997
France	12,475	13,442	14,387	15,218	16,137	17,127
Germany	19,449	21,345	23,292	25,183	26,944	28,494
Italy	7,342	8,310	9,238	9,966	10,666	11,310
Spain	6,155	6,591	7,108	7,651	8,136	8,668
UK	16,019	16,979	17,854	18,637	19,261	19,949
EU5	61,440	66,667	71,879	76,655	81,144	85,548
EU15	76,106	82,539	88,942	94,849	100,479	105,963
Western Europe	78,698	85,316	91,926	98,045	103,888	109,567
USA	68,481	73,565	78,125	82,445	86,533	90,875
Japan	31,320	32,531	33,744	34,970	36,037	37,422

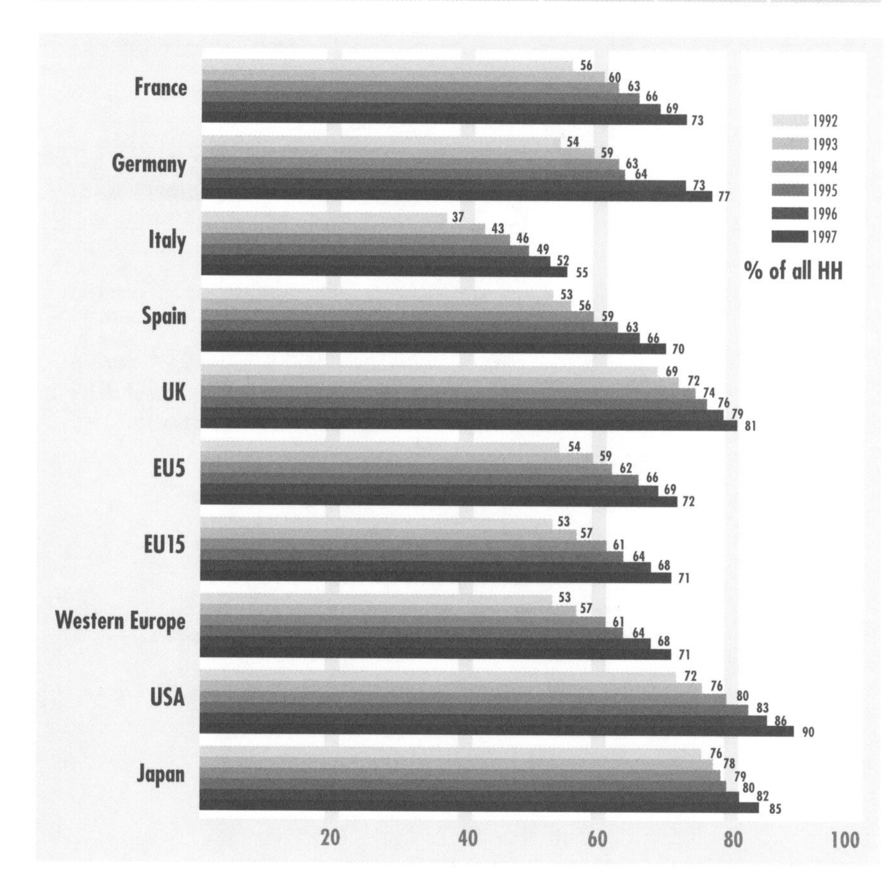

Sources

OBS/Screen Digest, Prognos calculations, no data for 1998 available.

6.2.6 Telephone Main Lines

per 100 inhabitants	1993	1994	1995	1996	1997
France	54	55	56	56	58
Germany	46	48	51	54	55
Italy	42	43	43	44	45
Spain	36	38	38	39	40
UK	47	49	50	52	54
EU5	46	47	49	50	51
EU15	46	47	49	50	52
Western Europe	46	48	49	51	52
USA	58	60	63	65	n.a.
Japan	47	48	49	49	n.a.

France has the highest density of telephone main lines within Europe, but the US figure was 15 % higher in 1996. On the other hand the modernisation of the telephone system is more advanced in Europe and in Japan. In Germany and Japan the conversion to digital lines had already been completed in 1997.

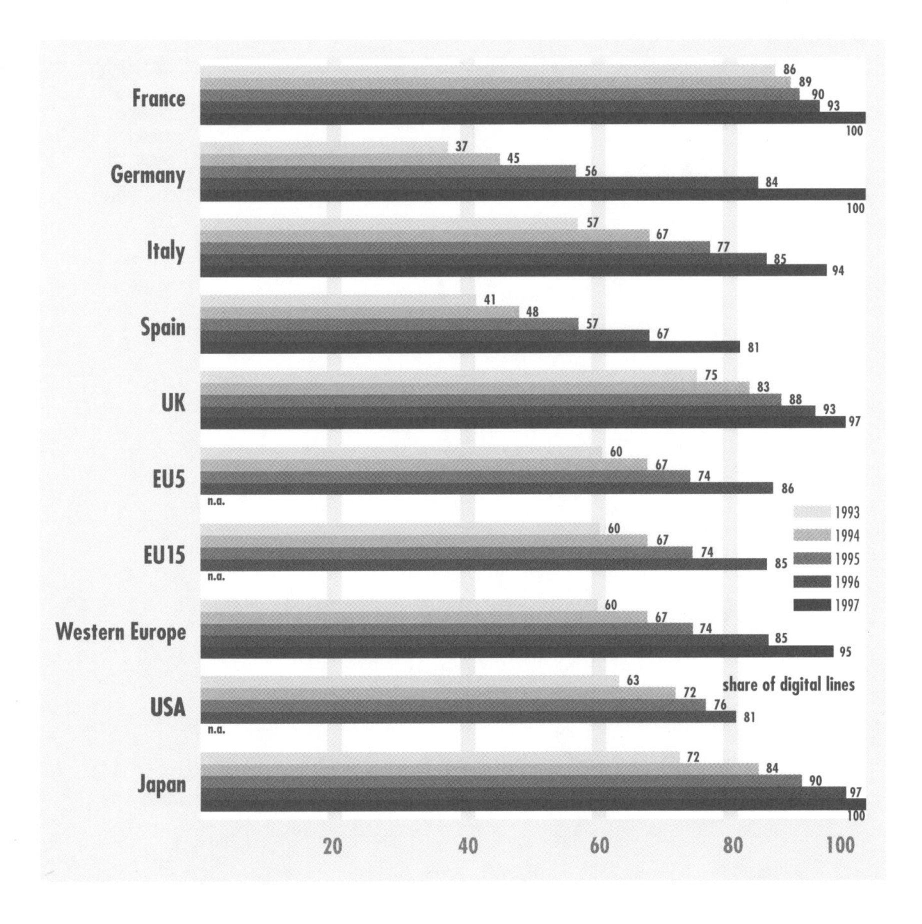

Sources

ITU-Database, EITO 1999, Prognos calculations.
Private and business sectors cannot be separated.

6.2.7 Mobile Phone Subscribers

The dynamic mobile phone market is still in a take-off-phase, the growth rates will increase further. Japan is both leading in saturation and growth rates. In Europe mobile phones are most popular in Italy, where the saturation rate has reached the US level.

per 100 inhabitants	1993	1994	1995	1996	1997
France	1	2	2	4	10
Germany	2	3	5	7	11
Italy	2	4	7	11	20
Spain	1	1	2	3	11
UK	4	7	10	12	14
EU5	2	3	5	8	n.a.
EU15	2	4	6	8	n.a.
Western Europe	2	4	6	9	13
USA	6	9	13	17	21
Japan	2	3	9	21	30

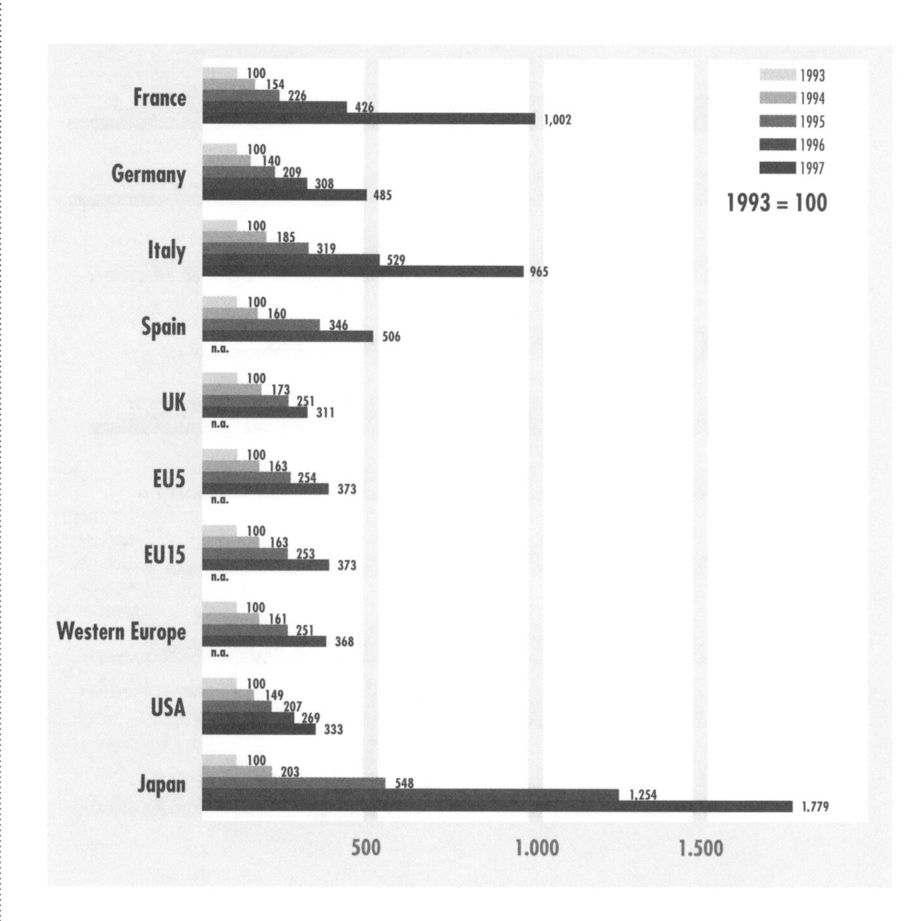

Sources

ITU-Database, EITO 1999, Progonos calculations. Private and business sectors cannot be separated.

6.2.8 Multimedia-PC Households

in 000s	1993	1994	1995	1996	1997
France	37	280	1,080	1,931	2,944
Germany	127	1,100	3,220	4,513	5,763
Italy	48	250	1,040	1,474	2,289
Spain	20	105	720	1,014	1,495
UK	84	509	1,200	2,270	3,646
EU5	316	2,244	7,260	11,202	16,137
EU15	441	2,643	8,424	13,468	n.a.
Western Europe	462	2,743	8,701	13,991	n.a.
USA	n.a.	n.a.	n.a.	22,300	26,500

% of all HH

	1993	1994	1995	1996	1997
France	0	1	5	8	13
Germany	0	3	9	12	16
Italy	0	1	5	7	11
Spain	0	1	6	8	12
UK	0	2	5	9	15
EU5	0	2	6	10	14
EU15	0	2	6	9	n.a.
Western Europe	0	2	6	9	n.a.
USA	n.a.	n.a.	n.a.	22	26

Within the last five years the share of European HH with a MM-PC increased from 0% to 10%, but this is still half the level of the USA. There are no significant differences between the large European countries.

Sources

Inteco, Prognos calculations.
Multimedia-PC: PC with CD-ROM-drive.

6.2.9 Households With Modem or Internet Access

In accordance with the high MM PC saturation, the share of households with Internet access in the USA is four times the European average. There is still a significant share of HH not using their modems for complete Internet access (e.g. only eMail). These figures do not include net access at the workplace, therefore the level is much lower than reported in other sources.

Modem	HH in 000s 1996	HH in 000s 1997	% of all HH 1996	% of all HH 1997
France	907	1,549	4	7
Germany	2,501	3,495	7	9
Italy	765	1,236	4	6
Spain	651	989	5	8
UK	1,601	2,544	7	10
EU5	6,425	9,813	5	8
USA	27,000	31,100	27	31

Internet access	HH in 000s 1996	HH in 000s 1997	% of all HH 1996	% of all HH 1997
France	281	735	1	3
Germany	1,404	2,315	4	6
Italy	271	633	1	3
Spain	270	523	2	4
UK	900	1,498	4	6
EU5	3,126	5,704	3	5
EU15	4,067	7,350	3	5
Western Europe	4,267	7,685	3	5
USA	12,800	19,800	13	20

6.2.10 Forecast: Internet Access

In the next five years, there will be in all European countries a significant growth in the number of households with Internet access. Between 55 and 75 percent of households in the different member states could be online by the year 2010.

in % of all HH	1997	2002	2010
France	3	20	55
Germany	6	25	60
Italy	3	18	48
Spain	4	19	50
UK	6	25	60
EU5	5	22	56
EU15	5	21	55
Western Europe	5	21	55
USA	20	60	75

Sources

6.2.9 Screen Digest/Inteco, Prognos calculation.
6.2.10 Forrester Research, Prognos estimates

Research and Development, Production and Supply

6.3.1 IT Patent Applications for Europe

from ...	1993	1994	1995	1996	1997	1998
France	235	191	231	203	216	268
Germany	343	350	387	438	575	699
Italy	70	63	70	80	78	89
Spain	13	9	8	16	15	19
UK	124	121	137	160	183	215
EU5	785	734	833	897	1,067	1,290
EU15	928	910	991	1,137	1,395	1,657
Western Europe	1,017	1,014	1,087	1,199	1,517	1,780
USA	1,298	1,256	1,446	1,497	1,817	1,825
Japan	666	593	627	709	914	992

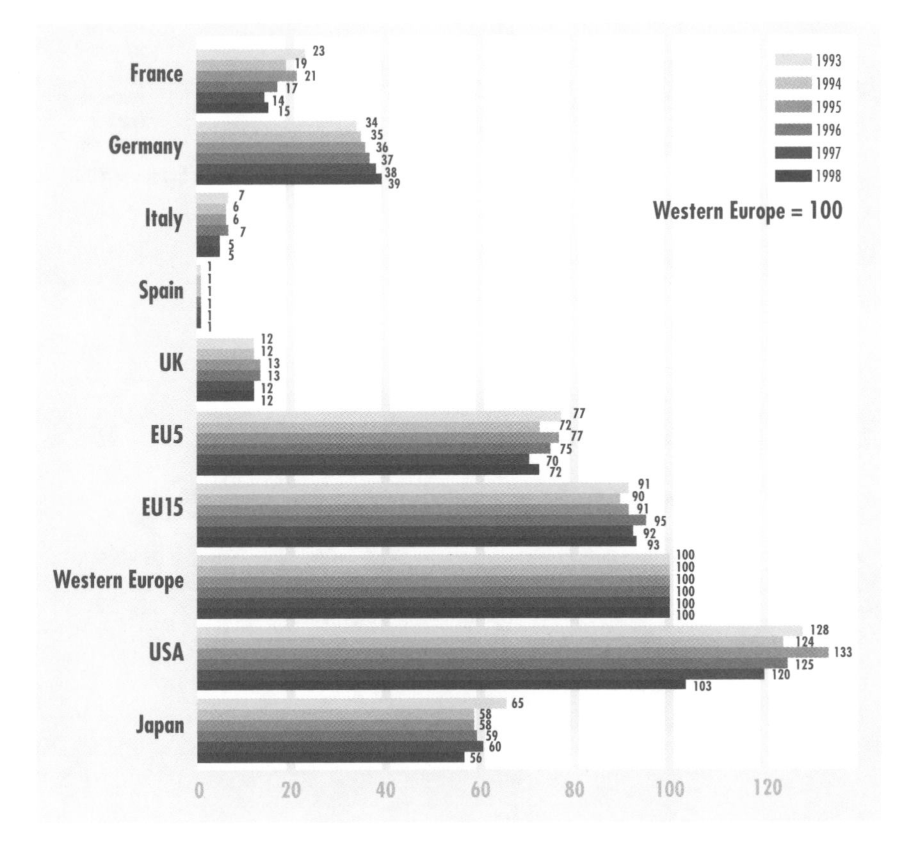

The leading role of the USA in the information technology sector is shown by the number of US patent applications for Europe, which is still 20% higher than the number of applications from all Western European countries for this area.

Sources

Patents in the information technology (IT) sector applicated for the area of the European Patent Office per country of origin.
European Patent Office EPO annual reports.
EU15: Finland from 1996, Western Europe without Norway.

6.3.2 TC Patent Applications for Europe

Compared to Japan and the USA, the European countries traditionally have competitive advantages in the telecommunications (TC) sector.
TC patent applications for Europe from European countries increased about 60% since 1993.

from ...	1993	1994	1995	1996	1997	1998
France	299	312	312	347	410	549
Germany	508	502	545	581	624	841
Italy	90	95	99	93	92	93
Spain	8	18	15	6	12	9
UK	219	263	216	244	301	391
EU5	1,124	1,190	1,187	1,271	1,439	1,883
EU15	1,501	1,617	1,651	2,022	2,424	3,013
Western Europe	1,542	1,659	1,712	2,065	2,482	3,073
USA	1,269	1,301	1,515	1,688	1,819	2,197
Japan	959	1,070	1,083	1,254	1,480	1,673

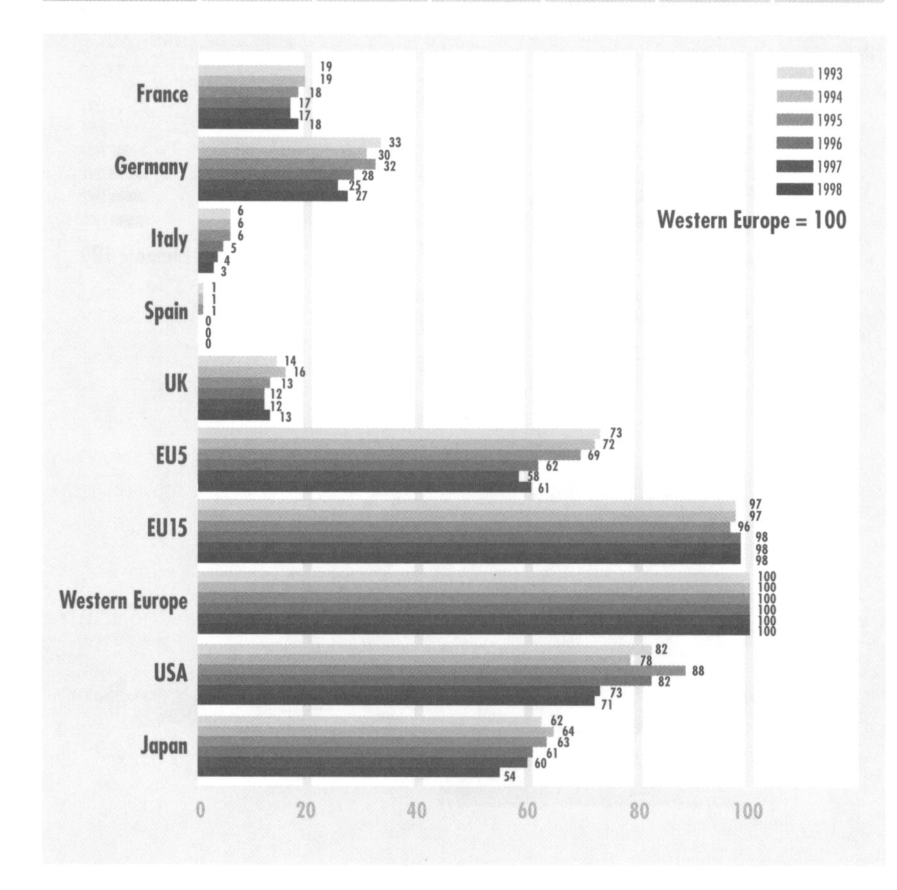

Sources

Patents in the telecommunication (TC) sector applicated for the area of the European Patent Office per country of origin.
European Patent Office EPO annual reports
EU15: Finland from 1996, Western Europe without Norway.

6.3.3 IT Production Value

in mill. ECU	1992	1993	1994	1995	1996
France	8,621	7,432	7,888	8,940	9,351
Germany	12,915	12,104	12,308	13,385	14,004
Italy	5,136	4,920	7,681	7,011	6,860
UK	27,296	27,524	35,455	36,124	39,578
USA	108,459	121,545	135,976	145,427	167,678
Japan	102,558	119,799	121,705	121,654	132,116

ECU per capita

	1992	1993	1994	1995	1996
France	150	128	136	153	160
Germany	159	149	151	164	170
Italy	90	86	134	122	119
UK	469	471	605	615	672
USA	421	466	517	551	627
Japan	823	960	973	969	1,051

0 200 400 600 800 1.000

IT production values show the performance of the IT production industry. Regarding the per capita figures, Japan is far ahead of all other countries. The highest growth since 1992 has been achieved by the USA. In Europe only the UK has reached a similar level.

Sources

ZVEI (IT=information technology), no data available for 1997.

6.3.4 TC Production Values

The Japanese TC production values per capita are twice the value of France which is the leading European country. The gap between Japan and the other countries is still increasing.

in mill. ECU	1992	1993	1994	1995	1996
France	8,406	7,962	8,369	9,156	9,998
Germany	10,989	10,246	10,014	11,488	12,572
Italy	4,495	3,236	4,007	3,728	4,665
UK	6,426	6,639	7,889	8,046	9,892
USA	39,335	43,537	48,558	47,674	55,779
Japan	26,560	32,195	35,359	37,041	43,708

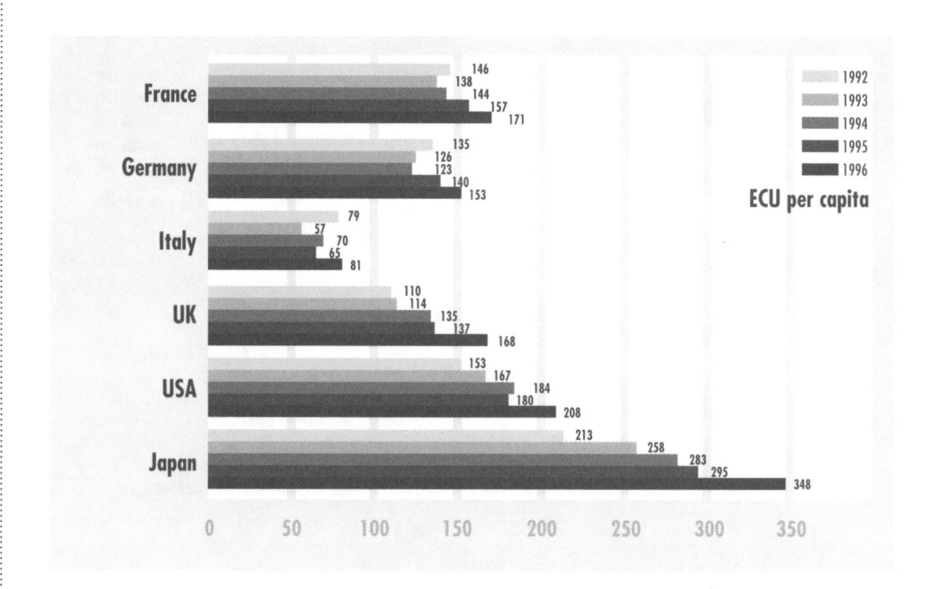

Sources

ZVEI (TC= telecommunication), no data available for 1997.

6.3.5 Feature Films Produced

	1993	1994	1995	1996	1997
France	101	89	97	104	130
Germany	67	60	63	64	61
Italy	106	95	75	99	87
Spain	56	44	59	91	80
UK	60	70	76	128	108
EU5	390	358	370	486	466
EU15	536	500	508	588	619
Western Europe	581	551	567	602	669
USA	440	410	370	421	n.a.
Japan	238	251	289	278	278

In regard to the sheer number of feature films produced, the gap between Europe and the USA is not as large as often expected. This says nothing about the poor commercial relevance of many European films. The main reason for the strong competition is the limited number of release dates.

Sources

EAO, European Audiovisual Observatory including Co-productions 1996 without Greece and Sweden USA and Japan: feature films released. 1998 data not available.

6.3.6 National TV Channels

As there are different ways to count pan-European channels, the total number of European TV channels cannot be calculated exactly. Although many channels are in a deficit situation, the trust in market opportunities seems to be unbroken.

	1992	1993	1994	1995	1996
France	11	14	14	16	18
Germany	8	14	14	18	19
Italy	11	12	12	13	13
Spain	13	13	13	13	13
UK	42	52	52	64	80
EU5	85	105	105	124	143
EU15	124	159	162	189	214
Western Europe	133	168	171	199	225

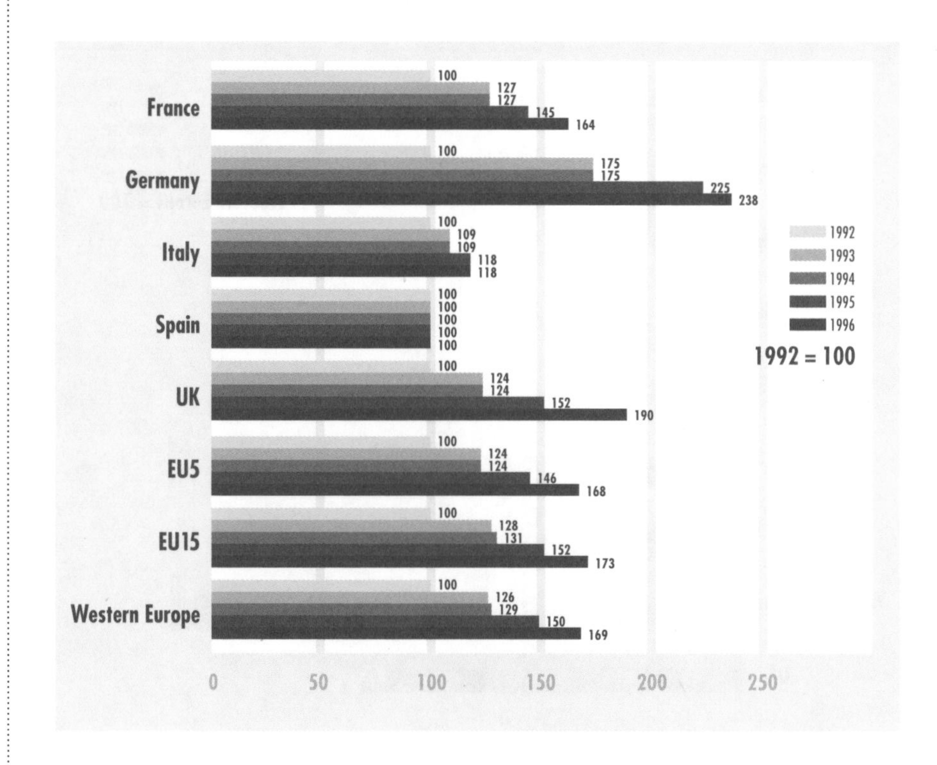

Sources

European Commission, national channels only.

6.3.7 Total Spending for TV Programming (Estimate)

in mill. ECU	1992	1993	1994	1995	1996
France	1,727	2,104	2,073	2,283	2,305
Germany	2,277	2,901	3,463	3,963	4,262
Italy	1,638	2,001	1,682	1,724	1,825
Spain	521	869	807	866	946
UK	2,796	3,592	3,243	3,372	3,642
EU5	8,959	11,467	11,268	12,207	12,980
EU15	10,385	13,163	12,886	14,146	14,983
Western Europe	10,719	13,592	13,351	14,643	15,494

These Screen Digest estimates are based on highly heterogeneous information from 43 main channels in 16 Western European countries. The growth rates in Germany and Spain are remarkable. Germany took over the leading role from the UK in 1994.

Sources

EAO, European Audiovisual Observatory/Screen Digest (estimate).

Based on the same information sources as for table 3.7, these figures should also be considered as first estimates.

6.3.8 Spending for TV Programme Purchases (Estimate)

in mill. ECU	1992	1993	1994	1995	1996
France	423	514	526	565	601
Germany	380	457	514	678	793
Italy	181	226	212	216	232
Spain	117	197	178	192	209
UK	290	450	431	499	625
EU5	1,392	1,845	1,860	2,149	2,461
EU15	1,616	2,119	2,141	2,472	2,797
Western Europe	1,700	2,235	2,297	2,617	2,949

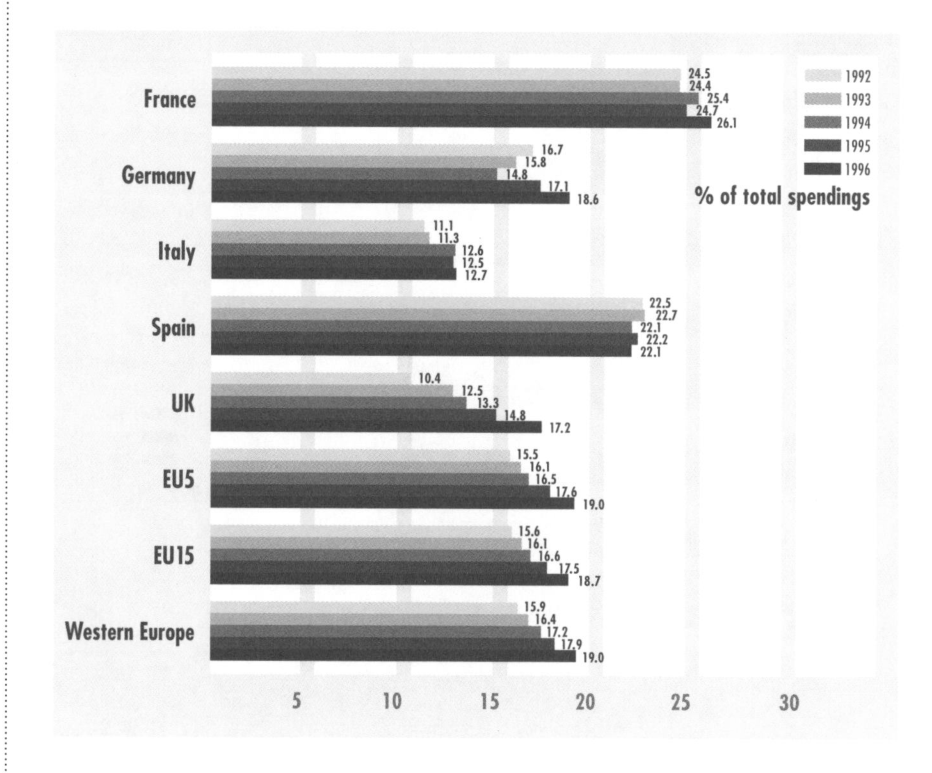

Sources

EAO, European Observatory/Screen Digest (estimate).

6.3.9 Daily Newspapers: Number of Titles

	1993	1994	1995	1996	1997
France	85	87	87	85	84
Germany	414	411	406	400	402
Italy	81	76	75	77	n.a.
Spain	126	125	126	126	128
UK	98	99	100	99	99
EU5	804	798	794	787	n.a.
EU15	1,153	1,118	1,135	1,122	n.a.
Western Europe	1,343	1,306	1,323	1,308	n.a.
USA	1,556	1,548	1,533	1,520	1,509
Japan	109	108	108	109	109

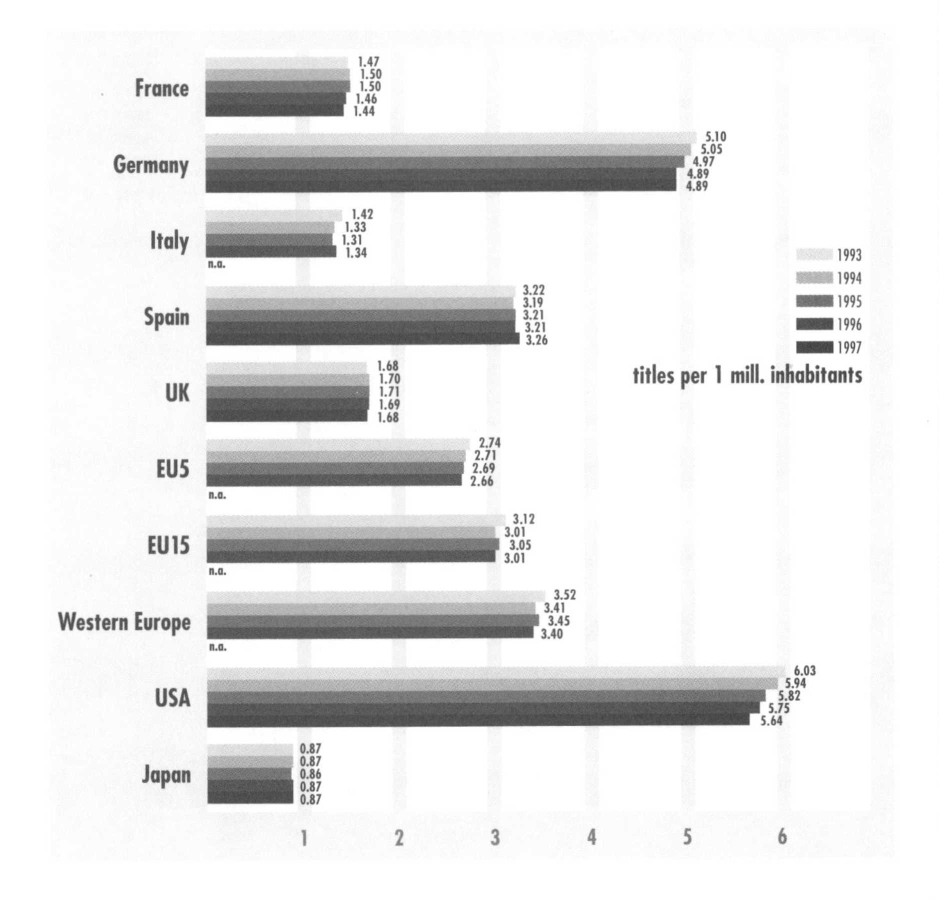

The number of daily newspapers per 1 million inhabitants differs in a wide range, from nearly six titles in the USA to less than one title in Japan. Within Western Europe, Germany has the highest density of titles.

Sources

World Association of Newspapers (WAN), Prognos calculations.

Despite of the small number of titles the sales of copies per HH are highest in Japan. The situation in the UK is similar: more copies per HH are sold than in the countries with the highest density of titles, Germany and the USA. With the exception of Spain the number of copies per HH is decreasing.

6.3.10 Daily Newspapers: Average Circulation

in 000s ...	1993	1994	1995	1996	1997
France	9,090	9,080	9,284	8,820	n.a.
Germany	25,902	25,757	25,467	25,217	25,038
Italy	6,438	6,208	5,988	5,904	5,920
Spain	3,950	4,100	4,237	4,180	4,265
UK	19,578	19,794	19,742	19,332	18,447
EU5	64,958	64,939	64,718	63,453	n.a.
EU15	84,189	83,566	83,226	82,181	n.a.
Western Europe	89,388	88,793	88,349	87,272	n.a.
USA	59,812	59,305	58,193	56,990	56,728
Japan	72,043	71,924	72,047	72,705	72,699

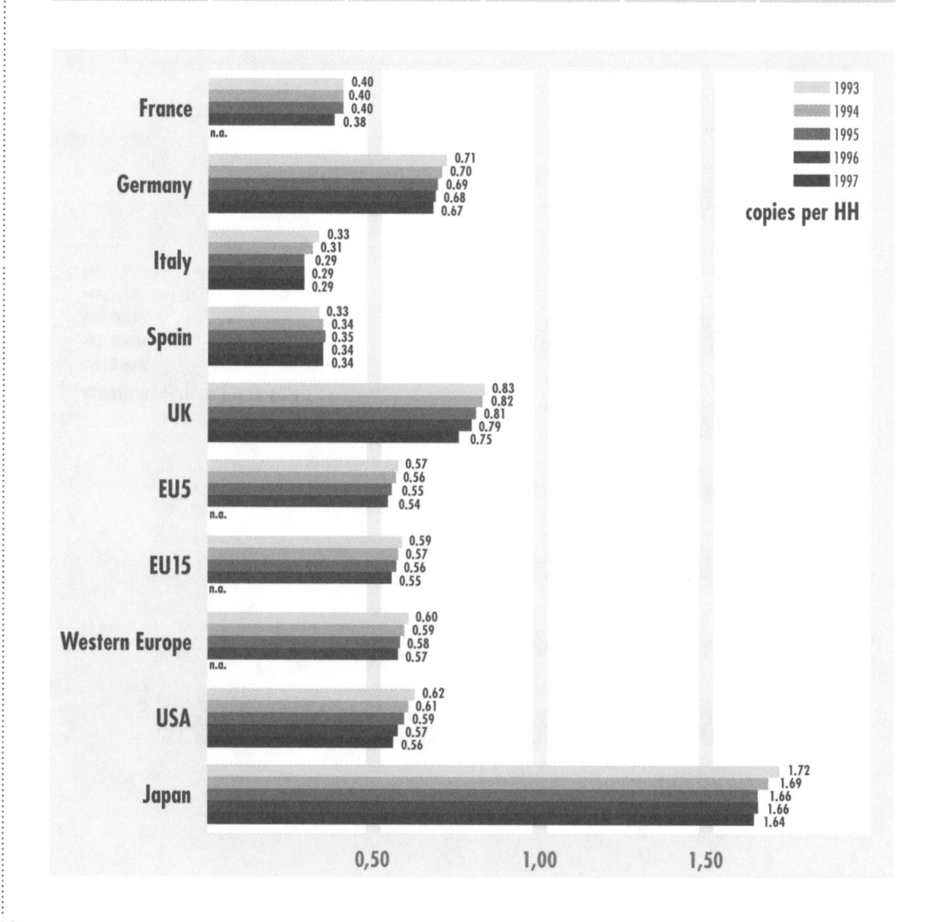

Sources

World Association of Newspapers WAN, Prognos calculations.
EU15 and Western Europe 1994 without Portugal.

6.3.11 Radio Stations 1997

	total number	private stations	market share in %
France	1,500	1,412	94
Germany	294	165	56
Italy	2,515	2,512	100
Spain	1,800	811	45
UK	230	183	80
EU5	6,339	5,083	80
EU15	9,559	7,909	83

	stations per 1 mill. inhabitants	inhabitants per Station
France	26	39,000
Germany	4	279,500
Italy	44	22,800
Spain	46	21,800
UK	4	255,900
EU5	21	46,700
EU15	26	39,100

The radio landscape across Europe is very heterogeneous. The number of stations per 1 million inhabitants differs from 4 to 46. It is not clear whether all stations listed are completely independent or run with commercial interests. The share of private stations is no indicator for their market relevance.

Sources

WorldDAB, Frankreich, Spanien: estimates includes national, regional and local services and networks. 1998 data not available.

The fast increasing number of hosts (Computers connected to the Internet) and webpages (Internet pages) shows the dynamic growth of net content according with the growth of Internet access opportunities (see table 2.9).

6.3.12 Internet Hosts and Webpages 1997

Hosts (000s)	July 1996	July 1997	% change 96/97	Hosts per 1,000 inhab. (1997)
France	190	292	54	5
Germany	548	876	60	11
Italy	114	212	86	4
Spain	62	122	95	3
UK	579	878	52	15
EU5	1,494	2,380	59	8
EU15	2,418	3,727	54	10
Western Europe	2,641	4,084	55	11
USA	8,224	11,829	44	44

webpages (000s)	July 1996	July 1997	% change 96/97 in %	pages per 1,000 inhab. (1997)
France	637	2,016	216	34
Germany	1,593	5,083	219	62
Italy	907	2,875	217	50
Spain	271	903	233	23
UK	1,605	5,132	220	87
EU5	5,014	16,010	219	54
EU15	7,471	23,485	214	63
Western Europe	8,243	25,764	213	67

Sources

European Commission DG XIII
no data for webpages in the USA available

Section 6.4

Spending on Information and Telecommunication Technologies

6.4.1 Total ICT Expenditure

in mill. ECU	1993	1994	1995	1996	1997	1998
France	42,294	46,059	49,237	53,974	58,596	64,374
Germany	66,770	66,202	72,430	77,782	82,867	89,040
Italy	24,575	29,648	31,774	36,152	39,264	44,038
Spain	11,152	12,267	13,488	16,522	17,952	19,678
UK	44,409	49,855	55,441	63,008	68,261	73,582
EU5	189,200	204,030	222,370	247,438	266,940	290,712
EU15	236,961	256,894	280,374	313,676	339,788	370,515
Western Europe	252,198	272,406	296,933	332,440	359,934	392,130

ECU per capita

	1993	1994	1995	1996	1997	1998
France	734	795	847	926	1.002	1.096
Germany	823	813	887	951	1.008	1.083
Italy	431	518	555	631	683	765
Spain	285	313	344	421	457	500
UK	763	854	946	1.073	1.160	1.245
EU5	645	694	754	838	901	979
EU15	641	693	754	842	909	989
Western Europe	661	712	774	865	934	1.015

200 400 600 800 1000 1200

Germany is the largest ICT market, but the highest per capita level was reached by the UK with an increase of 50 % in the last five years. Similar growth rates were performed by Italy and Spain on a lower level.

Sources

EITO, Prognos calculations, 1997 conversion rates. Total ICT expenditure consists of expenditure for IT information technology (hardware, software, services) and expenditure for TC telecommunication (hardware/net equipment and services). Private and business expenditure cannot be separated. Per capita values are only arithmetical.

6.4.2 IT Hardware Expenditure

The IT hardware expenditure shows the same pattern as total ICT expenditure: the UK reaches the highest per capita level, Spain and Italy come last.

in mill. ECU	1993	1994	1995	1996	1997	1998
France	8,485	9,850	10,487	11,115	11,899	12,513
Germany	16,135	15,973	17,536	18,237	19,115	21,110
Italy	4,622	5,540	5,885	5,949	6,246	6,563
Spain	2,535	2,832	3,047	3,285	3,600	3,822
UK	10,234	14,292	15,601	17,081	18,418	19,448
EU5	42,009	48,487	52,556	55,668	59,277	63,456
EU15	52,267	61,945	67,665	71,731	76,674	81,724
Western Europe	55,296	65,784	71,867	76,248	81,510	86,752

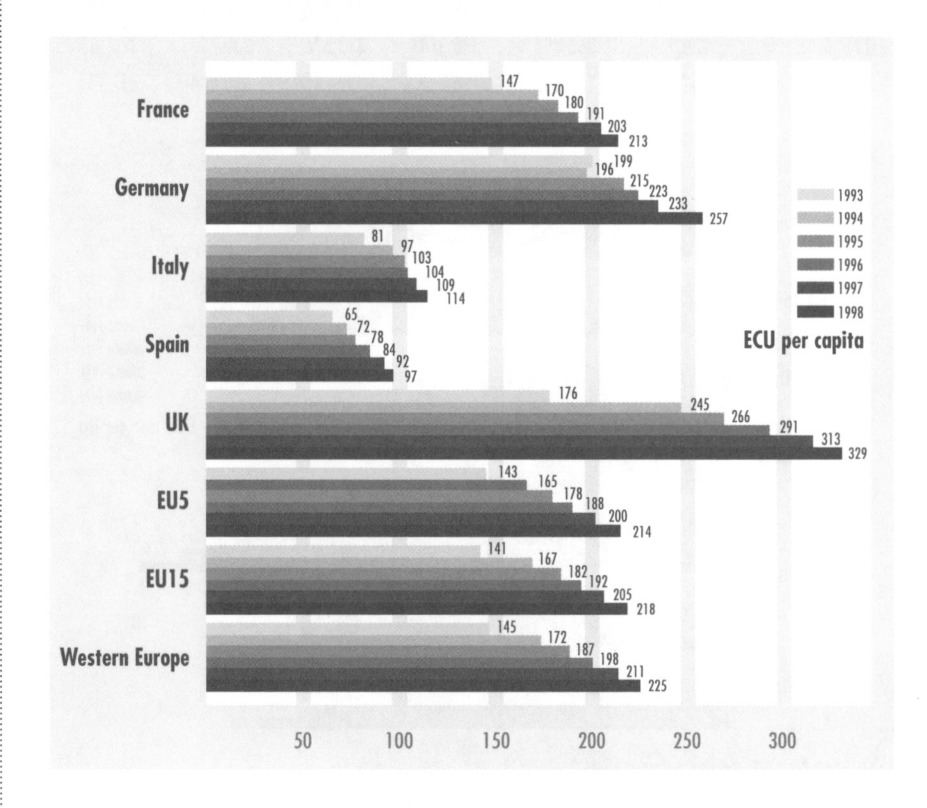

Sources

EITO, Prognos calculations, 1997 conversion rates
For further explanations see table 6.4.1

6.4.3 IT Software Expenditure

in mill. ECU	1993	1994	1995	1996	1997	1998
France	3,397	3,857	4,262	4,629	5,013	5,581
Germany	8,157	7,930	8,695	9,325	10,183	11,399
Italy	2,140	2,415	2,514	2,648	2,789	3,021
Spain	823	641	765	869	956	1,086
UK	4,347	5,115	5,666	6,212	7,025	7,946
EU5	18,864	19,958	21,902	23,683	25,966	29,033
EU15	23,598	24,662	27,009	29,268	32,197	36,125
Western Europe	25,167	26,115	28,611	31,020	34,127	38,262

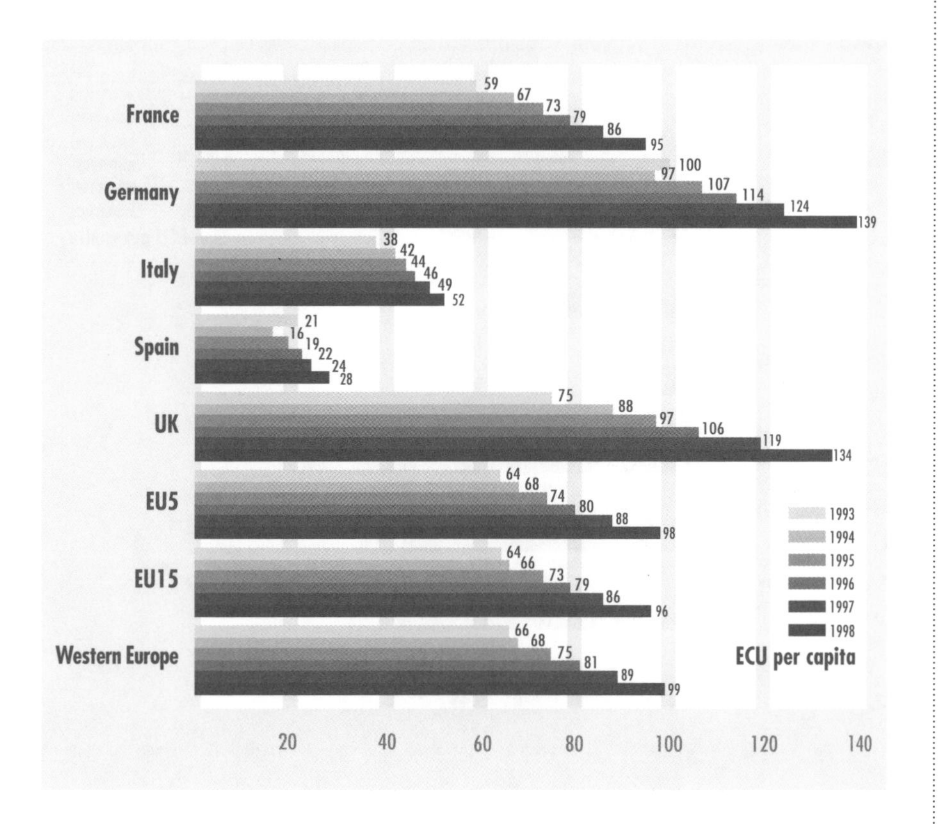

In this category Germany is still top both in market volume and per capita level, although in all countries except Spain the IT software expenditure grows faster.

Sources

EITO, Prognos calculations, 1997 conversion rates
For further explanations see table 6.4.1

6.4.4 IT Services Expenditure

The IT services expenditure is the only ICT category where France is the largest market with a significant lead.

in mill. ECU	1993	1994	1995	1996	1997	1998
France	10,017	11,857	12,278	13,020	14,245	16,074
Germany	10,435	10,410	10,924	10,233	11,079	12,365
Italy	4,587	4,965	5,199	5,602	6,064	6,756
Spain	1,479	1,644	1,714	1,848	2,074	2,353
UK	7,411	9,049	9,835	10,334	11,475	13,048
EU5	33,928	37,925	39,950	41,037	44,937	50,596
EU15	43,578	47,266	49,776	51,486	56,510	63,709
Western Europe	46,918	50,624	53,266	55,182	60,470	68,029

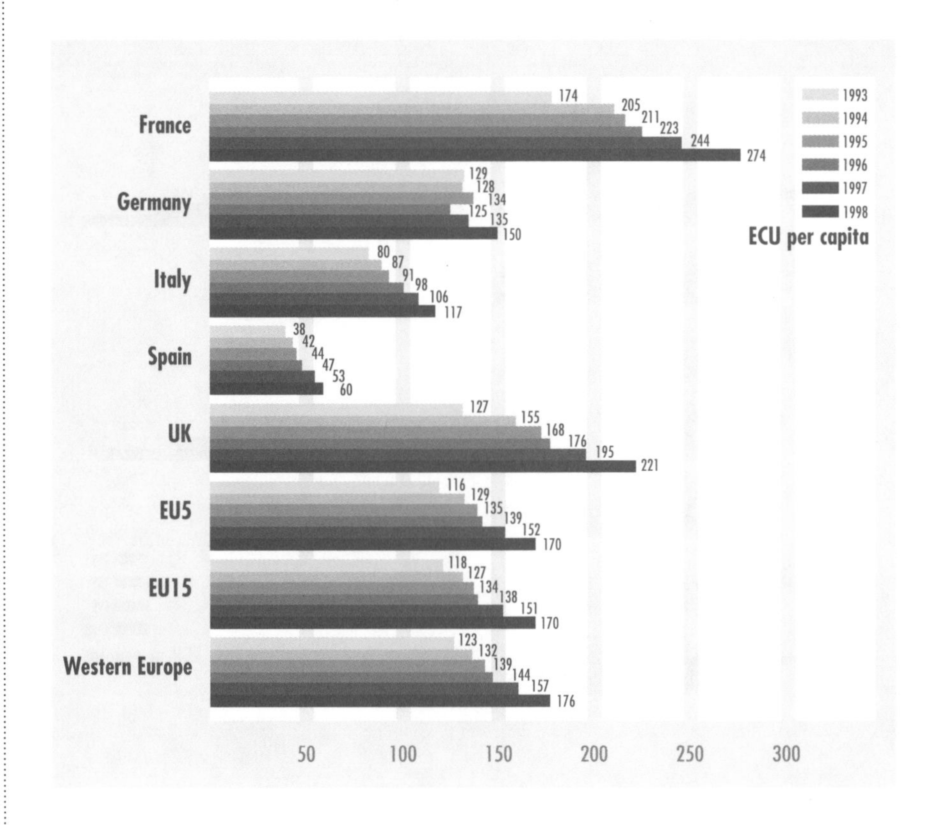

Sources

EITO, Prognos calculations, 1997 conversion rates
For further explanations see table 4.1

6.4.5 TC Hardware/Net Equipment Expenditure

in mill. ECU	1993	1994	1995	1996	1997	1998
France	2,611	2,351	2,570	2,800	3,089	3,357
Germany	3,298	3,554	3,866	4,120	4,377	4,617
Italy	1,285	1,588	1,694	1,921	2,056	2,192
Spain	705	548	579	685	690	793
UK	2,249	2,672	2,852	2,986	3,166	3,257
EU5	10,149	10,713	11,561	12,512	13,378	14,216
EU15	12,914	13,186	14,272	15,434	16,500	17,580
Western Europe	13,803	13,933	15,037	16,218	17,327	18,466

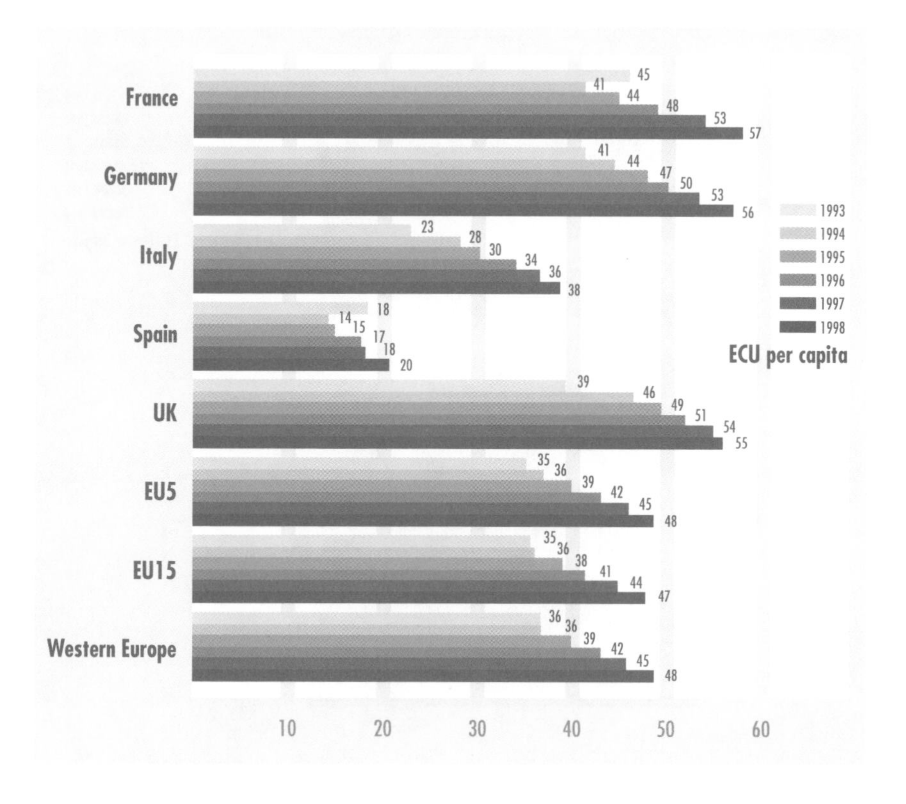

The expenditure for TC hardware/net equipment is on the same per capita level in France, Germany and the UK. The gap between Spain and the rest of Europe increased because of reduced investments between 1994 and 1996.

Sources

EITO, Prognos calculations, 1997 conversion rates
For further explanations see table 6.4.1

6.4.5 TC Hardware/Net Equipment Expenditure

With a share of 44 % the expenditure for TC services is the largest component of the total ICT expenditure. Once more, the UK reaches the highest per capita level, with Germany remaining the largest market.

in mill. ECU	1993	1994	1995	1996	1997	1998
France	17,784	18,145	19,640	20,731	22,610	24,733
Germany	28,746	28,334	31,408	31,995	34,058	36,354
Italy	11,940	15,140	16,483	18,236	20,233	23,102
Spain	5,610	6,600	7,383	8,897	9,583	10,468
UK	20,169	18,728	21,487	24,421	26,159	27,730
EU5	84,250	86,947	96,401	104,280	112,643	122,387
EU15	104,604	109,834	121,652	132,749	144,323	157,277
Western Europe	111,014	115,949	128,151	139,913	152,079	165,629

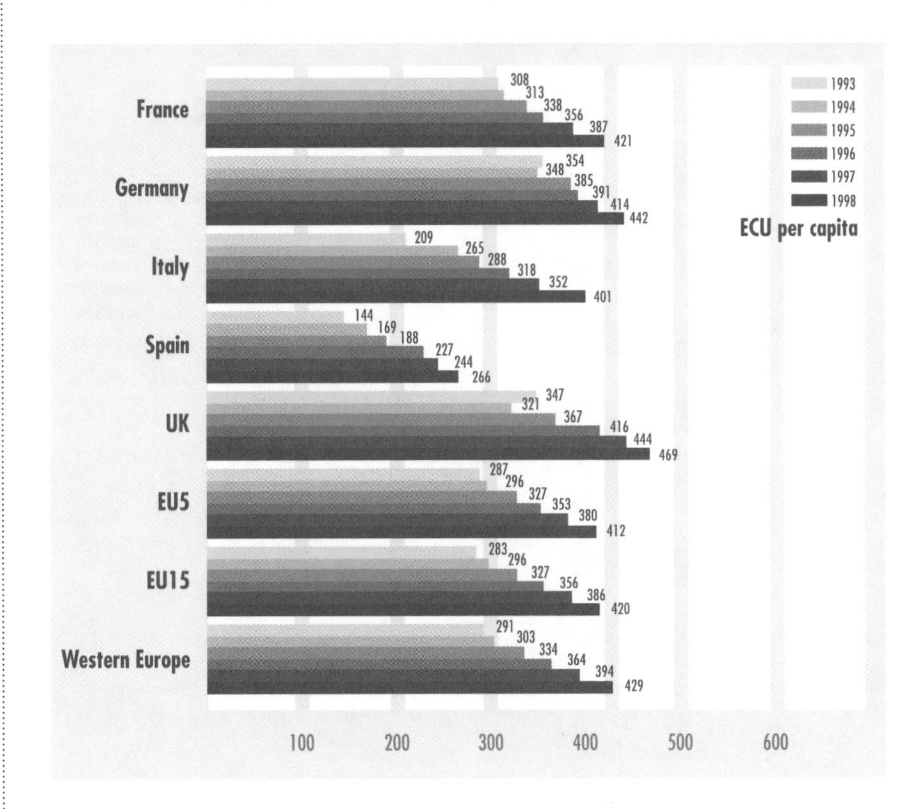

Sources

EITO, Prognos calculations, 1997 conversion rates.
For further explanations see table 6.4.1

Advertising Markets

6.5.1 Total Advertising Expenditure

in mill. ECU	1993	1994	1995	1996	1997
France	6,989	7,343	7,644	7,873	8,172
Germany	14,141	15,107	16,025	16,279	16,922
Italy	4,056	4,176	4,377	4,798	5,186
Spain	3,376	3,429	3,552	3,669	3,899
UK	9,780	10,822	11,728	12,629	13,864
EU5	38,343	40,876	43,326	45,248	48,043
EU15	46,996	50,424	53,910	56,015	55,734
Western Europe	49,519	53,122	56,852	58,979	56,738
USA	68,411	74,698	79,619	86,307	92,603
Japan	25,256	25,570	26,854	28,765	29,764

ECU per capita

	1993	1994	1995	1996	1997
France	121	127	131	135	140
Germany	174	186	196	199	206
Italy	71	73	76	84	90
Spain	86	88	91	94	99
UK	168	185	200	215	236
EU5	131	139	147	153	162
EU15	127	136	145	150	158
Western Europe	130	139	148	153	159
USA	265	287	302	327	346
Japan	203	205	215	229	237

Total advertising expenditure shows significant differences between countries. The USA per capita value is more than twice the European average. With an average growth of 9 % p.a. in the last five years, the UK has reached the Japanese per capita level, whilst Germany remains the largest advertising market in Europe.

Sources

The Advertising Association, Advertising Information Group.
The data have been adjusted in order to facilitate comparison. They include agency commission and press classified advertising, but exclude production costs. 1997 Data for Denmark, Netherlands and Switzerland are not included.

6.5.2 Newspaper Advertising Expenditure

The share of newspaper adspend of total adspend varies between 21 % in Italy to 48 % in Germany. The European average is 37 %.

in mill. ECU	1993	1994	1995	1996	1997
France	1,747	1,833	1,883	1,924	1,959
Germany	7,306	7,624	7,889	7,897	8,157
Italy	846	899	944	1,005	1,064
Spain	1,040	1,091	1,119	1,154	1,220
UK	4,239	4,632	4,905	5,158	5,615
EU5	15,178	16,080	16,741	17,139	18,014
EU15	19,321	20,517	21,525	22,098	21,180
Western Europe	20,833	22,080	23,220	23,774	21,766
USA	26,825	28,778	30,420	32,167	34,904
Japan	7,696	7,784	8,098	8,594	8,776

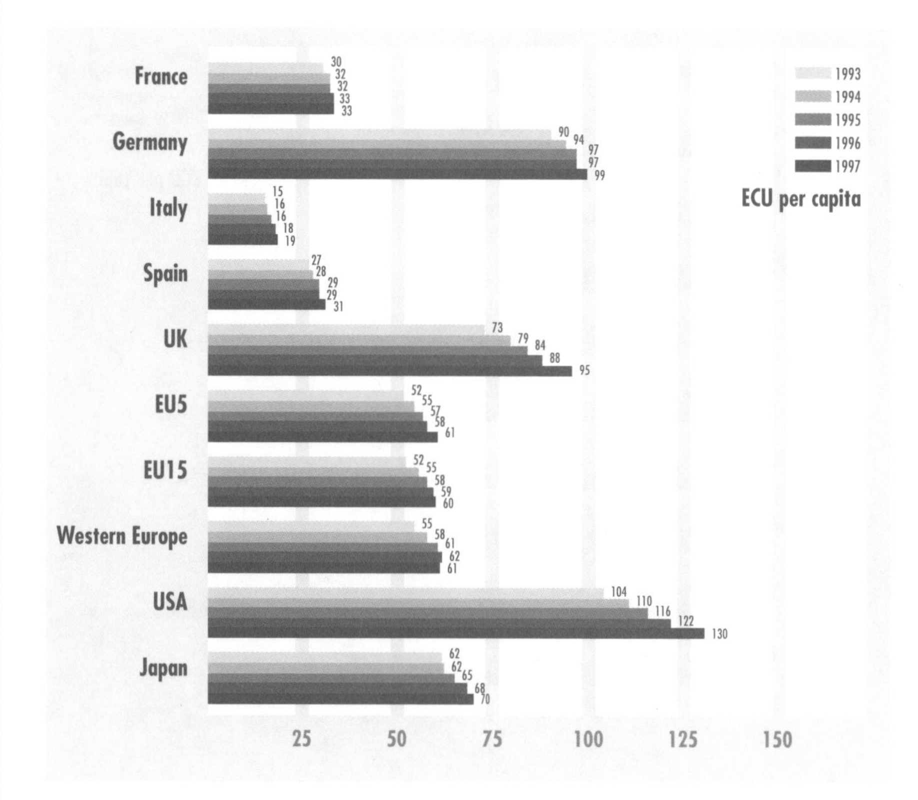

Sources

The Advertising Association, Advertising Information Group.
The data have been adjusted in order to facilitate comparison. They include agency commission and press classified advertising, but exclude production costs. 1997 Data for Denmark, Netherlands and Switzerland are not included.

6.5.3 Magazine Advertising Expenditure

in mill. ECU	1993	1994	1995	1996	1997
France	1,645	1,701	1,743	1,800	1,888
Germany	2,915	2,978	3,177	3,072	3,152
Italy	743	709	737	788	829
Spain	619	554	563	571	598
UK	1,678	1,856	2,067	2,313	2,551
EU5	7,601	7,798	8,286	8,543	9,018
EU15	9,174	9,440	10,066	10,450	10,337
Western Europe	9,608	9,884	10,534	10,932	10,444
USA	8,911	9,464	10,189	10,758	11,693
Japan	2,334	2,371	2,553	2,779	2,998

ECU per capita

	1993	1994	1995	1996	1997
France	29	29	30	31	32
Germany	36	37	39	38	38
Italy	13	12	13	14	14
Spain	16	14	14	15	15
UK	29	32	35	39	43
EU5	26	27	28	29	30
EU15	25	25	27	28	29
Western Europe	25	26	27	28	29
USA	35	36	39	41	44
Japan	19	19	20	22	24

Germany represents the largest magazine advertising market in Europe. Nevertheless, the UK has the highest per capita adspend for magazines, ranking at the same level as the USA.

Sources

The Advertising Association, Advertising Information Group.
The data have been adjusted in order to facilitate comparison. They include agency commission and press classified advertising, but exclude production costs. 1997 Data for Denmark, Netherlands and Switzerland are not included.

6.5.4 Television Advertising Expenditure

The most important increase of television adspend was registered in Germany with an average growth of 11 % p.a. during the period from 1993 to 1997. Therefore, the television share of the total German adspend increased to 25 % in 1997, but didn't reach the European average of 32 %.

in mill. ECU	1993	1994	1995	1996	1997
France	2,185	2,345	2,526	2,640	2,780
Germany	2,730	3,185	3,587	3,901	4,207
Italy	2,275	2,394	2,508	2,721	2,982
Spain	1,232	1,281	1,327	1,382	1,492
UK	3,204	3,546	3,848	4,083	4,499
EU5	11,626	12,750	13,797	14,727	15,962
EU15	13,680	15,281	16,728	17,441	18,210
Western Europe	13,915	15,594	17,069	17,793	18,396
USA	24,317	27,184	29,020	32,592	34,154
Japan	10,089	10,432	11,140	12,161	12,745

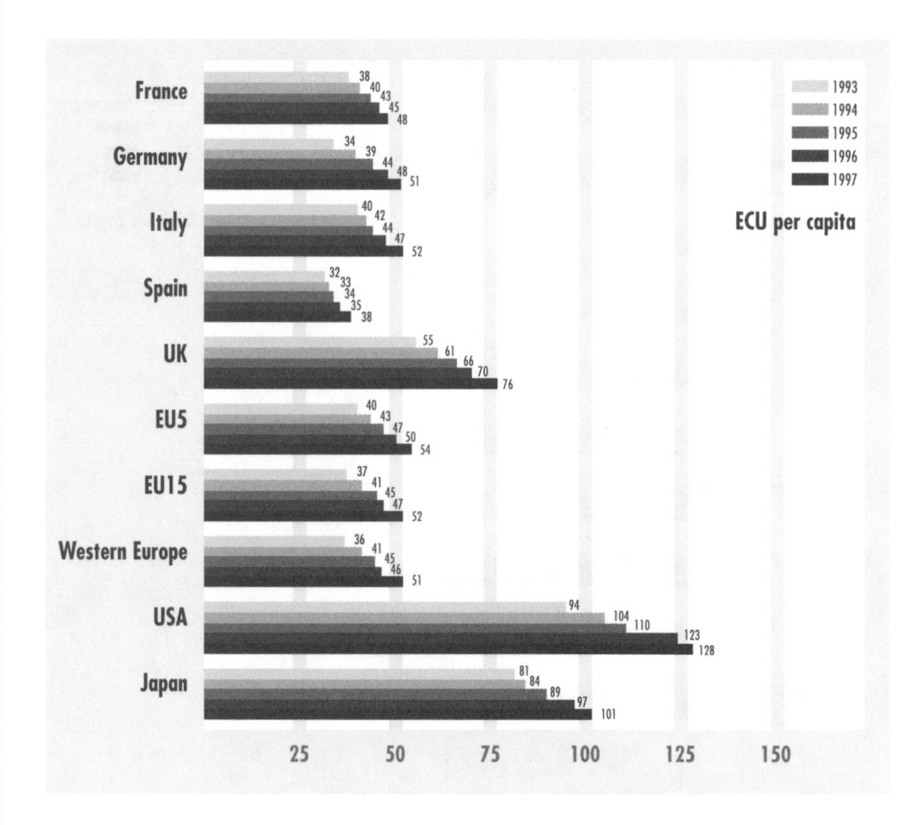

Sources

The Advertising Association, Advertising Information Group.
The data have been adjusted in order to facilitate comparison. They include agency commission and press classified advertising, but exclude production costs. 1997 Data for Denmark, Netherlands and Switzerland are not included.

6.5.5 Radio Advertising Expenditure

in mill. ECU	1993	1994	1995	1996	1997
France	538	560	567	550	539
Germany	569	623	638	652	665
Italy	55	58	73	159	183
Spain	311	320	347	361	378
UK	256	319	389	452	517
EU5	1,729	1,879	2,012	2,173	2,283
EU15	2,151	2,349	2,543	2,744	2,746
Western Europe	2,224	2,447	2,669	2,881	2,848
USA	7,589	8,449	9,098	9,845	10,826
Japan	1,401	1,350	1,379	1,444	1,488

France 9 10 10 9 9
Germany 7 8 8 8 8
Italy 1 1 1 3 3
Spain 8 8 9 9 10
UK 4 5 7 8 9
EU5 6 6 7 7 8
EU15 6 6 7 7 8
Western Europe 6 6 7 7 8
USA 29 32 35 37 40
Japan 11 11 11 12 12
10 20 30 40
1993
1994
1995
1996
1997
ECU per capita

In the radio advertising market the per capita level of the USA is five times the European average. The share of radio advertising amounts to 12 % compared to only 5 % in Western Europe.

Sources

The Advertising Association, Advertising Information Group.
The data have been adjusted in order to facilitate comparison. They include agency commission and press classified advertising, but exclude production costs. 1997 Data for Denmark, Netherlands and Switzerland are not included.

6.5.6 Outdoor/Transport Advertising Expenditure

Outdoor/transport is the only advertising market where the European per capita level is higher than the US level.

in mill. ECU	1993	1994	1995	1996	1997
France	839	869	888	917	959
Germany	473	541	567	587	567
Italy	136	116	116	123	128
Spain	147	155	161	167	179
UK	347	404	436	493	578
EU5	1,942	2,085	2,167	2,287	2,411
EU15	2,364	2,517	2,645	2,822	2,865
Western Europe	2,609	2,773	2,933	3,112	2,883
USA	769	824	891	944	1,026
Japan	3,735	3,633	3,677	3,772	3,757

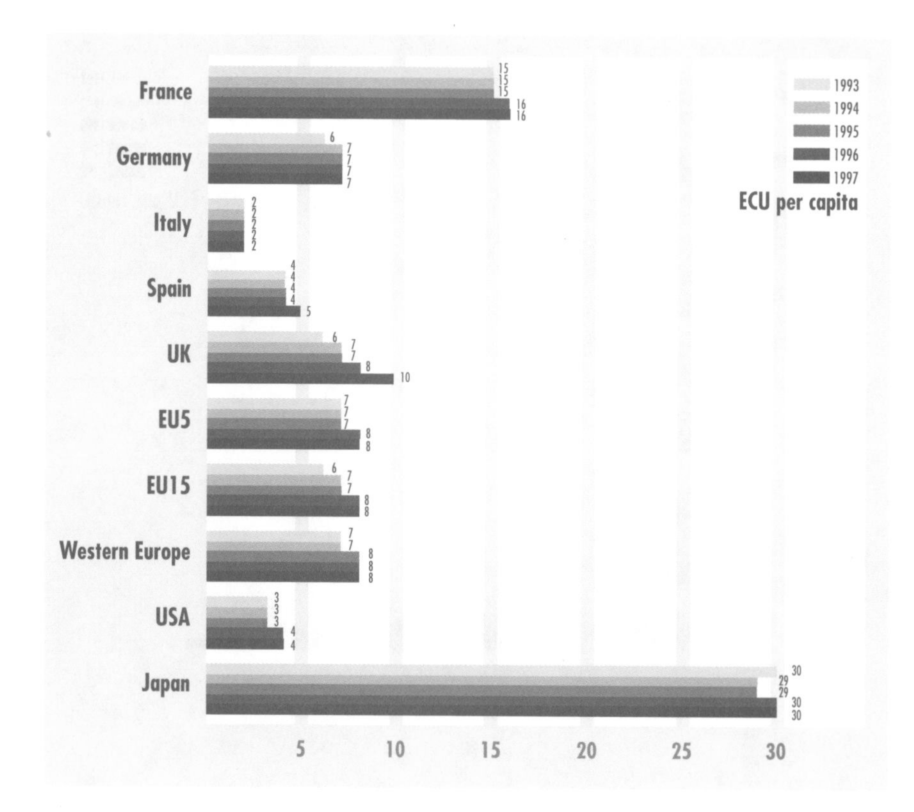

Sources

The The Advertising Association, Advertising Information Group
The data have been adjusted in order to facilitate comparison. They include agency commission and press classified advertising, but exclude production costs. 1997 Data for Denmark, Netherlands and Switzerland are not included.

6.5.7 Forecast: Online Advertising Expenditure

in mill. ECU	1997	2002	2010
France	5	84	423
Germany	26	230	944
Italy	3	36	230
Spain	3	43	230
UK	26	214	867
EU5	62	607	2,694
EU15	71	663	3,112
Western Europe	74	679	3,179
USA	561	3,469	5,969

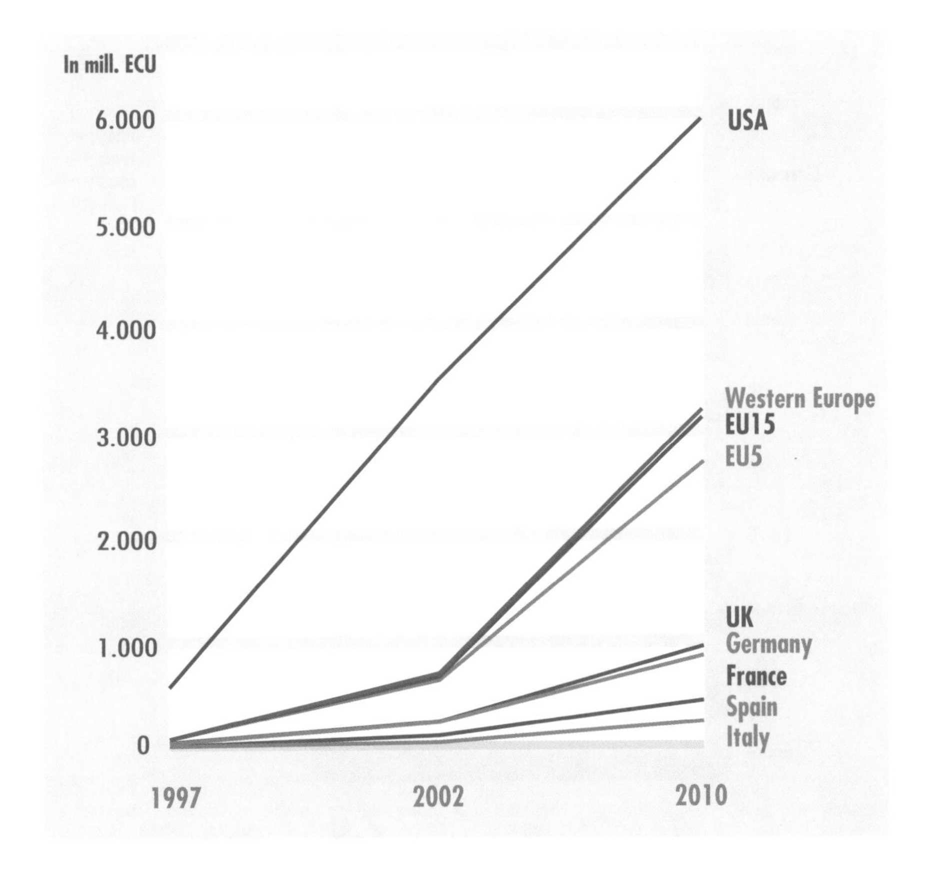

Sources

Prognos estimates; advertising expenditure only, 1997 conversion rates. Online advertising expenditure is expected to grow from 561 ECU in the year 1997 to 3.5 billion ECU by the year 2002.

Selected Media Markets

6.6.1 Daily Newspapers: Sales Revenues

Sales revenues' contribution to total revenues of daily newspapers is 36 % in Western Europe, 14 % in the USA and more than 50 % in Japan. Regarding the high number of copies per HH in Japan (table 6.3.10), its leading position is not surprising.

in mill. ECU	1993	1994	1995	1996	1997
France	1,920	1,936	2,041	1,993	n.a.
Germany	2,900	3,027	3,155	3,292	3,309
Italy	1,065	1,103	1,247	1,304	1,319
Spain	627	735	844	916	965
UK	2,609	2,570	2,609	2,775	n.a.
EU5	9,121	9,371	9,895	10,280	n.a.
EU15	11,748	12,111	12,698	13,320	n.a.
Western Europe	12,204	12,594	13,181	13,802	n.a.
USA	5,031	5,156	5,297	5,430	5,492
Japan	8,718	9,163	9,287	9,411	n.a.

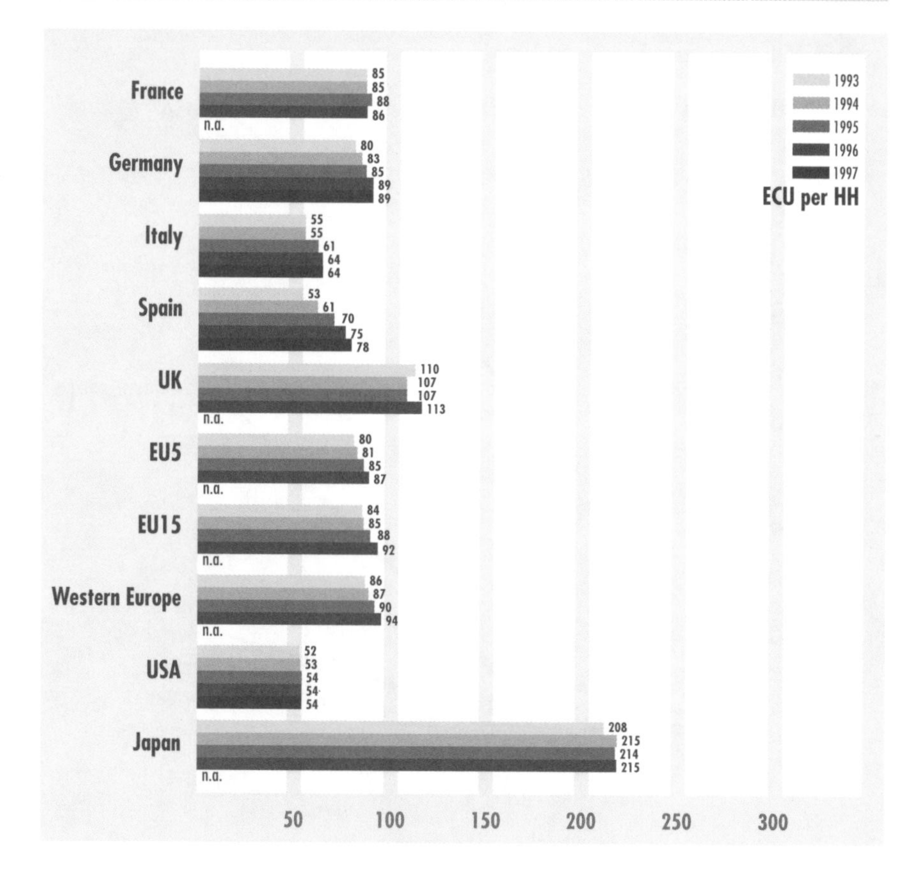

Sources

World Association of Newspapers WAN, Prognos calculations.
EU15 without Austria, Western Europe without Switzerland. 1997 conversion rates

6.6.2 TV Licence Fee Revenues

in mill. ECU	1994	1995	1996
France	1,010	1,084	1,625
Germany	2,811	2,865	2,852
Italy	1,166	1,224	1,249
Spain	0	0	0
UK	2,432	2,633	2,778
EU5	7,419	7,806	8,503
EU15	9,487	9.826	n.a.
Western Europe	9,750	10,112	n.a.
Japan	3,980	4,038	n.a.

Licence fee revenues depend on the amount of fees and the number of licence fee accounts. Estimates of the licence fee evasion rate range between 5 % (Belgium) and 15 % (Italy) in 1996. The tremendous increase in France in 1996 is questionable because no reasons could be found.

Sources

Eurostat. No licence fees in Spain, Portugal, Greece. Western Europe without Switzerland. 1997 conversion rates.

The gradual decrease of cinema admissions in Europe since 1984 has come to an end. In all European countries the figures in 1996/1997 look better than in the years before. With 5 admissions per capita and year the USA remain the cinema eldorado.

6.6.3 Cinema Admissions

mill.	1993	1994	1995	1996	1997
France	133	124	130	137	148
Germany	131	133	124	133	143
Italy	92	98	91	97	100
Spain	88	89	95	104	105
UK	114	124	115	124	139
EU5	557	568	555	594	635
EU15	661	676	659	703	760
Western Europe	689	705	686	731	788
USA	1,244	1,213	1,222	1,265	1,310
Japan	131	123	127	120	141

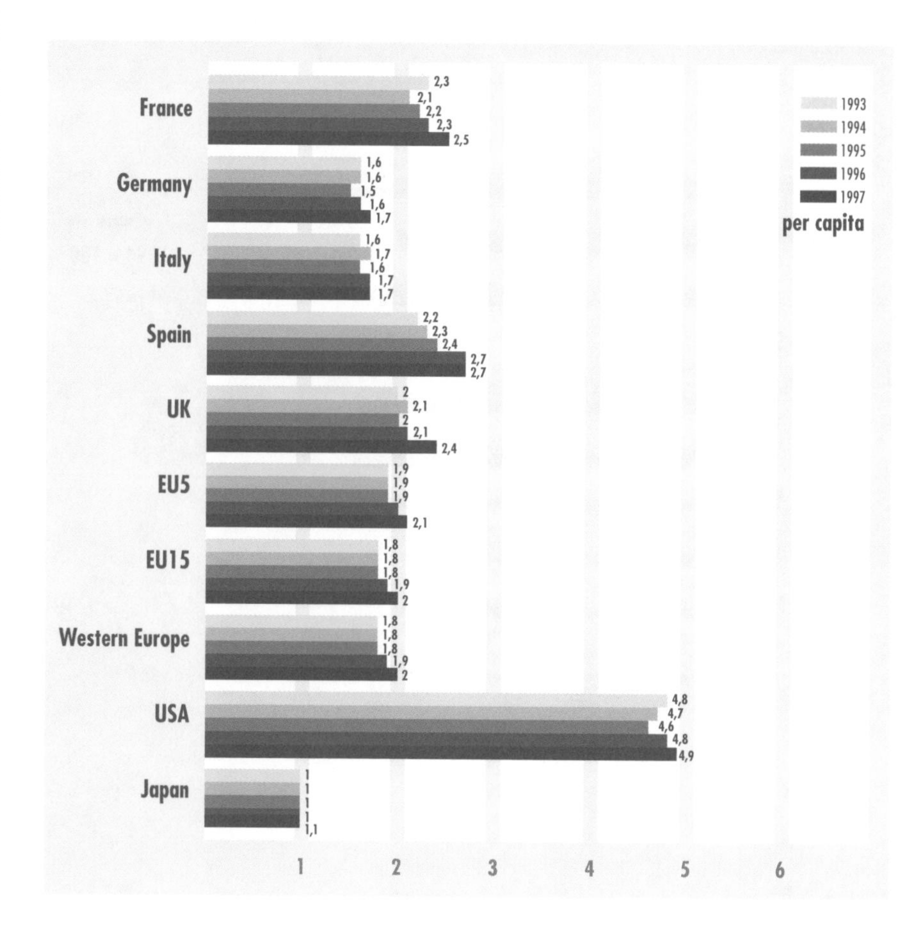

Sources

European Audiovisual Observatory EAO

6.6.4 Cinema Gross Box Office

in mill. ECU	1993	1994	1995	1996	1997
France	683	648	685	718	778
Germany	596	625	602	669	749
Italy	393	427	413	454	479
Spain	245	263	277	333	354
UK	514	526	556	636	734
EU5	2,431	2,489	2,534	2,810	3,095
EU15	2,936	3,021	3,002	3,280	3,730
Western Europe	3,104	3,202	3,181	3,465	3,925
USA	4,391	4,629	4,721	5,081	5,634
Japan	1,194	1,120	1,152	1,086	1,291

ECU per capita

	1993	1994	1995	1996	1997
France	38	40	43	45	48
Germany	34	39	44	48	51
Italy	40	42	44	47	52
Spain	32	33	34	35	38
UK	55	61	66	70	76
EU5	40	43	47	50	54
EU15	37	41	45	47	52
Western Europe	36	41	45	46	51
USA	94	104	110	123	128
Japan	81	84	89	97	101

25 50 75 100 125 150

In the last five years the cinema gross box office increased even more than the admissions. The new multiplexes attracted a new audience in spite of higher ticket prices, the success of national films is the other factor. As cinema tickets are cheaper in the USA, the gap between Europe and the USA is not as large as could be expected from per capita admissions.

Sources

European Audiovisual Observatory EAO. 1995 without Greece and Portugal, 1996 without Greece, Portugal and Ireland. 1997 conversion rates.

The boom in unit sales to the consumer is continuing because the reduction in average prices leads to an increasing number of purchased pre-recorded videos per VCR home. As with cinema admissions the USA seem to be an exceptional case: US VCR homes buy more than twice videocassettes than the average European home. One main reason is pricing policy: new blockbuster videos are offered for discount prices in a certain period.

6.6.5 Video Purchases

in mill. units	1992	1993	1994	1995	1996	1997
France	26,0	26,0	50,0	55,0	55,0	50,5
Germany	21,0	28,5	35,0	41,7	44,0	40,0
Italy	15,3	20,9	28,7	25,0	28,5	21,9
Spain	3,3	8,2	10,9	12,1	12,7	14,5
UK	48,0	60,0	66,0	73,0	79,0	87,0
EU5	113,6	143,6	190,6	206,8	219,2	213,9
EU15	129,8	168,5	220,9	241,6	254,9	252,6
Western Europe	132,2	171,9	225,9	247,5	261,9	259,8
USA	386,8	326,9	398,4	470,9	600,1	633,1
Japan	31,0	22,7	29,4	31,4	40,4	59,8

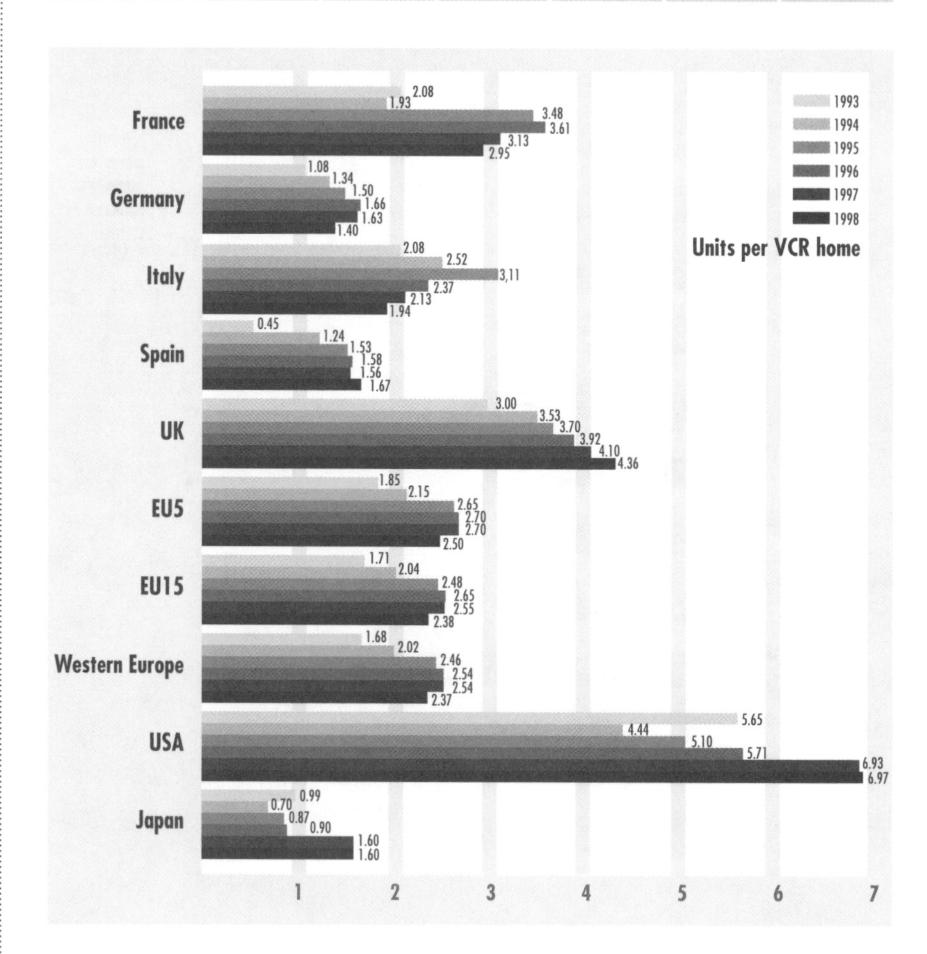

Sources

European Audiovisual Observatory EAO, Screen Digest.

6.6.6 Video Sales Turnover

in mill. ECU	1992	1993	1994	1995	1996	1997
France	665	681	922	965	998	999
Germany	295	423	484	529	535	471
Italy	177	232	286	224	235	294
Spain	65	106	157	175	168	176
UK	731	929	1,008	1,140	1,160	1,244
EU5	1,934	2,371	2,857	3,033	3,096	3,183
EU15	2,169	2,723	3,305	3,659	3,769	3,731
Western Europe	2,223	2,803	3,392	3,755	3,879	3,831
USA	3,493	4,152	4,847	5,523	6,477	6,696
Japan	1,021	1,016	720	878	1,253	n.a.

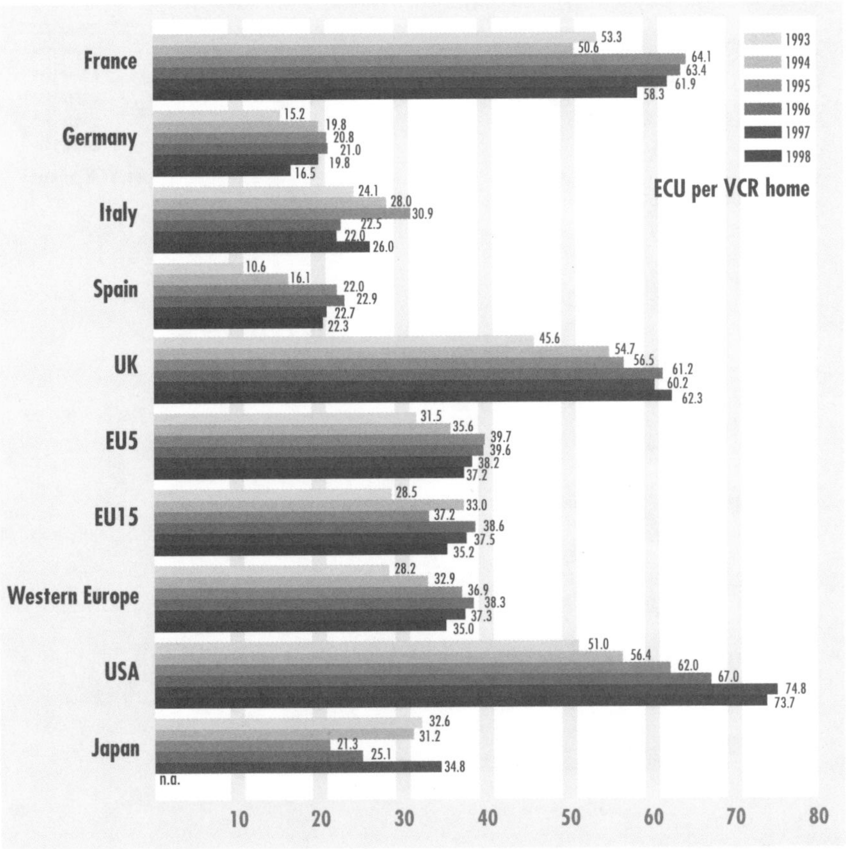

ECU per VCR home

1993 1994 1995 1996 1997 1998

France
53.3
50.6
64.1
63.4
61.9
58.3

Germany
15.2
19.8
20.8
21.0
19.8
16.5

Italy
24.1
28.0
30.9
22.5
22.0
26.0

Spain
10.6
16.1
22.0
22.9
22.7
22.3

UK
45.6
54.7
56.5
61.2
60.2
62.3

EU5
31.5
35.6
39.7
39.6
38.2
37.2

EU15
28.5
33.0
37.2
38.6
37.5
35.2

Western Europe
28.2
32.9
36.9
38.3
37.3
35.0

USA
51.0
56.4
62.0
67.0
74.8
73.7

Japan
32.6
31.2
21.3
25.1
34.8
n.a.

10 20 30 40 50 60 70 80

Despite a slow drop-off in prices the sales turnover was increasing in all European countries until 1995. Since then, the average sales per VCR home have stagnated. The average price of a prerecorded video is lowest in Italy and Germany, resulting in little spending per VCR home.

Sources

European Audiovisual Observatory EAO, Screen Digest

The video rental market still suffers from the growing sales of prerecorded videos, but in most countries the dramatic decline seemes to have stopped now. With the growing number of VCR homes in most countries the average rental transactions per VCR home are falling.

6.6.7 Video Rental Turnover

in mill. ECU	1992	1993	1994	1995	1996	1997
France	34	150	160	166	170	178
Germany	458	377	367	377	397	357
Italy	117	106	106	100	102	99
Spain	151	95	86	86	72	78
UK	738	660	633	660	709	588
EU5	1,699	1,387	1,350	1,389	1,449	1,301
EU15	2,149	1,830	1,777	1,815	1,883	1,740
Western Europe	2,276	1,958	1,895	1,929	1,990	1,844
USA	8,595	8,780	8,391	7,994	8,133	7,942
Japan	2,957	2,618	2,445	2,490	2,748	2,762

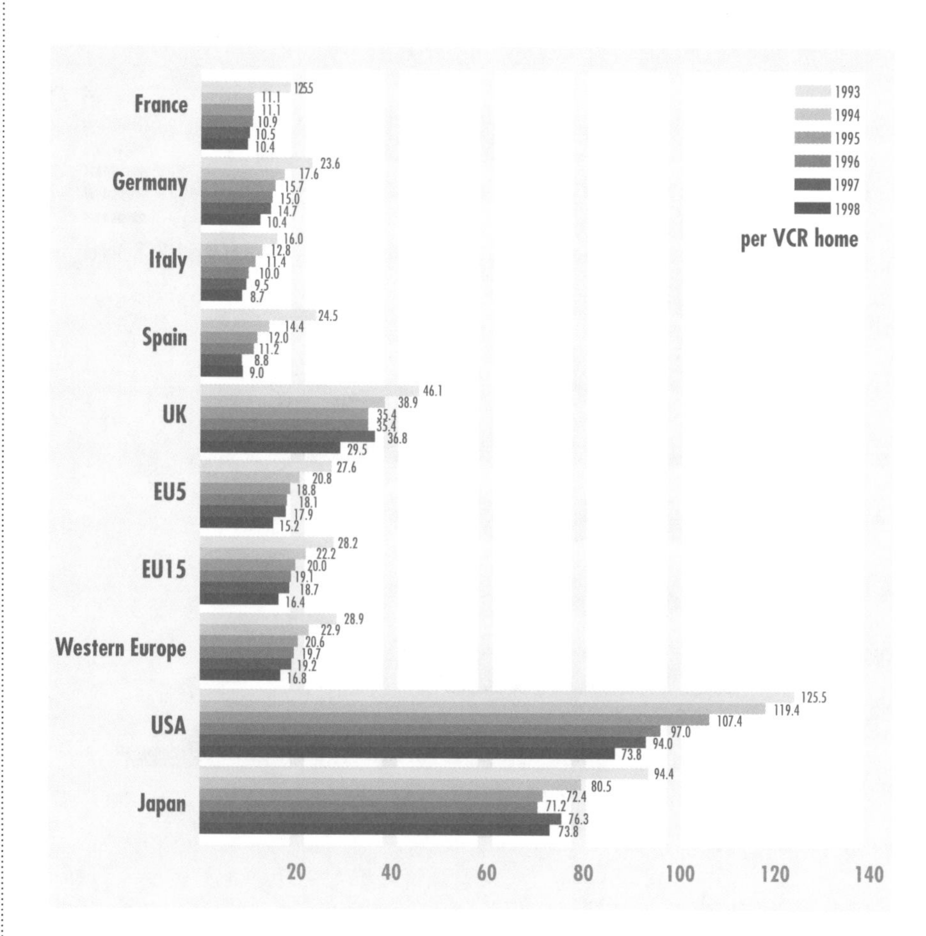

Sources

European Audiovisual Observatory EAO, Screen Digest

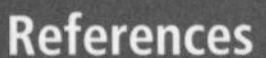

References

Adstead, Stephen; **McGarvey,** Patrick (1997): Convergence in Europe. The New Media Value Chain. London, UK: Financial Times Management Report

Alden, Christopher (1998): Kingmaker. Yahoo is becoming an online service. Internet: http://www.redherring.com/mag/issue57, August 1998

Aley, James (1996): The Theory That Made Microsoft. Internet: http://www.pathfinder.com/fortune/archives/htm, 29.4.96

Alison, Alexander; **Owers,** James; **Carveth,** Rod (Ed.) (1998): Media Economics – Theory and Practice. Mahwah, N.J.: Lawrence Erlbaum Associates (2nd. ed.)

Altmeppen, Klaus-Dieter (Ed.) (1996): Ökonomie der Medien und des Mediensystems. Grundlagen, Ergebnisse und Perspektiven medienökonomischer Forschung. Opladen: Westdeutscher Verlag

Anon. (1997a): Der Preis der Freiheit. In: Manager Magazin, Nr. 8/97, pp. 94-106

Anon. (1997b): Neues Spiel: Der Monopolist lernt den Wettbewerb. In: Manager Magazin, Nr. 8/97, pp. 108-109

Anon. (1997c): Platzangst im All. In: Manager Magazin, Nr. 8/97, pp. 112-121

Anon. (1997d): Assembling the new economy. In: The Economist, 13.9.97, pp. 77-83

Anon. (1997e): Unentgeltlich telefonieren - in Schweden ein gutes Geschäft. In: Frankfurter Allgemeine Zeitung, 6.10.97, p. 8

Anon. (1997f): Telekom-Konkurrenten dürfen Preise vergleichen. In: Spiegel Online, 3.12.97

Anon. (1998a): In the shark pond. In: The Economist, 3.1.98, pp. 59-60

Anon. (1998b): Fernsehen, Computer und Internet sollen verschmelzen. In: Frankfurter Allgemeine Zeitung, 6.1.98, S. 16

Anon. (1998c): Microsoft's contradiction. In: The Economist, 21.1.98, pp. 71-72

Anon. (1998d): Zwei ungleiche Rivalen. In: Die Zeit, 22.1.98, pp. 18-19

Anon. (1998e): Compaq kauft Digital Equipment für 9,6 Mrd. Dollar. In: Frankfurter Allgemeine Zeitung, 27.1.98, p. 14

Anon. (1998f): Sun and Microsoft. A simmering dispute. In: The Economist, 21.2.98, pp. 81-82

Anon. (1998g): Schneller als ISDN - die Technik bei Highspeed-Übertragung. In: Connect, Nr. 2/98, p. 69

Anon. (1998h): Netz des Vertrauens – das World Wide Web muß verläßlicher und intelligenter werden (Interview mit Tim Berners-Lee). In: Manager Magazin, MM-Spezial Internet, Nr. 3/98, pp. 214-222

Anon. (1998i): Grundig und Primus-Online starten Internet-TV fürs Wohnzimmer. In: Spiegel Online, 1.4.98

Anon. (1998j): Mangel an Kundenorientierung in der High-Tech Industrie. In: Blick durch die Wirtschaft, 19.5.98, p. 3

Anon. (1998k): Deutschland ist Weltmeister im Wettbewerb. In: Handelsblatt, 16.5.98, p. 13

Anon. (1998l): ADSL macht das Rennen. In: Net investor, Nr. 6/98, p. 10

Anon. (1998m): A bid too far? In: The Economist, 16.5.98, pp. 73f.

Anon. (1998n): The Rebirth of IBM - Blue is the colour. In: The Economist, 6.6.98, pp. 77-80

Anon. (1998o): Microsoft Narrows Netscape's Lead, Internet: http://www.wired.com, (Headline News), 13.7.98

Anon. (1998p): Regulierer zieht positive Bilanz. In: Handelsblatt, 16.7.98, p. 1

Anon. (1998q): Intel - Zukunftsaussichten gelten als positiv. In: Handelsblatt, 16.7.98, p. 17

Anon. (1998r): Intel server chip coming soon. Internet: http://www.news.com/News/Item/04 24514,00html, CNet News.com, 23.7.98

Arrow, Kenneth (1962): Economic Welfare and the Allocation of Resources for Invention. In: Nelson R.R. (Ed.): The Rate and Direction of Inventive Activity: Economic and Social Factors, New York, N.Y.: Princeton University Press

Artopé, Alexander; **Zerdick,** Axel (1995): Die Folgen der Media-Mergers in den USA. Die neue Ausgangssituation auf dem deutschen und europäischen Fernsehmarkt. München: MGM MediaGruppe München

Bachofer, Michael (1998): Wie wirkt Werbung im Web? Blickverhalten, Gedächtnisleistung und Imageveränderung beim Kontakt mit Internet-Anzeigen. (Ed.) STERN-Bibliothek, Hamburg: STERN Anzeigenabteilung

Bakos, Yannis; Brynjolfsson, Erik (1996): Bundling Information Goods: Pricing, Profits and Efficiency, SIMS Working Paper, Boston, MA.: MIT Sloan School of Business

BancAmerica Robertson Stephens (Ed.) (1998): E-Tailing Update. Internet Research. (see also: Internet: http://www.rsco.com), San Francisco, CA., 1998

Barlas, Pete (1998): Web Content Firm's Status: Little Profits, Fewer Investors. Internet: http://www.investorsbusiness.com, Investors Business Daily, 27.7.98

Bayers, Charles (1998): The Promise of One to One (A Love Story). In: Wired, 5/98, pp.130-134; 184-187

Booz Allen & Hamilton (Ed.) (1997): Zukunft Multimedia. Grundlagen, Märkte und Perspektiven in Deutschland. Frankfurt a.M.: Institut für Medienentwicklung und Kommunikation (4. Aufl.)

Browning, John; **Reiss,** Spencer (1998): Encyclopedia of the New Economy (Part I). In: Wired, 3/98, pp. 105-114

Brynjolfsson, Erik; **Hitt,** Lorin M. (1996): Productivity, Business Profitability, and Consumer Surplus: Three Different Measures of Information Technology Value. In: MIS Quarterly, 7/96, pp. 121-142

Burrows, Peter (1998): Cheap PCs - The Model T of the Digital Age. In: Business Week, 23.3.1998, pp. 28-32

Cairncross, Frances (1998): The Death of Distance. How the Communications Revolution Will Change Our Lives. Boston, MA.: Harvard Business Press

Carlton, Jim (1997): Apple. The Inside Story of Intrigue, Egomania and Business Blunders. New York, N.Y.: Times Business Books

Cerf, Vinton G.; **Kahn,** Robert (1974): A Protocol for Packet Network Interconnection. In: IEEE Transactions on Communications, vol. 22(5), pp. 637-648

Chandler, Alfred (1977): The Visible Hand. The Managerial Revolution in American Business. Cambridge, MA.: Belknapp Press

Clevé, Bastian (1997): Wege zum Geld. Film-, Fernseh- und Multimedia-Finanzierungen. Gerlingen: Bleicher-Verlag

Clevé, Bastian (Ed.) (1998) Investoren im Visier. Film- und Fernsehproduktionen mit Kapital aus der Privatwirtschaft, Gerlingen: Bleicher-Verlag

Collis, David J.; **Bane,** William P.; **Bradley,** Stephen P. (1997): The Converging World of Telecommunication, Computing, and Entertainment. In: Yoffie, David B. (Ed.): Competing in an Age of Digital Convergence, Boston, MA.: Harvard Business School Press, pp. 159-200

Cringely, Robert X. (1996): Accidental Empires. How the Boys of Silicon Valley Make their Millions, Battle Foreign Competition, And Still Can't Get A Date. New York, N.Y.: Harper Collins (2nd. ed.)

Cringely, Robert X. (1998): Sega, the once and future king of video games, has a plan to dominate personal computing, too, and it just might work. Internet: http://www. pbs.org/cringely/archive/may2898_text.html; PBS Online, Vol. 1.61, 28.5.1998

Databank Consulting (Ed.) (1997): Evolution of the Internet and WWW in Europe. Report for the European Commission, DG XII, with the support of IDATE/TNO, (see also: Internet: http://www.echo.lu), Brussels

Dehn, Peter (1998): Als Düsenjäger über die Datenrennstrecke jagen. In: Der Tagesspiegel, 12.7.98, p. 30

De Thier, Peter (1998): AT&T sprengt das Monopol auf US-Ortsgespräche. In: Berliner Zeitung, 27./28.6.98, p. 27

Diamond, David (1998): Building the Future Proof Telco. In: Wired, 5/98, pp. 124-127, 178-183

Diller, Hermann (1991): Preispolitik. Stuttgart: Kohlhammer-Verlag (2.Aufl.)

Downes, Larry; **Mui,** Chunka (1998): Unleashing the Killer App. Digital Strategies for Market Dominance. Harvard, MA.: Harvard Business School Press

dti (Ed.) (1998): Regulating communications: Approaching Convergence in the Information Age. London: Department of Trade and Industry; Department for Culture, Media and Sport

EAO (Europäische Audiovisuelle Informationsstelle) (Ed.) (1998): Statistisches Jahrbuch. Film, Fernsehen, Video und Neue Medien in Europa 1998. Strasbourg

Economides, Nicholas (1996): The Economics of Networks. In: International Journal of Industrial Organization, vol. 14, no.2, März 1996, pp. 12-35 (See also: http://edgar.stern.nyu.ed/networks)

Economides, Nicholas (1998): U.S. Telecommunications today. In: Business Economics, April 1998, Vol. XXXIII, Number 2, 1998, pp. 74-93

Economides, Nicholas (1999): The Telecommunications Act of 1996 and its Impact. In: Japan and the World Economy, forthcoming; quoted from the manuscript.

Edstrom, Jennifer; **Eller,** Marlin (1998): Barbarians Led by Bill Gates - Microsoft from the Inside. New York, N.Y.: Henry Holt and Company

Eimeren, Birgit van; **Gerhard,** Heinz; **Oehmichen,** Ekkehart; **Schröter,** Christian (1998): ARD-Online-Studie 1998: Onlinemedien gewinnen an Bedeutung. In: Media Perspektiven, Nr. 8/98, pp. 423-435

Eimeren, Birgit van; **Oehmichen,** Ekkehart; **Schröter,** Christian (1997): ARD-Online-Studie 1997: Onlinenutzung in Deutschland. In: Media Perspektiven, Nr. 10/97, pp. 548-557

EITO (Ed.) (1998): European Information Technology Observatory 98, Mainz: Eggebrecht-Presse

EITO (Ed.) (1999): European Information Technology Observatory 99, Frankfurt: EITO

EITO (Ed.) (2000): European Information Technology Observatory 2000, Frankfurt: EITO (See also: http://www.eito.com)

Elstrom, Peter; Reinhardt, Andy; Jackson, Susan; Yang, Catherine (1998): The New Trailblazers. Special Report Telecommunications, In: Business Week, 6.4.1998, pp. 48-56

Enquete-Kommission (1998): Enquete-Kommission Zukunft der Medien in Wirtschaft und Gesellschaft, Deutschlands Wege in die Informationsgesellschaft. Bonn : ZH Zeitungs-Verlag Service 1998 (and additional volumes of specialised reports)

Europäische Kommission, DG XIII/A4 (Ed.) (1997): Grünbuch zur Konvergenz der Branchen Telekommunikation, Medien und Informationstechnologie und ihren ordnungspolitischen Auswirkungen. Brüssel, 3. Dezember 1997

Europäische Kommission, DG X (Ed.) (1998a): Arbeitsdokument der Kommission, Zusammenfassung der Ergebnisse der öffentlichen Konsultation zum Grünbuch zur Konvergenz der Branchen Telekommunikation, Medien und Informationstechnologie und Themen für weitere Überlegungen. Brüssel, 29. Juli 1998

European Commission, Directorate General XIII/E (Ed.) (1998b): Condrinet. Content and Commerce Driven Strategies in Global Networks; A Study by Gemini Consulting Commissioned by the European Commission. Internet: http://www2.echo.lu/condrinet/, 10.10.98

European Communication Council - ECC (Ed.) (1997): Exploring the Limits. Europe's Changing Communication Environment. Berlin; Heidelberg; New York: Springer

European Communication Council - ECC (1999): Die Internet-Ökonomie: Strategien für die digitale Wirtschaft. European Communication Council. Von Axel Zerdick, Arnold Picot, Klaus Schrape; Alexander Artopé, Klaus Goldhammer, Ulrich T. Lange, Eckart Vierkant; Esteban López-Escobar, Roger Silverstone - Berlin; Heidelberg; New York; Barcelona; Hongkong; London; Mailand; Paris; Singapur; Tokio: Springer

Fries, Cornelia (1997): Nutzerkompetenz als Determinante der Diffusion multimedialer Dienste. Diskussionsbeitrag Nr. 184, Bad Honnef: Wissenschaftliches Institut für Kommunikationsdienste - WIK (Ed.)

Froitzheim, Ulf J. (1997): Reif für den Härtetest. In: Global Online, 9/97, pp. 40-45

Gates, Bill (1995): Der Weg nach vorn. Die Zukunft der Informationsgesellschaft. Hamburg: Hoffmann&Campe

Geirland, John (1998): The New Mousketeer. Disney's Online Jake Winebaum on Walt, the Web, and the future of couch potatoes. In: Wired, 6/98, p. 155

Gilder, George (1997): Feasting on the Giant Peach. Will the Internet Collapse? No Way! Internet: http://www.discovery.org/Giler/gilderaug96.html, 24.8.97

Gillmor, Dan (1998): Mail order is unfair dodge of sales tax. In: San Jose Mercury News, 6 October 1998; http://www.mercurycenter.com/columnists/gillmor

Goldhammer, Klaus (1998): Hörfunk und Werbung. Berlin: Vistas Verlag

Goldhammer, Klaus; **Zerdick,** Axel (1999): Rundfunk Online. Entwicklung und Perspektiven des Internets für Hörfunk- und Fernsehanbieter. Berlin: VISTAS Verlag

Grove, Andy S. (1996): Only The Paranoid Survive. How to exploit the crisis points that challenge every company and career. New York, N.Y.: Doubleday

Gründler, Ansgar (1997): Computer und Produktivität: das Produktivitätsparadoxon der Informationstechnologie. Wiesbaden: Gabler Verlag

Haas, Karl-Gerhard (1997): Schöne Aussichten - das TV von morgen. In: Stern, Nr. 36/1997, pp. 24-32

Haas, Michael H.; **Frigge,** Uwe; **Zimmer,** Gert (1991): Radio-Management. Ein Handbuch für Radio-Journalisten, München: Ölschläger

Hachmeister, Lutz ; **Rager,** Günther (1997): Wer beherrscht die Medien? Die 50 größten Medienkonzerne der Welt. C.H.Beck: München

Hagel III, John (1996): Spider versus Spider. In: The McKinsey Quarterly, 1/96, S. pp-18

Hagel III, John; Armstrong, Arthur G. (1997): Net Gain. Expanding Markets Through Virtual Communities. Boston, MA.: Harvard Business School Press

Heinrich, Jürgen (1994): Medienökonomie. Band 1: Mediensystem, Zeitung, Zeitschrift, Anzeigenblatt. Opladen: Westdeutscher Verlag

Hohensee, Matthias (1998): Fernsehen über Telefonleitung. In: Wirtschaftswoche, 23/98, Spezial Telekommunikation, p. 148

Jacob, Frank (1995): Produktindividualisierung - ein Ansatz zur innovativen Leistungsgegestaltung im Business-to-Business Bereich. Neue Betriebswirtschaftliche Forschung, Bd. 144, Wiesbaden: Gabler Verlag

Jacobides, Michael G. (1997): Information, Technology and Coordination in Firms and Markets. Working Paper, Boston, MA; Wharton School, University of Pennsylvania

Jupiter Communications (Ed.) (1997): Online Marketplace. Digital Commerce Monthly 12/1997, New York, N.Y.

Jupiter Communications (Ed.) (1998): 1998 Online Advertising Report. New York, N.Y.: Jupiter Online Advertising Group (www.jup.com)

Kanellos, Michael (1998): Cheap computers: A scary business. In: CNET NEWS.COM, Internet: http://www.news.com/News/Item/0,4,24293,00.html, 16.7.98

Karepin, Rolf (1998): Anbieter forschen nach dem Pay Value. Info-Provider und Verlage wollen kostenpflichtige Web-Abrufdienste etablieren - aber die User sträuben sich. In: Horizont, 23.7.98, S. 64

Kelly, Kevin (1997): New Rules for the New Economy. Twelve dependable principles for thriving in a turbulent world. In: Wired, 9/97, pp.140-144; 186-197

Kelly, Kevin (1998): New Rules for the New Economy. 10 Radical Strategies for a Connected World. New York, N.Y.: Viking Press

Kirzner, Israel M. (1973): Competition and Entrepreneurship. Chicago, Ill.: University of Chicago Press

Kleinaltenkamp, Michael (1997): Kundenintegration, In: Zeitschrift für wirtschaftswissenschaftliches Studium, Heft 7, Juli 1997, pp. 350-354

Kotler, Philip; Bliemel, Friedhelm (1992): Marketing-Management – Analyse, Planung, Umsetzung und Steuerung. Stuttgart: Schäffer-Poeschl

Krieger, David J. (1996): Einführung in die allgemeine Systemtheorie. München: W. Fink

Kruse, Jörn (1994): Die amerikanische Dominanz bei Film- und Fernsehproduktionen. In: Rundfunk und Fernsehen, 42. Jhrg., Nr. 42/94, pp. 184-199

Künstner, Thomas (1997): Finanzierung & Marktchancen von Net- und Online-TV. Unveröffentlichtes Manuskript zu einer Präsentation auf den Medientagen München am 16.10.1997. Booz • Allen & Hamilton: Düsseldorf

Lambert, Patrick (1998): The Latest Web Rankings. Internet: http://www.businessweek.com, Business Week Online, 23.6.98

Lange, Ulrich Th. (1998): Wohnwert Multimedia. Unveröffentlichte Vorstudie für ein Auftragsgutachten im Rahmen eines Berliner Feldversuches. Berlin: o. Verl.

Laube, Helene (1998): Daten an die Macht. In: Manager Magazin, Nr. 8/98, pp. 80-84

Laube, Helene; **Preissner,** Anne; **Rieker,** Jochen (1998): Rette sich, wer kann. In: Manager Magazin, Nr. 3/98, mm-spezial Internet, pp. 204-213

Leo, Angelika (1998): AOL mit über 100 Werbemillionen. In: Horizont, 12.3.1998, pp. 61

Lewis, Theodore G. (1997): The Friction-Free Economy. Marketing Strategies for a Wired, World. New York, N.Y.: Harper Collins

Loizos, Constance (1998): Feeling the Burn: as the Costs of Online Publishing continue to exceed ad revenues, web publishers are devising increasingly elaborate ways to make money. In: The Red Herring, April 1998, pp.34-38

Luhmann, Niklas (1984): Soziale Systeme, Grundriss einer allgemeinen Theorie. Frankfurt/Main: Suhrkamp

Madden, Andrew P. (1997): Data and Voice Come Together on the Net. Internet: http://www.redherring.com/mag/issue49/top/1.html, Red Herring Online, December 1997

Madden, Andrew P. (1998a): The Lawgiver Gordon Moore. In: The Red Herring, April 1998, pp. 64-69

Madden, Andrew P. (1998b): The network computer's strange evolution. Internet: http://www.redherring.com/mag/issue57/nc.html, The Red Herring Online, August 1998

Malone, T. W.; **Yates,** J.A.; **Benjamin,** R.I. (1987): Electronic Markets and Electronic Hierarchies. In: Communications of the ACM, 6/87, pp. 484-497

Mandel, Michael J. (1998): The New Economy Starts to Hit Home. In: Business Week, 23.3.1998, p. 34

Mankiw, Gregory N. (1998): Principles of Economics. Fort Worth (TX): Dryden Press

Mankiw, Gregory N. (1998): The Microsoft quick quiz. In: Financial Times, 26.5.98, p. 14

McKnight, Lee W.; **Bailey,** Joseph P. (Ed.) (1997): Internet Economics. Cambridge, MA.: MIT Press

Mertz, Andreas (1998): High-Speed über Kupferdoppelader. xDSL-Techniken im Aufwind. In: Gateway, 6/98, pp. 94-97

Mieszkowski, Katharine (1998): The Best Things in Life Are Free. In: Fast Company, April-May 1998, pp. 38-40

Moore, Geoffrey A.; **Johnson,** Paul; **Kippola,** Tom (1998): The Gorilla Game. An Investor's Guide to Picking Winners in High Technology. New York, N.Y.: Harpers Business

Morgan Stanley Dean Witter (1997): Internet Quarterly: The Business of the Web..., Internet: http://www.ms.com/link24.htm, 23.9.97

Moschella, David C. (1997): Waves of Power. The Dynamics of Global Leadership 1964-2010. New York, N.Y.: AMACOM

Mougar, Walid (1998): Opening Digital Markets - Battle Plans and Business Strategies for Internet Commerce. New York, N.Y.: McGraw Hill (2nd ed.)

Negroponte, Nicholas (1995): Being Digital. New York, N.Y.: Vintage Books

Negroponte, Nicholas (1997): Reintermediated. In: Wired, 9/97; p. 208

Negroponte, Nicholas (1998): Telekommunikation und Neue Medien: Business-Strategien der Zukunft. Protokoll der Rede zur Tagung des Institute for International Research am 27.1.98

Negroponte, Nicholas (1998b): Taxing Taxes. In: Wired, 5/98; p. 210

Network Wizards (1998): Growth of the Internet Hosts. Internet: http://www.nw.com, Juli 1998

Noam, Eli M.; **Singhal,** Anjali (1996): Supra-national regulation for supra-national telecommunication carriers?. In: Telecommunications Policy, 10/96; pp. 769-787

Ohmae, Kenicho (1986): Japanische Strategien, Hamburg

Owen, Bruce M.; **Wildman,** Steven S. (1992): Video Economics. Cambridge, MA: Harvard University Press

PaineWebber (Ed.) (1997): Presentations from the 25th Annual Media Conference. New York, N.Y.: Communications Research Paine Webber

PaineWebber (Ed.) (1998): Entertainment Focus Quarterly Review – Time for a breather. New York, N.Y.: Communications Research PaineWebber

Pavlik, John V. (1999): New Media Technology: Cultural and Commercial Perspectives. Needham Heights, MA: Allyn & Bacon 2nd ed.

Peppers, Don; Rogers, Martha (1997): Enterprise One to One. Tools for Competing in the Interactive Age. New York, N.Y.: Currency/Doubleday

Pethig, Rüdiger; **Blind,** Sofia (1998). Fernsehfinanzierung. Ökonomische, rechtliche und ästhetische Perspektiven. Opladen/Wiesbaden: Westdeutscher Verlag

Picot, Arnold (1998a): Zusammenhänge zwischen Innovation und Marktentwicklung durch Telekommunikation. In: Picot, Arnold (Ed.): Telekommunikation im Spannungsfeld von Innovation, Wettbewerb und Regulierung. Tagungsband des Münchner Kreises, Heidelberg: Hüthig, pp. 77-98

Picot, Arnold (1998b): Die Transformation wirtschaftlicher Aktivitäten unter dem Einfluß der Informations- und Kommunikationstechnik, Vortrag anläßlich der Verleihung der Ehrendoktorwürde durch die Fakultät für Wirtschaftswissenschaften der Technischen Universität Bergakademie Freiberg am 23.1.98 (unveröffentlichtes Manuskript)

Picot, Arnold; **Reichwald,** Ralf; **Behrbohm,** Peter (1985): Menschengerechte Arbeitsplätze sind wirtschaftlich! Das Vier-Ebenen-Modell der Wirtschaftlichkeitsbeurteilung. In: Schriftenreihe des RKW, Eschborn

Picot, Arnold; **Reichwald,** Ralf; **Wigand,** Rolf T. (1998): Die grenzenlose Unternehmung. Information, Organisation und Management, Wiesbaden: Gabler Verlag (3. Aufl.)

Porter, Michael E. (1980): Competitive Strategy. Techniques for Analyzing Industries and Competitors. New York, N.Y.: The Free Press

Porter, Michael E. (1985): Competitive Advantage. New York, N.Y.: The Free Press

Quittner, Joshua (1998): Netscape's Survival Kit. How Jim Barksdale learned to stop worrying and love the monopoly. In: Wired, 4/98, pp. 154-158, 182-184

Rayport, J.E.; **Sviokla,** J.J. (1994): Managing in the Marketspace. In: Harvard Business Review, 11-12/94, pp.141-150

Rich, Laura (1998): Lucent Sprouts In-House Ventures. In: The Industry Standard, 24.8.98

Rupp, Hans-Björn (1996): Ein Preissystem für das Internet. Diskussionsbeitrag Nr. 164, Wissenschaftliches Institut für Kommunikationsdienste - WIK (Ed.), Bad Honnef

Sahal, Devendra (1981): Patterns of Technological Innovation. Reading, MA.: Addison-Wesley Publishing

Schrape, Klaus (1997): Strategien und Szenarien zur erfolgreichen Implementierung von DVB. Vortrag anläßlich des BLM-Rundfunk-Kongresses 1997 in München. Basel

Schrape, Klaus; **Seufert,** Wolfgang (1995): Künftige Entwicklung des Mediensektors. Gutachten im Auftrag des Bundesministers für Wirtschaft (Ed.), Berlin/Basel

Schrape, Klaus; **Seufert,** Wolfgang (1997): The Economics of New Information and Communication Technologies. In: European Communication Council - Report 1997 (Hrsg): Exploring the Limits. Europe's Changing Communication Environment. Heidelberg/Berlin: Springer Verlag, pp. 69-109

Schuler, Thomas (1996): Im Netz bleibt kaum etwas hängen. In: Süddeutsche Zeitung, 30.11.96, S. VII

Schürmann, Hans (1997): Daten huckepack übers Stromnetz transportieren. In: Handelsblatt, 5.11.97, S. 47

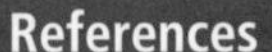

Sennewald, Nicola (1997): Der Markt für Massenmedien: Wettbewerbswandel durch das Internet. Eine Analyse anhand der Pressebranche. Inaugural-Dissertation an der Ludwig-Maximilians-Universität München, München

Shapiro, Carl; **Varian,** Hal (1998): Information Rules: A Strategic Guide to the Network Economy, Boston, MA: Harvard Business School Press

Sherman, Barry L. (1995). Telecommunications Management. Broadcasting/Cable and the New Technologies. International Edition: McGraw-Hill.

Steinberg, Steve G. (1998): Living Room LAN. In: Wired, 8/98, p. 79

Steiner, Peter O. (1952): Program Patterns and the Workability of Competition in Radio Broadcasting. In: Quarterly Journal of Economics, 66, p. 194

Sydow, Jörg (1991): Strategische Netzwerke – Evolution und Organisation. Wiesbaden: Gabler

Sydow, Jörg et al. (1995): Organisation von Netzwerken. Opladen: Westdeutscher Verlag

Tapscott, Donald (1996): The Digital Economy. Promise and Peril in the Age of Networked Intelligence, New York, N.Y.: McGraw-Hill

Tewes, Daniel (1997): Chancen und Risiken netzunabhängiger Service Provider. Diskussionsbeitrag Nr. 179, Wissenschaftliches Institut für Kommunikationsdienste – WIK (Ed.), Bad Honnef

Tongue, Carole (1998): Public interest issues in European policies. In: Media in Europe, The Yearbook of the European Institute for the Media 1998. Düsseldorf: European Institute for the Media 1998; pp. 70-79

U.S. Department of Commerce (Ed.) (1998): The Emerging Digital Economy. Internet: http://www.ecommerce.gov

Wallis, John J.; **North,** Douglas C. (1986): Measuring the Transaction Sector in the American Economy, 1870-1970. In: Engerman, Stanley L.; Gallmann, Robert E. (Ed.): Long-Term Factors in American Economic Growth, Chicago/London: University of Chicago Press, 1986, pp. 95-161

Wayner, Peter (1998): Plugging In to the Internet: Many Paths, Many Speeds. In: New York Times On the Web, Internet:http://www.nytimes.com/library/tech/98/07/circuits/howitworks/02plug.html, 2.7.98

Weiber, Rolf; Adler, Jost (1995): Informationsökonomisch begründete Typologisierung von Kaufprozessen. In: Zeitschrift für betriebswirtschaftliche Forschung (zfbf), Heft 1/1995, pp. 43-63

Weishaupt, Georg (1997): Richter brechen eine Lanze für den Wettbewerb im Telekommunikationsmarkt. In: Handelsblatt, 21.8.97, p. 2

Weiß, Michael (1998): Electronic Commerce im Einzelhandel. Erfolgsfaktoren und Implikationen einer neuen Distributionsform. Diplomarbeit im Fach Kommunikationswirtschaft der Fakultät Verkehrswissenschaft, TU Dresden

Wenger, Albert (1998): Information Technology and Firm Size. (mimeo), MIT Sloan School of Management, Cambridge (MA)

Werbach, Kevin (1997): Digital Tornado: The Internet and Telecommunications Policy. OPP Working Paper No. 29; Office of Plans and Policy, Federal Communications Commission, Washington, DC, March 1997

Wigand, Rolf; **Picot,** Arnold; **Reichwald,** Ralf (1997): Information, Organization and Management. Expanding Markets and Corporate Boundaries. Chichester, UK: John Wiley & Sons

Wilke, Hartmut (1993): Systemtheorie I: Grundlagen. Stuttgart: Lucius & Lucius (4. Aufl.)

Winsbury, Rex (1998): Why a pan-European regulator for telecoms is only a matter of time. In: Intermedia, 10/1998; pp. 33-35

Zerdick, Axel (1995): Markt- und Unternehmenstransparenz im Multimedia-Zeitalter. In: Wolfgang Hoffmann-Riem und Heide Simonis (Ed.): Chancen und Regelungsbedarf im Übergang zum Multi-Media-Zeitalter, Dokumentation zum Medienworkshop am 21. August 1995. Kiel: Landesregierung Schleswig-Holstein 1995, pp. 72-87